CHICAGO PUBLIC LIBRARY
HAROLD WASHINGTON LIBRARY CENTER
R0021974298

TECHNOLOGICAL SUBSTITUTION

TECHNOLOGICAL SUBSTITUTION

Forecasting Techniques and Applications

Edited by
Harold A. Linstone
Devendra Sahal
Portland State University, Portland, Oregon

ELSEVIER
New York/Oxford/Amsterdam

AMERICAN ELSEVIER PUBLISHING COMPANY, INC.
52 Vanderbilt Avenue, New York, NY 10017

ELSEVIER SCIENTIFIC PUBLISHING COMPANY
335 Jan Van Galenstraat, P.O. Box 221
Amsterdam, The Netherlands

Library of Congress Cataloging in Publication Data
Main entry under title:

Technological substitution.

Bibliography: p.
Includes index.
1. Substitution (Technology)–Addresses, essays, lectures. 2. Technological forecasting–Addresses, essays, lectures. I. Linstone, Harold A. II. Sahal, Devendra.
T173.2.T4 658.4'01 75-39768
ISBN 0-444-00181-6

Manufactured in the United States of America

To the pioneers in forecasting research
who grasped the nettle

Contents

Biographical Data

The Editors

HAROLD A. LINSTONE received his B.S. degree in 1944 from the City College of New York and his Ph.D. in Mathematics from the University of Southern California in 1954. He has been a member of the RAND Corporation, a Senior Scientist at Hughes Aircraft Company, and Associate Director of Corporate Development Planning at Lockheed Aircraft Corporation. From 1965 to 1970, he was also Adjunct Professor of Industrial and Systems Engineering at the University of Southern California. There he introduced courses in "Technological Forecasting" and "Planning Alternative Futures." The latter won USC's Dart Award for Innovation in Teaching in 1970. He has presented numerous seminars on Technological Forecasting and Long-Range Planning in the United States and Europe. Since 1970, he has been Professor of Systems Science at Portland State University and Director of its new interdisciplinary Ph.D. Program in this field. He is Senior Editor of the international journal, *Technological Forecasting and Social Change*, and recently co-edited a book on *The Delphi Method* with Murray Turoff.

DEVENDRA SAHAL received the B.S. degree in Mechanical Engineering from Banares Hindu University, the M.Phil. degree in Nuclear Physics from Helsinki University, and the Licentiate and Doctor of Technology degrees in Industrial Engineering from Helsinki University of Technology. He has worked as a mechanical engineer and operations research analyst for many leading North European corporations. He has also carried out and directed research in irradiation resistance of high polymers. Currently he is a post-doctoral Research Associate in the Systems Science Ph.D. Program at Portland State University. His publications include more than thirty technical papers in fields such as Cybernetics, General Systems, Theory of Evolutionary Processes, Industrial Engineering, and Technological Forecasting.

The Contributors

ROBERT U. AYRES is Vice President of International Research and Technology Corporation and has worked for many years in the field of technological forecasting and technology assessment. He has also conducted studies on environmental and resource economics and transportation technology, among other topics.

A. WADE BLACKMAN, JR. is Deputy Assistant Administrator for Planning and Analysis at the Energy Research and Development Administration in Washington, D.C.

M. BUNDGAARD-NIELSEN is an Associate Professor in chemical engineering economics at the Technical University of Denmark. He is currently serving as chairman of a government committee on the long range planning of higher education.

STEPHEN K. CONVER is a Systems Analyst with the Division of Safeguards, U.S. Nuclear Regulatory Commission, Bethesda, Maryland. He received the M.S. degree from Ohio State University in 1970, and has extensive experience with mathematical modelling and computer simulation.

P. FIEHN is currently completing his doctoral dissertation. He is employed with the Hempel Company in Copenhagen.

HAROLD E. GOELLER is a senior engineer with Oak Ridge National Laboratory. His areas of specialization have included the nuclear fuel cycle, nuclear desalination and agro-industrial complexes, and more recently general studies on nonrenewable resources.

CHOWDHURY KABIR is an Executive Engineer of a grain silo project in Bangladesh. He has an M.S. degree from Purdue University. He received his doctorate in 1975 from the Asian Institute of Technology.

HYDER LAKHANI is an Economist with the Maryland Department of Economic and Community Development, Division of Research, Annapolis, Maryland. In the past, he was a Professor of Economics at K. C. College of the University of Bombay, India. He has written a book, *Problems of Economic Development of India,* as well as a few journal articles.

JOSEPH P. MARTINO is a Research Scientist at the University of Dayton Research Institute, Dayton, Ohio. He received his doctorate from the Ohio State University in 1961, and has extensive experience in operations analysis and technological forecasting.

EDWARD J. SELIGMAN is Senior Mathematician at United Technologies Research Center in East Hartford, Connecticut. He is a Ph.D. candidate in mathematical statistics at the University of Connecticut and an Associate of the Society of Actuaries. His field is numerical analysis, investment analysis and statistical consulting.

ADELE SHAPANKA does substitution analysis and technological-economic analysis at Resources For the Future. She is currently involved in a fifty-year forecast of the U.S. economy.

M. NAWAZ SHARIF is Associate Professor of Industrial Engineering and Management at the Asian Institute of Technology. He has an M.E.A. degree from The George Washington University and a Ph.D. degree in Industrial Engineering from the Texas A&M University.

G. C. SOGLIERO is a Senior Research Engineer at United Technologies Research Center in East Hartford, Connecticut. She received her Ph.D. in mathematical statistics at the University of Connecticut in 1970. She investigates stochastic processes, performs time-series analysis and does statistical counsulting.

EARL STAPLETON prepared the article in this book while a graduate student in the Systems Science Doctoral Program at Portland State University.

M. O. STERN Senior Staff Member of International Research and Technology Corporation, is a physicist and systems analyst. He has specialized in the application of mathematical modeling techniques to urban, technological, and environmental problems. His interests have centered on the economics and mechanics of urban land use, transportation, and exploitation of natural resources.

ALVIN M. WEINBERG is Director of the Institute of Energy Analysis, a post he assumed July 1, 1975. He was, for 17 years, Director of Oak Ridge National Laboratory.

Introduction

Substitution has been an instrument of man's material progress since prehistoric days. The combination of two uniquely human characteristics–intelligence and adaptability–has impelled man to substitute wheels for muscles, agriculture for hunting, aircraft for surface vehicles, and preventive inoculations for disease treatment. Substitution proceeded at a modest pace until the eighteenth century when John Locke's philosophy of liberalism and the surge of applied science ushered in the Industrial Revolution. Over the last two hundred years technological substitution has become a vivid measure of the pace of this era. It took about 75 years to substitute steam for sail (10% to 90% of gross tonnage) and 42 years for the open hearth to replace the Bessemer steel manufacturing process. More recently it took a mere 18 years for synthetic to replace natural tire fibers and less than nine years to substitute detergents for natural soap in the United States (in all cases 10% to 90% period).

Although there have been occasional studies of the innovation process, e.g., Colum Gilfillan's work,[1] in-depth analysis of technological substitution[2] began only in the 1950's. The push came from two groups–economists and technological forecasters. Among the former Zvi Griliches and Edwin Mansfield exerted a major influence.[3] The latter were themselves a product of a perceived need. We recall that World War II connotes a milestone in military research and development. Prior to this time the R&D budget was measured in millions of dollars, subsequently it jumped to billions. The United States Air Force plunged into advanced technology systems more rapidly than any other organization in the world. It created not only the RAND Corporation, but its own in-house groups to assist in the enormous planning task. In this context Ralph Lenz was a pioneer in the development of technological forecasting methods. He early perceived an analogy between functional technological and biological growth patterns.[4] In 1925 the

---

[1] S. C. Gilfillan, *The Sociology of Invention,* originally published in 1935, reprinted by the M.I.T. Press, 1970.

[2] The term "technological substitution" refers to the adoption of an innovation and should not be confused with the production function studies in econometrics.

[3] Z. Griliches, Hybrid corn: An exploration in the economics of technical change, *Econometrica* **25** (1957), pp. 501–522; E. Mansfield, Technical change and the rate of imitation, *Econometrica* **29** (1961), pp. 741–766.

[4] R. C. Lenz, Jr., *Technological Forecasting,* 2nd Ed., USAF Aeronautical Systems Division, Tech. Rept. ASD–TR–62–414, June 1962.

American biologist Raymond Pearl[5] had observed that the growth in weight of a pumpkin, the growth of a culture of yeast cells, and other biological phenomena could be represented by the S-shaped curve now known as the Pearl or logistic curve (see Fig. 1). This growth curve has provided one starting point for fruitful analyses of the technological substitution process. The harvest of the diverse activities of recent years is presented in the following chapters.

In the two centuries spanned by the Age of Technology, the American per capita gross national product has grown from $200 to $7,000 while the world gross product has zoomed from less than $0.1 trillion to more than $5 trillion (1975 dollars). But, as the saying goes in business, "there is no such thing as a free lunch." Technological substitution creates new problems as it solves old ones. Most obvious of all new problems is the depletion of familiar forms of energy and material resources. We are warned that economically accessible reserves of petroleum, copper, natural gas, mercury, zinc, and silver will disappear within the next hundred years.

As any student of technological forecasting knows, the basis for innovation is inevitably an amalgam of a new technological capability and a perceived new need. In the era of the Technological Age the igniting element for the technological substitution process has usually been (1) a technological achievement, (2) a military threat, or (3) a business strategy to increase profitability (e.g., competitive pressure, consumer conditioning to planned obsolescence).

By contrast, the primary trigger for technological substitution in the future may well be a perceived depletion of resources. And the ability to plan, initiate, time, and implement this process can prove decisive in managing the difficult transformation we face today. Herman Kahn, Jonas Salk, Dennis Meadows, John Platt, and others see the United States at a historic inflection point on the logistic curve representing its material growth (Fig. 1). The past two hundred years correspond to the portion preceding this inflection point. It has been a period of accelerating economic growth, characterized by free competition, and virtually unlimited resources. At this point the sign of the second derivative of the curve changes from positive to negative, denoting slowing growth

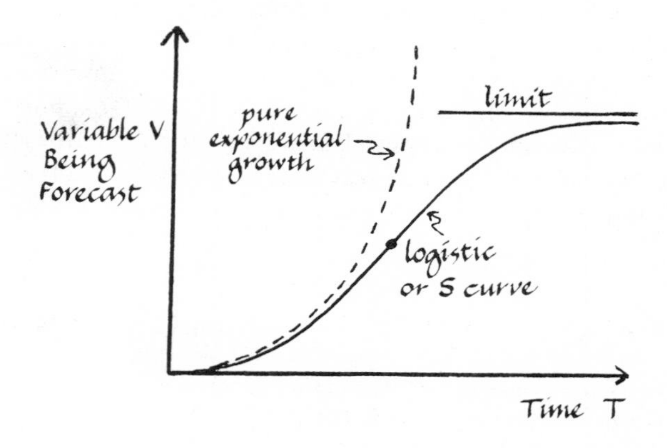

Fig. 1.

[5] R. Pearl, *The Biology of Population Growth,* Alfred A. Knopf, New York, 1925.

presumably characterized by cooperation and growing resource constraints.[6] In this situation technologists confront an unprecedented challenge to facilitate the transition with minimal dislocations and disruptions. The help of policy sciences is also vital—a major question is whether the incentives for implementation will be provided by the institutional arrangements, i.e., by government to business. If these efforts are effective, physical resources will cease to be a primary issue. In the words of Goeller and Weinberg, we will asymptotically approach the "Age of Substitutability."[7]

Another new factor in substitution is the consequence of the secondary impacts of the massive use of technology, effects that could be neglected in the past but that should prove very significant in the small "Spaceship Earth" of the future. High mobility has weakened the family, loosened social roots and has led to alienation and loss of security for the individual. Telecommunications appears to create a profusion of stimuli with a profound impact on the psyche that we have barely begun to recognize. Pollution of the environment endangers health in diverse and subtle ways. Substitution of safe for dangerous products is occurring with increasing frequency. Organic pesticides have largely replaced DDT, minimal polluting automobile engines will replace current engines, resilient "safe-fail" ecosystem design must replace vulnerable (albeit efficient) "fail-safe" systems.

A third interesting new characteristic of technological substitution in the coming decades is the increasing impact of "intermediate technology." Technological substitution need not imply greater integration, centralization, and complexity. As E. F. Schumacher has so cogently argued, economic growth in the poor world need not, must not, follow in the footsteps of today's advanced societies. Business has surprised itself by its ability to develop "intermediate technology" substitutions for current products and processes far more suitable to the needs of the developing regions than the mere transfer of advanced technology.

These brief comments portend a significantly intensified concern with technological substitution on the part of planners in government and industry. Most of the articles in this book were initially written for the journal *Technological Forecasting and Social Change*. In the present format they should be conveniently accessible to a wider group of researchers and practitioners. We hope that the material will stimulate the former to improve the methodology and the latter to enhance the effectiveness of their planning.

---

[6] The possibility that the entire earth's population can achieve an economic standard of living comparable to, or better than, that of the United States today should not be dismissed as long as technological substitution is permitted to proceed. However, most of the world's nations do not see themselves as close to the inflection point. They must continue to seek accelerating economic growth. The question for them is this: does slowing growth for an advanced society permit or impede exponential growth for a poor country? If slowing economic growth has an adverse effect on the developing world's economies the likelihood of deadly conflict between rich and poor itself increases exponentially.

[7] See H. E. Goeller and A. M. Weinberg, Part 6, Chapter 15.

Part 1.
Basic Models

Part 1. Basic Models

Introduction

Very few innovations gain immediate universal acceptance. Far more typically, the substitution of a technology over time (expressed in terms of the percentage of the potential users who have adopted the innovation, the percentage of the market acquired by the new product, etc.) tends to take the form of an S-shaped curve (see Fig. 1 of Introduction on p.xiv).The process of substitution is characterized by a slow initial rise followed by a period of more rapid growth and tapering off toward a fixed saturation value, the final level of adoption of the innovation.

The assertion that the path of substitution tends to take the form of an S-shaped curve is generally supported by the empirical evidence. Fisher and Pry have analyzed seventeen cases of substitution (see Table 1) and found good support for an S-shaped curve corresponding to a version of the so-called logistic function.[1] Since a number of studies in this book refer to their work, we will describe it in some detail.

The underlying hypothesis of the substitution model used by Fisher and Pry is that the rate of adoption of a new product or process is proportional to the fraction of the old one still in use. Mathematically we express this relationship as follows: if f is the fraction of the market captured by the innovation, t the time, and b a constant, then

$$\frac{1}{f}\left(\frac{df}{dt}\right) = b(1-f). \tag{1}$$

Now let t_0 denote the time at which the fractional substitution reaches its midpoint, i.e., $f = \frac{1}{2}$. Integration of Eq. (1) yields the familiar logistic curve

$$f = \frac{1}{1 + \exp b(t - t_0)} \tag{2}$$

This function, shown in Fig. 1(a), is expressed alternatively in the form

$$\frac{f}{(1-f)} = \exp b(t - t_0). \tag{3}$$

The constant b may be interpreted as the initial rate of adoption of the innovation. Equation (3) permits representation in linear form on semilog paper (Fig. 1(b)).

Substitution data and their fit to this model for a number of illustrative cases are shown in Fig. 2. It can be seen that the data in the form of $f/(1-f)$ do indeed form a straight line when plotted against time on semilog paper–irrespective of the type of

[1] Fisher, J. C., and Pry, R. H., A simple substitution model of technological change, *Technol. Forecast. Soc. Change* 3 (1971), American Elsevier Publishing Co., NY, pp. 75–88.

Table 1

Takeover Times (Δt) and Substitution Midpoints, t_0, for a Number of Substitution Cases

Substitution	Units	Δt Years	t_0 Year
Synthetic/Natural Rubber	Pounds	58	1956
Synthetic/Natural Fibers	Pounds	58	1969
Plastic/Natural Leather	Equiv. Hides	57	1957
Margarine/Natural Butter	Pounds	56	1957
Electric-Arc/Open-Hearth Speciality Steels	Tons	47	1947
Water-Based/Oil-Based House Paint	Gallons	43	1967
Open-Hearth/Bessemer Steel	Tons	42	1907
Sulfate/Tree-tapped Turpentine	Pounds	42	1959
TiO_2/PbO-ZnO Paint Pigments	Pounds	26	1949
Plastic/Hardwood Residence Floors	Square Feet	25	1966
Plastic/Other Pleasure-Boat Hulls	Hulls	20	1966
Organic/Inorganic Insecticides	Pounds	19	1946
Synthetic/Natural Tire Fibers	Pounds	17.5	1948
Plastics/Metal Cars	Pounds	16	1981
BOF/Open-Hearth Steels	Tons	10.5	1968
Detergent/Natural Soap (U.S.)	Pounds	8.75	1951
Detergent/Natural Soap (Japan)	Pounds	8.25	1962

Note: Δt is the time from 10% to 90% takeover. (See footnote 1 for source.)

products and processes considered. Normalization of the time scale by use of the term $2(t - t_0)/\Delta t$ collapses all seventeen cases of substitution in Table 1 into the single curve shown in Fig. 3. Blackman and Floyd have presented similar models based on the assumption that the substitution of technology over time tends to take the form of an S-shaped curve.[2,3] The empirical results of their studies provide further evidence that technological substitution evinces such a pattern.

The three articles in this section provide us with further extension and evaluation of the models based on extrapolation of the S-shaped trend. The chief attraction of the methods presented in all three works lies in their simplicity. They are ideally suited to cases where data on economic variables are not readily available. In the first article, Sharif and Kabir show how the considerations of the effective life span of the new product and the availability of data may be taken into account so as to improve the reliability of the basic trend extrapolation model. A number of applications are presented including those of substitution of diesel and electric locomotives in U.S. railroads, and substitution of metal for wood in the U.S. merchant marine market.

However, it must be borne in mind that forecasting based on trend extrapolation implies the *ceteris paribus* assumption. This can be a serious handicap in the case of a significant change in the exogenous factors that the influence the course of substitution. The recent oil crisis is one such example that has perhaps invalidated all forecasts of the

[2] Blackman, A. W., The market dynamics of technological substitutions, Part 2, Chap. 6.

[3] Floyd, A., Trend forecasting: A methodology for figure of merit, in *Technological Forecasting for Industry and Government* (J. Bright, ed.) Prentice-Hall, Englewood Cliffs, NJ, 1968.

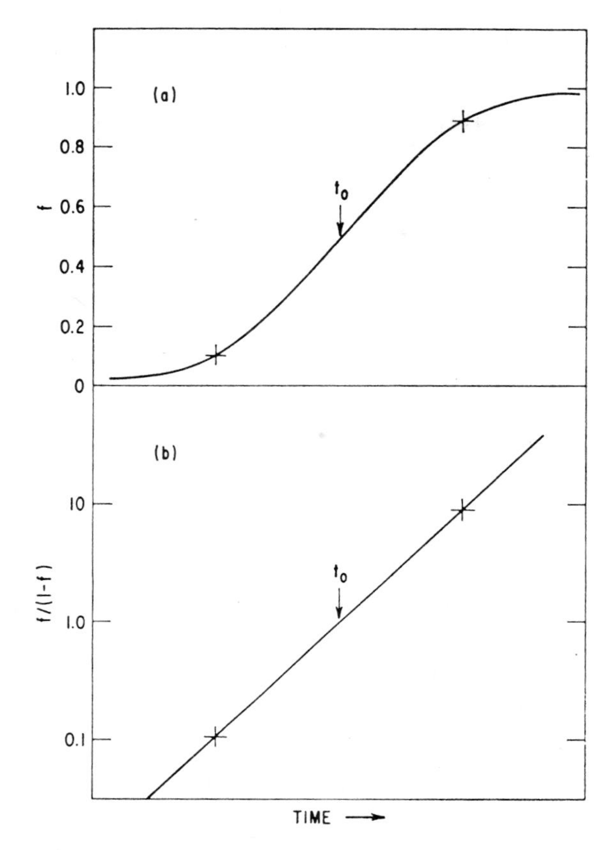

Fig. 1. General form of the substitution model function of Fisher and Pry. (See footnote 1 for source.)

substitution of natural fibers by oil-based synthetic fibers. A practical forecasting model, therefore, has to be responsive to exogenous influences. There is a need for some simple (requiring less data), dynamic (allowing parametric variations) and adaptive (to changes in the environment) models for forecasting technological substitution. In their second paper Sharif and Kabir show how the "system dynamics" technique can be utilized to build such a model. The methodology they present also permits forecasting in cases where a particular product or technology is replacing an older one while at the same time being replaced by a newer one—a multilevel substitution process. In one such application of the methodology, the three way substitution problem among piston, turbo-prop and turbo-jet aircraft is studied. There is yet another application presented in this paper that is of current interest and considerable importance—the substitution of jute by polypropylene.

Despite their consideration of exogenous changes, the models of Sharif and Kabir are of a predominantly deterministic nature. They do not consider the role of random elements—at least not explicitly. In contrast, Stapleton in the third paper presents a probabilistic model of technological substitution based on the assumption of a normal distribution. The empirical evidence presented in this paper does indeed support the assumption that the fractional growth of the new product when plotted on normal

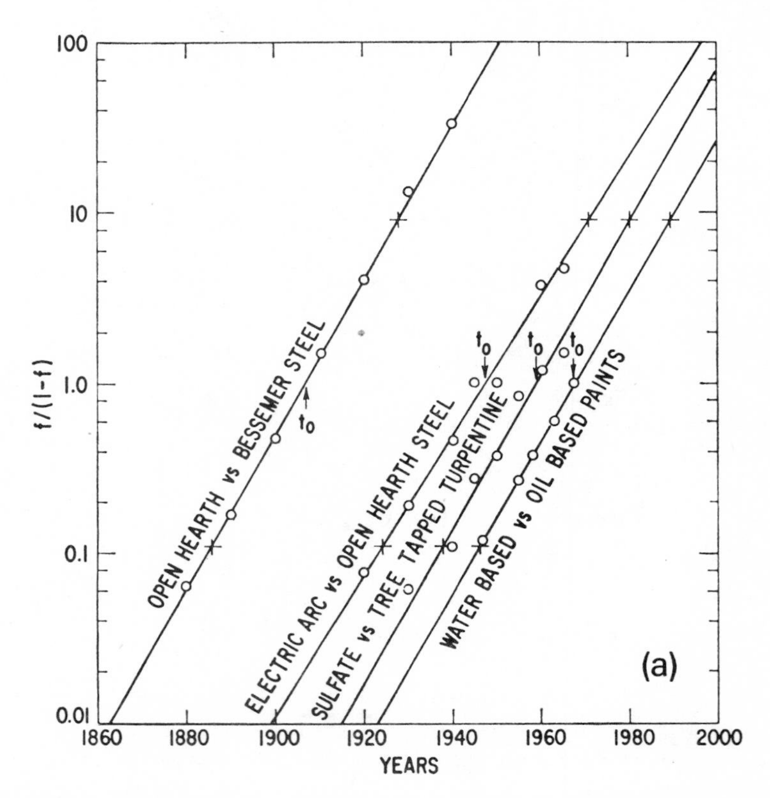

Fig. 2. Substitution data and their fit to model for a number of products and processes. (See footnote 1 for source.)

probability paper will result in a straight line. This work may therefore be regarded as a further verification of the hypothesis that technological substitution follows as S-shaped pattern.

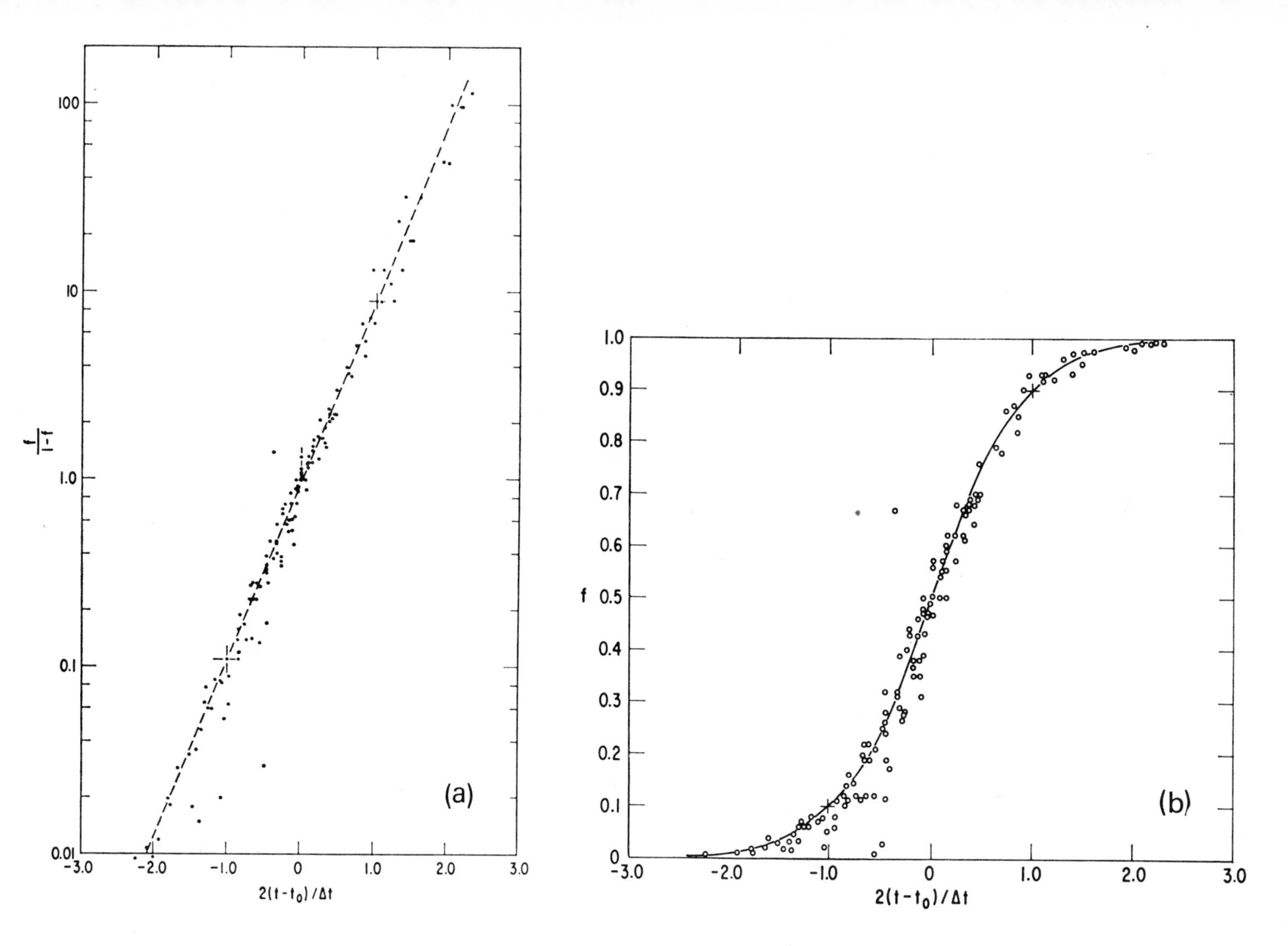

Fig. 3(a,b). Fit of substitution model function to substitution data for all 17 cases vs. normalized units of time. (See footnote 1 for source.)

A Generalized Model for Forecasting Technological Substitution

M. NAWAZ SHARIF and CHOWDHURY KABIR

ABSTRACT

This paper presents a generalized mathematical model for forecasting technological substitution under a wide variety of circumstances. Some of the existing models are shown as special cases of the generalized model. Methods are also suggested for improving the reliability of the model by taking corrective measures on the available data and following a step-wise forecasting procedure.

Introduction

In the present era of rapid technological advancement, the accuracy of forecasts regarding technological substitution of new products of techniques for old are becoming more and more critical for minimizing the risk of the prohibitive cost of error in decision-making. This study is an attempt towards refinement of some of the existing models. The modifications are suggested on the basis of experience gained through working with available mathematical models of trend extrapolation for forecasting technological substitution under different circumstances.

Working with existing models for forecasting the market share of different technology-based products of wide variety, the authors identified many factors that significantly influence the accuracy of the forecasts. One important factor is the realization of the necessity for adapting existing models to suit a particular purpose. An elegant adapting procedure from a generalized model is presented in this paper.

Some other factors that were considered and are presented in this paper for increasing the reliability of a forecast are: (i) the effect of scatterness of the historical data; (ii) the effect of data extent both in terms of number of observations and in terms of time and magnitude interval; (iii) the effect of the last data point; and (iv) the effect of technological characteristics of the product life cycle.

Model Formulation

The three existing models of technological forecasting by trend extrapolation that form the background for this study are—Blackman's model [1], the Fisher-Pry model [4], and Floyd's model [5]. The mathematical formulation of these models are as follows:

This article appeared in *Technological Forecasting and Social Change*, Vol. 8, No. 4, 1976.

Blackman:

$$\ln \frac{f}{F-f} = C_1 + C_2 t, \tag{1}$$

Fisher-Pry:

$$\frac{f}{1-f} = \exp\ 2\alpha(t-t_0)\ , \tag{2}$$

Floyd:

$$\ln \frac{f}{F-f} + \frac{F}{F-f} = C_1 + C_2 t, \tag{3}$$

where

f = market share of a product at time t;
F = upper limit of the market share;
t = time;
t_0 = time at which substitution is half complete;
2α = slope of the regression line; and
C_1, C_2 = constants.

It can be observed from the above equations that the Fisher-Pry model is a very special case of the substitution model presented by Blackman. Blackman's model considers an upper limit of the market share that the new product can capture in the long run whereas in the Fisher-Pry model this upper limit is taken to be 100% substitution. However, in a recent paper [2], Blackman seems to agree with the Fisher-Pry assumption that "if the new product is economically viable after it has gained a small market share, it is likely to become more competitive as time progresses, and once a substitution has begun, it is highly probable that it will eventually take over the available market". Therefore, there is hardly any difference between these two models, but they are different from Floyd's model. The form of the Floyd's model that is referred to here and used in this paper is the one for noncompeting technology.

Problems of overestimation and underestimation discussed in [8], and the analysis of historical data utilizing Floyd's and Blackman/Fisher-Pry models under the assumption of "possible full substition" as presented in [1, 2, 4 and 9] reveal that except for the cases where substitution has reached near completion at the time of forecast, Floyd's model gives an underestimation of the forecast while Blackman/Fisher-Pry model gives an overestimation of the forecast and the realistic forecast is likely to lie in the region bounded by these two extremes. One possible method for adaption within this region is by branching off from the Blackman/Fisher-Pry curve incorporating a reduced rate of substitution from a breakoff point corresponding to a significant environmental change. However, this procedure gives a curve that is not of the form of the intuitively acceptable S-shape.

A smooth S-shaped curve within the region bounded by the Blackman/Fisher-Pry and Floyd curves can be obtained by combining these two models as follows:

$$(1-\sigma)\left[\ln \frac{f}{F-f}\right] + \sigma\left[\ln \frac{f}{F-f} + \frac{F}{F-f}\right] = C_1 + C_2\, t, \tag{4}$$

where

f = market share at time t;
t = time;
C_1, C_2 = constants; and
σ = dimensionless factor, $0 \leqslant \sigma \leqslant 1$.

It can be observed from Eq. (4) that for $\sigma = 1$, we have Floyd's model; for $\sigma = 0$, we have Blackman's model; and for $\sigma = 0$, $F = 1$, we have the Fisher-Pry model.

Simplifying Eq. (4) and taking the value of $F= 1$ as acceptable for practical purposes, the general model for forecasting technological substitution can be expressed as

$$\ln \frac{f}{1-f} + \sigma \frac{1}{1-f} = C_1 + C_2 t. \tag{5}$$

The second term on the left-hand side of Eq. (5) is in fact a "delay factor" and σ is a "delay coefficient". Since σ can take a value between zero and one, a set of smooth S-shaped curves can be obtained by a decision-maker ranging from most optimistic to a pessimistic forecast. A set of guidelines are given in the next section for the selection of a proper delay coefficient.

Estimating the Delay Coefficient

In order to determine an appropriate value for σ to be used in the general model given by Eq. (5) many factors need to be considered. Although the judgment of the decision-maker to take care of the exogenous factors is the most important consideration in the selection of the value of σ, a preliminary estimation can be made by utilizing the following functional relationship:

$$\sigma = \phi\ [DS, DE, fl, ELS], \tag{6}$$

where

DS = data scatterness;
DE = data extent;
fl = last value of f (market share at the time of forecast);
ELS = effective life span (time required for f to go from 0.1 to 0.9).

If the substitution process is in its early stage, the effect of data scattering (DS) and the value of the market share at the time of the forecast (fl) are quite significant in determining the trend of the forecasting curve: this is demonstrated in Fig. 1. In order to minimize discrepancy, the smoothing of historical data by moving average (taking 3–5 at a time) is recommended. The moving average procedure quite obviously introduces serial correlation problems. The forecast would nevertheless work so long as serial correlation continues to exist. It is not proper to apply ordinary least squares in the presence of a serially correlated disturbance term. However, in this paper all diagrams except Fig. 1 are based on smoothed data.

Figure 2 shows the effect of data extent (DE) on the model predictions. As the rate of substitution in the very early stage is not significantly altered for quite a long period,

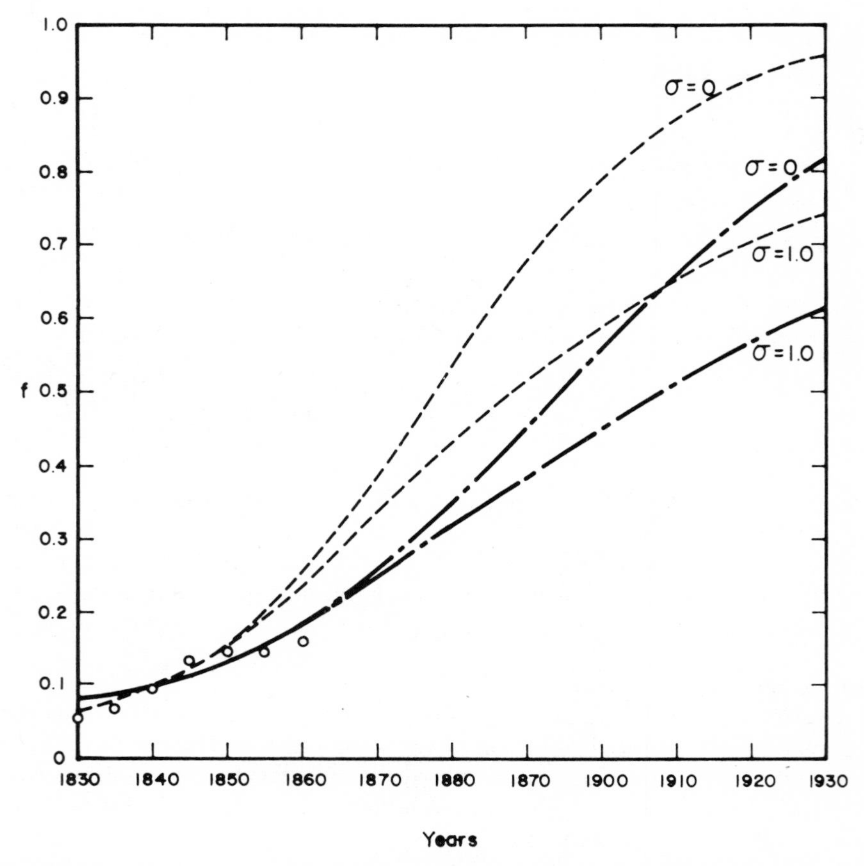

Fig. 1. Substitution of power for sail in the U.S. merchant marine market. ○, historical data [6]; ---, 1850 forecast (upper curve for $\sigma = 0$ and lower curve for $\sigma = 1.0$); ____, 1860 forecast (upper curve for $\sigma = 0$ and lower curve for $\sigma = 1.0$).

use of a large number of data points in this region is likely to make the forecast more pessimistic. One way to minimize this tendency is to disregard data points corresponding to the very early stage of substitution, and to fix an initial value of the market share f beyond which all data points are used for forecasting. It should be noted here that the selected value of x will depend on the level at which the initial value of f is set. In this presentation the initial value of f is arbitrarily set at 0.1.

Another factor that influences the selection of the delay coefficient is the effective life span of a product (*ELS*), defined as the time required for the market share to go from 0.1 to 0.9. It has been observed that better estimation of the value of σ is possible when the effective life span is also considered along with other corrective measures suggested in the preceding paragraphs. Figure 3 provides an empirical curve that can be used for predicting the effective life span. This curve has been obtained by simply observing the trend and a few data points from existing case studies. The value of C_2 is to be determined from Eq. (5) by regression analysis taking the value of $\sigma = 0$ and using data for f between 0.1 and 0.2.

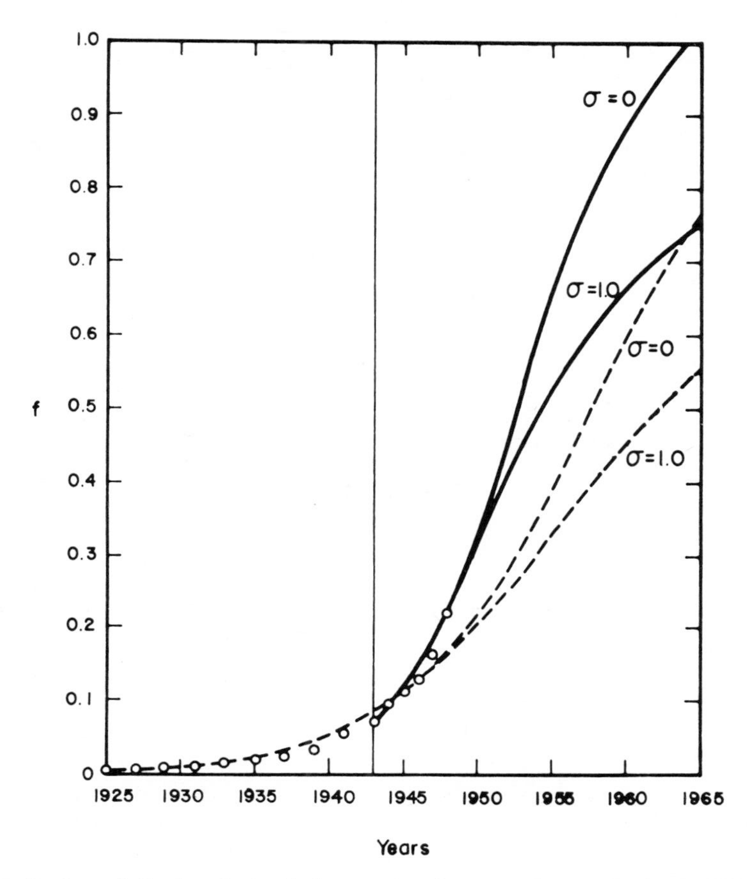

Fig. 2. Substitution of diesel and electric locomotives for steam locomotive in the U.S. railroad. ○, historical data [6]; – – –, model predictions at $f = 0.2$ with available data from $f = 0.006$ (upper curve for $\sigma = 0$ and lower curve for $\sigma = 1.0$); ____, model predictions at $f = 0.2$ with available data from $f = 0.08$ (upper curve for $\sigma = 0$ and lower curve for $\sigma = 1.0$).

Figure 4 shows three empirical curves corresponding to a short, medium and long effective life span. The values of σ to be used in the general model are estimated from these curves by selecting the proper life cycle curve (i.e., *ELS*) and the corresponding value of the market share at the time of forecast (*fl*). Attention is again drawn to the fact that the curves depicted in Fig. 4 are to be used only when data points for $f < 0.1$ are discarded and when smoothing of data is performed.

To incorporate a time dependent σ that correspondingly makes C_2 a function of time rather than invarient with time would be a logical extension of the work presented here.

An Illustrative Example

An illustration of the generalized model prediction using the delay coefficient estimated from the empirical curves as shown in the previous section is given in Fig. 5. The case used in this example is the substitution of metal for wood in the U.S. merchant marine market. The time of forecast is 1895 when approximately 20% of the market has been captured by metal. The procedure used for obtaining the forecast is as follows:

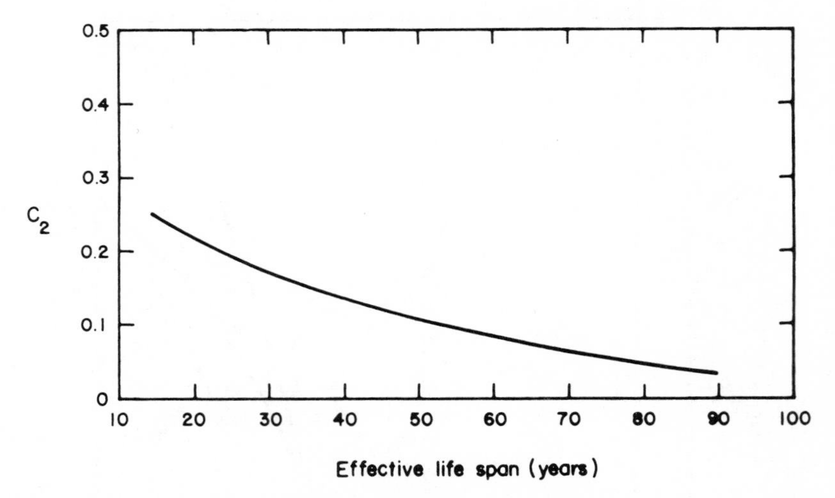

Fig. 3. Empirical curve to predict "effective life span" of the product being substituted.

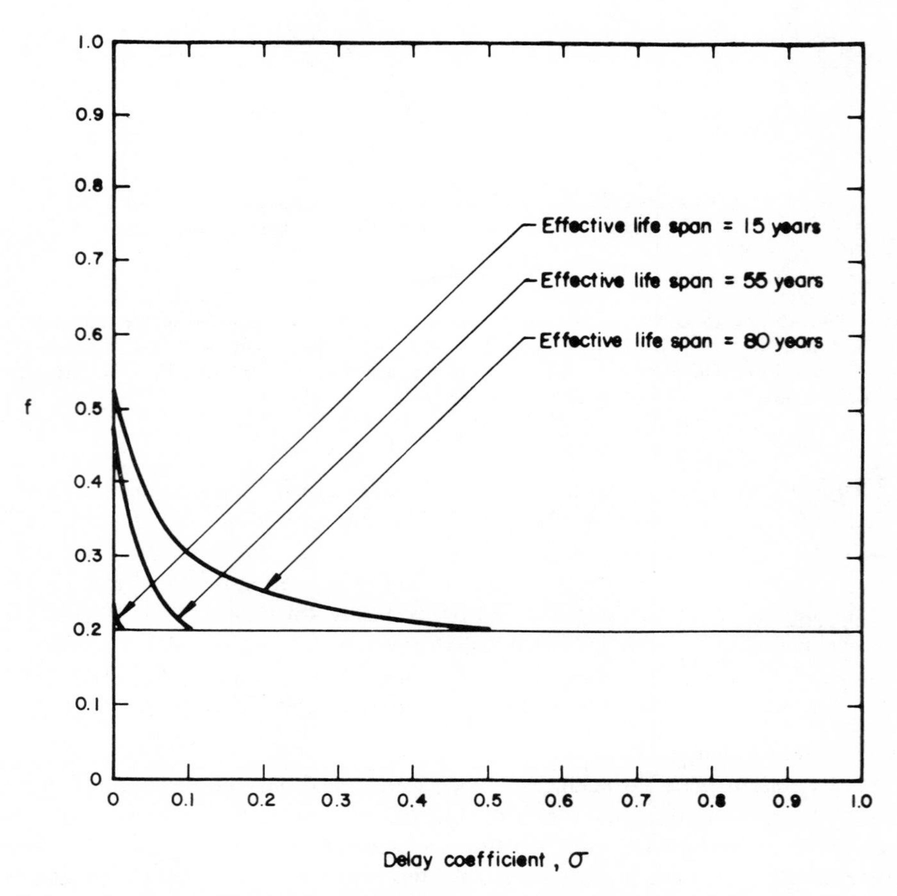

Fig. 4. Empirical curve to find the value of "delay coefficient".

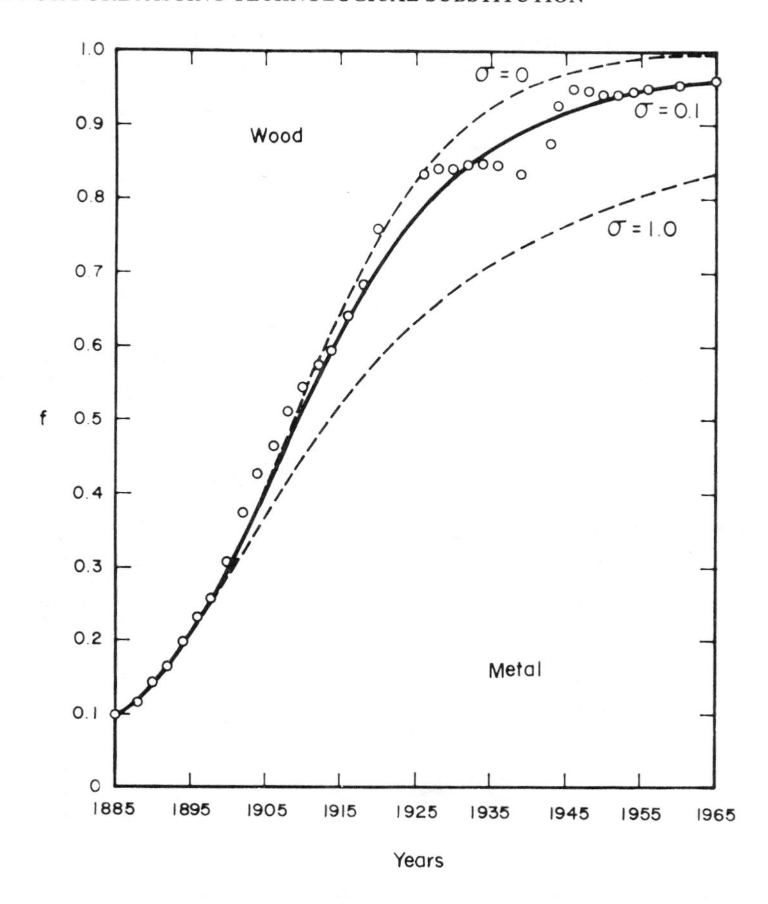

Fig. 5. Substitution of metal for wood in the U.S. merchant marine market. ○, historical data [6, 10]; – – –, model predictions at $f = 0.2$ (upper curve for $\sigma = 0$ and lower curve for $\sigma = 1.0$); ____, model predictions at $f = 0.2$ with $\sigma = 0.1$; $\ln [f/(1-f)] + 0.1 [1/(1-f)] = -2.14813 + 0.09578t$; f = market share of metal in merchant marine industry; t = year; correlation coefficient = 0.99903.

1. Moving average (taking three data points at a time) is used to smooth the historical data.
2. All data points corresponding to $f<0.1$ were eliminated.
3. Remaining data points are (1886, 0.104), (1888, 0.122), (1890, 0.141), (1892, 0.168) and (1894, 0.198).
4. Through regression analysis the value of C_2 is found to be 0.094.
5. *ELS* obtained from Fig. 3 for $C_2 = 0.094$ is 55 years.
6. Value of σ estimated from Fig. 4 corresponding to $ELS = 55$ and $fl = 0.2$ is 0.1.
7. Equation (5) is used to obtain the forecast as shown in Fig. 5.

Step-wise Procedure

While each substitution process can be depicted by some form of the S-shaped curve, the variations are too many. It has been observed in many cases that if the complete substitution process is considered to be a combination of many S-shaped curves instead of

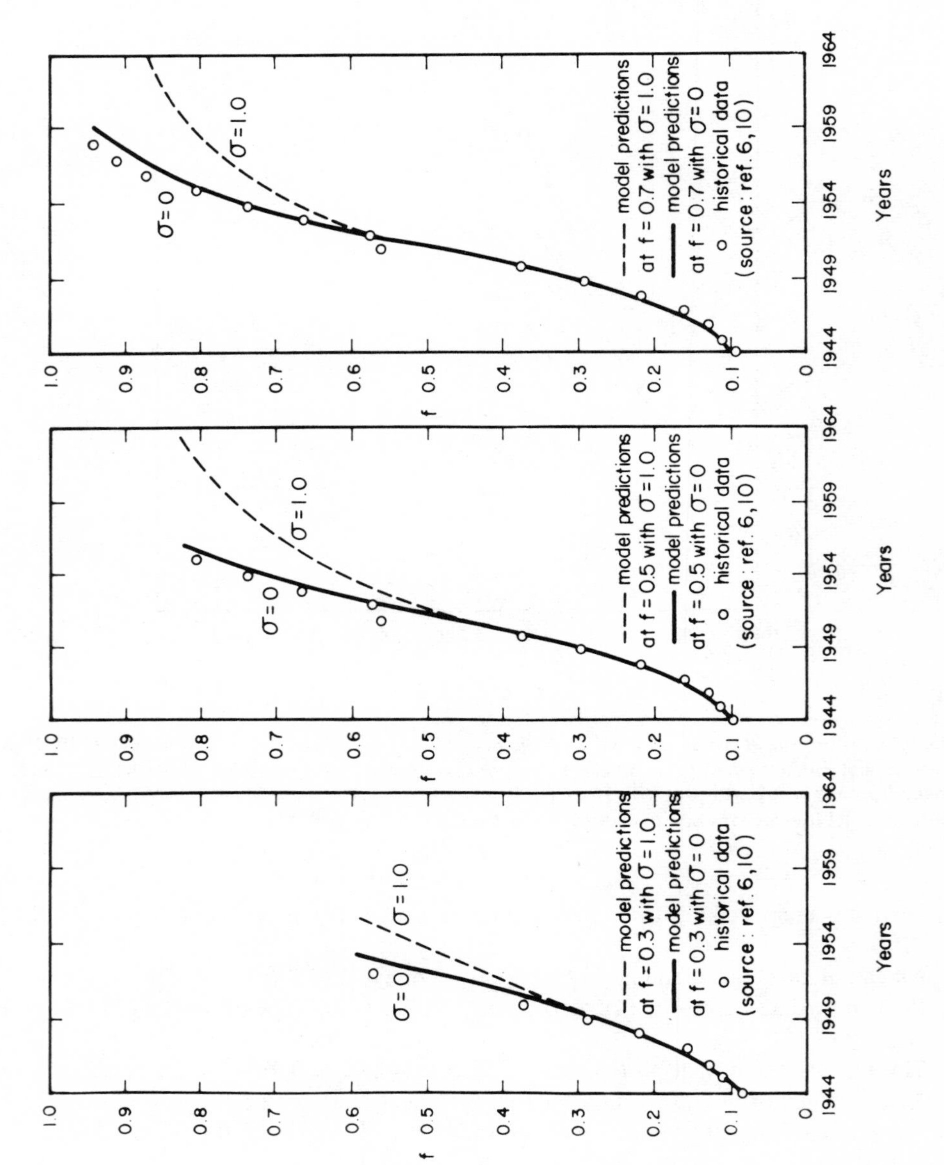

Fig. 6. Substitution of diesel and electric locomotives for steam locomotive in the U.S. railroad.

just one a better representation is possible. Moreover, there is universal agreement that the farther in the future that one tries to forecast the higher is the probability of making errors of greater magnitude. Therefore, it is recommended that the analyst should follow a step-by-step forecasting procedure. Limiting the forecast period to one quarter of the effective life span is the suggested rule of thumb.

Three illustrations are included here to demonstrate the usefulness of this procedure. In all of these examples the considerations were identical to that of Fig. 5 as far as the model and corrective measures are concerned; the only difference being the length of forecast limited to one quarter of *ELS* at a time. Table 1 gives all the pertinent data for the three examples shown in Fig. 6, 7 and 8.

Conclusion

The model presented in this study is hopefully general enough to find wide application under different circumstances. The proposed model is best suited when data on economic variables are not available. Several points have been noted while making this study and are worth mentioning here.

The value of the market share at the time of forecast influences the trend of the substitution curve quite significantly when the number of historical data is small. The influence decreases with the increase in data points. However, use of data points corresponding to a very early stage of substitution is likely to produce a pessimistic forecast when the market share at the time of forecast is also relatively small.

The accuracy of a forecast is highly dependent on the effective life span. For a very short effective life span, full substitution forecast is likely to be less accurate compared with the case when the effective life span is relatively long.

Forecasting technological substitution using the generalized model presented in this paper will be more meaningful if the period of forecast is limited to about one quarter of the effective life span and following a step-wise procedure.

TABLE 1

Relevant Information for Figures 6, 7 and 8

Figure	Effective life span (years)	Market share at the time of forecast	Delay coefficient (σ)	Period of forecast[a] (years)	Model constants C_1	C_2	Correlation coefficient
6(a)	15	0.30	0	4	−2.24308	0.29390	0.99485
6(b)	15	0.50	0	4	−2.32259	0.34226	0.99376
6(c)	15	0.70	0	4	−2.35634	0.35653	0.99649
7(a)	55	0.20	0.10	14	−2.19382	0.09521	0.99939
7(b)	55	0.40	0.01	14	−2.19298	0.09452	0.99951
7(c)	55	0.60	0	14	−2.18691	0.09387	0.99969
8(a)	80	0.25	0.20	20	−2.00846	0.04045	0.99741
8(b)	80	0.40	0.04	20	−2.24369	0.04055	0.99914

[a]Period of forecast = .25 effective life span (*ELS*) in term of years.

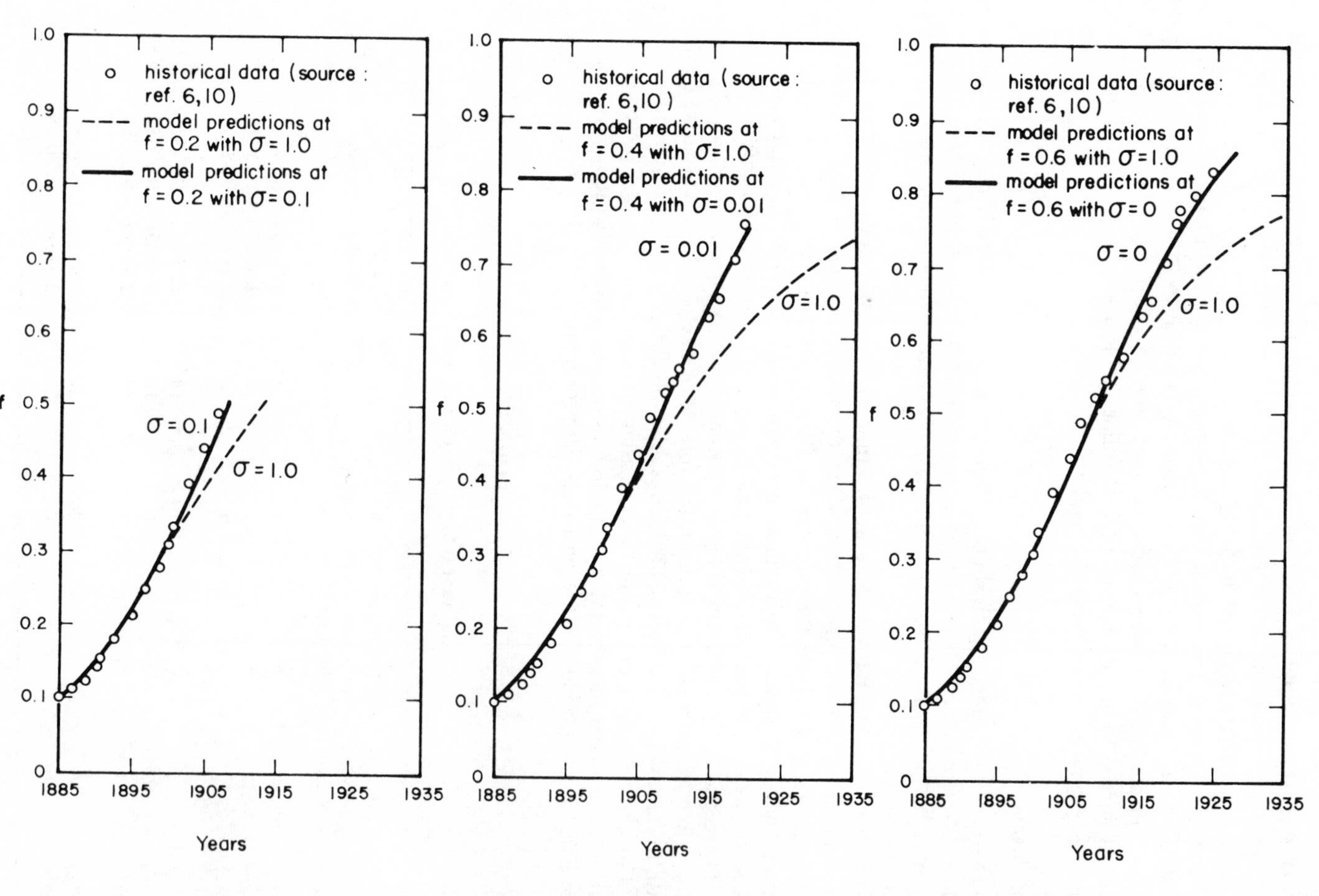

Fig. 7. Substitution of metal for wood in the U.S. merchant marine market.

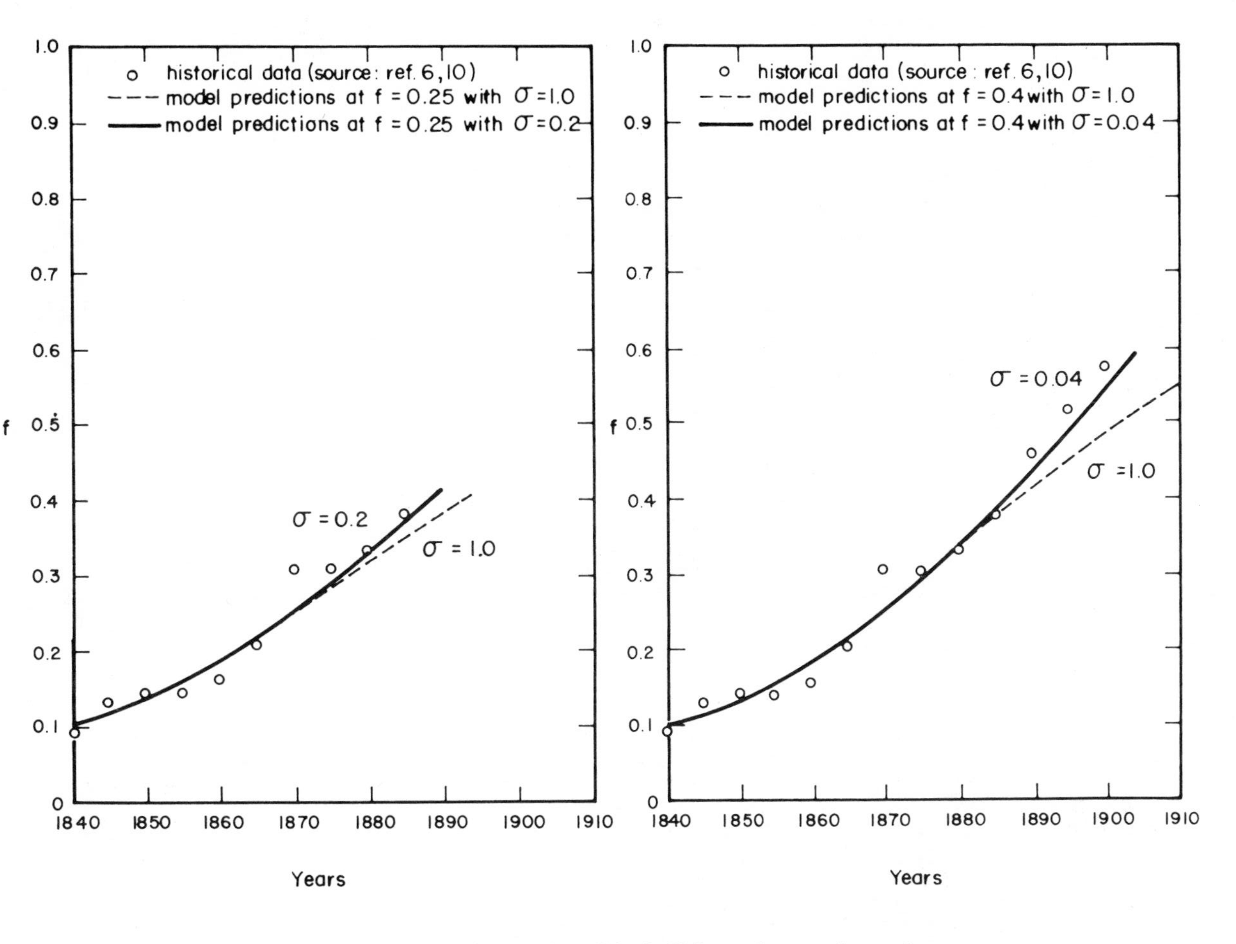

Fig. 8. Substitution of power for sail in the U.S. merchant marine market.

The empirical curves for determining the effective life span and the delay coefficient are only to provide some guides. The decision-maker still has the option to use his judgement in selecting a particular value by considering other environmental factors.

References

1. Blackman, A. W., A mathematical model for trend forecasts, *Technol. Forecast. Soc. Change* **3**, 441–452 (1972).
*2. Blackman, A. W., The market dynamics of technological substitutions, *Technol. Forecast. Soc. Change* **6**, 41–63 (1974).
3. Bright, J. E., *A Brief Introduction to Technological Forecasting,* Concepts and exercises, 3rd edition, Pemaquid Press, Texas, 1974.
4. Fisher, J. C., and Pry, R. H., A simple substitution model of technological change; *Technol. Forecast. Soc. Change,* **3**, 75–88 (1971).
5. Floyd, A., Trend Forecasting: A methodology for figure of merit, in *Proceedings of the First Annual Technology and Management Conference,* (J. Bright, Ed.), Prentice-Hall, Englewood Cliffs, New Jersey, 1968.
6. *Historical Statistics of the United States-Colonial Times to 1975,* A Statistical Abstract Supplement Prepared by the U.S. Department of Commerce, Bureau of the Census.
7. Lenz, R. C., and Lanford, H. W., The substitution phenomenon, *Business Horizons* **15** (1), 63–68 (1972).
8. Martino, J. P., *Technological Forecasting for Decision Making,* American Elsevier Publishing Co., New York, 1972.
9. Sharif, M. N., and Uddin, G. A., A procedure for adapting technological forecasting models, *Technol. Forecast. Soc. Change,* **7**, 99–106 (1975).
10. *Statistical Abstract of the U.S. Department of Commerce,* 93rd edition (1972) and earlier editions.

**This article also appears in Part 2 of this book.*

System Dynamics Modeling for Forecasting Multilevel Technological Substitution

M. NAWAZ SHARIF and CHOWDHURY KABIR

ABSTRACT

The existing mathematical models of trend extrapolation for forecasting technological substitution deal primarily with cases where a new product or technology replaces over a period of time the old one in at least some portion of the market. However, in the present era of rapid technological change, one can think of many examples where a particular product or technology is replacing an older one while at the same time being replaced by a newer one—a multilevel substitution process. Moreover, any forecast made at a given point in time needs to be revised as varying circumstances influence the elements of the forecast. This paper presents a systematic methodology (i) for forecasting multilevel technological substitution, and (ii) for incorporating various forms of time dependent parameters in the existing trend extrapolation models. The methodology is based on the "system dynamics" technique and combines the exploratory and the normative approaches to technological forecasting. In addition to the forecasting of market share, the models can also predict the actual size of the market of each of the competing technologies or products under different assumptions about the expansion of the joint market based on past trend as well as future anticipations.

Introduction

More and more the industrialists of the present world of technological revolution are finding themselves in a position where their recently introduced products that are supposed to capture over a period of time a substantial share of the market from existing older products very soon start to lose their own market to subsequently introduced newer products. Many examples can be cited from the electronics and other manufacturing industries where this kind of multilevel substitution of products or technologies are taking place; such examples are tube–transitor–integrated circuit in electronic devices, tape–cassette–cartridge players for sound reproduction, steam–diesel–electric locomotives, wood–steel–aluminium–plastic as building materials, piston–turboprop–jet engines in aircraft, etc. As a result of simultaneous substitution the effective life spans of the technology-intensive products are becoming very short and the costs of miscalculation of technological obsolescences are increasing exponentially. The need for appropriate models for forecasting multilevel technological substitution is, thus, very obvious. In addition, the increasing interaction of the elements of market dynamics calls for forecasting models that are adaptive to changing circumstances.

Technological forecasting on the basis of trend extrapolation implies the acceptance of the assumption that things are going to change in the future more or less the same way as

This article appeared in *Technological Forecasting and Social Change*, Vol. 8, No. 4, 1976.

it did in the past. This assumption is normally valid, because there exist a very large number of diversified factors, all interacting and jointly contributing to the process of change—some factors accelerating the process, others retarding it, such that in a normal situation the overall tendency deviates very little from the extrapolated central trend. However, when there is a significant change in the circumstances influencing the elements of the forecast, the assumption of "everything remaining the same" needs to be modified. Otherwise, there is very little chance that the forecast and reality will coincide. The recent "oil crisis" is one such example that has perhaps made invalid all forecasts regarding the substitution of natural fibers by oil-based synthetic fibers. A practical forecasting model, therefore, has to be dynamic in nature and responsive to exogeneous influences.

There are many mathematical models available for forecasting the substitution of one technology or product for another on the basis of trend extrapolation. However, the authors are not aware of any model that deals with the multilevel substitution process. Discussions on the simpler methods, namely those of Ayres-Noble-Overly [4], Blackman [7], Fisher-Pry [14], Floyd [16], and Mansfield [24] are available in most of the Technological Forecasting literature (see Ayres [3], Bright [9], Bright-Shoeman [10], Cetron-Ralph [12], Jantsch [23], and Martino [26]). All of these models use a simple curve-fitting technique to project a function having the appropriate S-shaped curve, using historical data to determine the parameters of the function. Sharif-Kabir [29] proposed a generalized version of these simple substitution models and suggested certain measures for improving the reliability of a forecast. A more elaborate market substitution model has been put forward by Stern [30] in which the total market has been subdivided on the basis of different uses of the product and price-utility factors were also incorporated. As the author himself has pointed out, the applicability of the Stern model is considerably limited by its demand for a good deal of technical as well as economic data.

Besides the models being for one-to-one substitution, the parameters used in the models mentioned in the previous paragraph are mostly static in nature. Although, Blackman [8] and Mansfield [24] allow for parametric variations, these models are not capable of incorporating systematic deviation over time due to some significant change in the exogeneous factors that influence the course of substitution. Therefore, there is a need for some simple (requiring less data), dynamic (in terms of parameters) and adaptive (to changes in the environment) models for forecasting technological substitution.

The purpose of this paper is twofold. Firstly, to describe the market dynamics of a technology-based product that is substituting one or more older products and at the same time being substituted by one or more newer products, all of which satisfy a particular consumer need. Secondly, to develop a systematic methodology for incorporating the dynamic as well as adaptive features in a general model of trend extrapolation.

Because of its simplicity as well as generality, the Sharif-Kabir model [29] is taken as the starting point for this research. When the model is extended to fully describe the rates and patterns of substitutions under the multilevel substitution conditions, and when the parameters of the model is made to vary with time, the derivations become complex. Computer based simulation methods in general and Forrester's "system dynamics" technique [17, 18] in particular can be of much help for such complicated situations. Blackman has introduced the system dynamics approach in the field of Technological Forecasting [5, 6]. However, he did not specifically work on substitution models. In this paper the system dynamics technique is used for the development of forecasting procedures.

The discussions that follow are divided into two parts. In the first part, the multilevel substitution case is discussed; this includes conceptualization, procedure development, and an illustrative example. The second part includes the development of a dynamic simulation procedure for making forecasting models adaptive to exogeneous influences. The procedure combines the exploratory and the normative approaches to technological forecasting. An illustrative example is also cited in this part.

1. Multilevel Substitution

Before getting involved in the forecasting procedure, a clear conceptualization of the problem is necessary. Figure 1 illustrates different phases of a technological substitution process. Let the outer circles represent 100% of the total market corresponding to the satisfaction of a particular consumer need at a particular time. With the passage of time the total market changes both in terms of volume and constituents. Now, suppose at time $t = t_1$ there is one product P1 in the market to satisfy the need. After some time, when $t = t_2$ (where $t_2 > t_1$), suppose that a second product P2 comes into the market that can satisfy the same need with some added advantages. Since we know that if the new product is economically viable after it has gained a small market share, it is likely to become more competitive as time progresses; therefore, once a substitution has begun, it is highly probable that it will eventually take over the available market. But before P2 completely replaces P1, suppose that at time $t = t_3$ (where $t_3 > t_2 > t_1$) a third product P3 enters the market. Now the situation is a little complex. Product P1 will lose its market to both P2 and P3 as they are newer than and technically superior to P1. P3, on the other hand, will gain market from both P1 and P2, because of its improved characteristics. The intermediate product P2 is likely to continue to gain market from P1 (note that P1's position is also changing because of P3) while at the same time losing its own market to P3. This is a case of multilevel substitution. Figure 1 also shows the case where there are four products in the market. In general, there can be N number of products in the market.

If there are N products in the market at a time t to satisfy a particular consumer need with the products moving into the market in a definite order starting from P1 to PN corresponding to the stages of technological development, then the principles of substitution can be stated as follows:

1. The oldest product in the market P1 being technologically least advanced will lose its market share to all other products (P2 to PN).
2. The newest product in the market PN being technologically most advanced will substitute over a period of time all other products (P1 to PN−1).
3. Any intermediate product PX (where $2 \leqslant X \leqslant N-1$) will substitute older products (P1 to PX−1) while at the same time being substituted by newer products (PX+1 to PN).

FORECASTING PROCEDURE FOR MULTILEVEL SUBSTITUTION

Under the multilevel substitution conditions, the products for which market share forecasts are desired can be classified into three groups corresponding to the principles of substitution as outlined in the previous section. For the two cases of P1 and PN the procedure is very simple. The case of P1 vs Others as well as the case of Others vs PN can be considered as cases of one-to-one substitution and forecasts can be obtained by using any of the existing models. The authors suggest the use of Sharif-Kabir model [29] for its adaptability under various circumstances. The model is given by

$$\ln \frac{f}{1-f} + \sigma \frac{1}{1-f} = C_1 + C_2 t$$

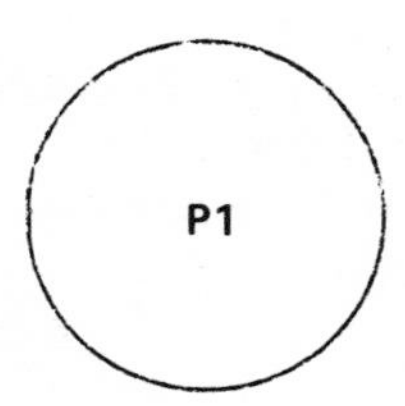

t = t_1 : A single product P1 is satisfying a particular consumer need.

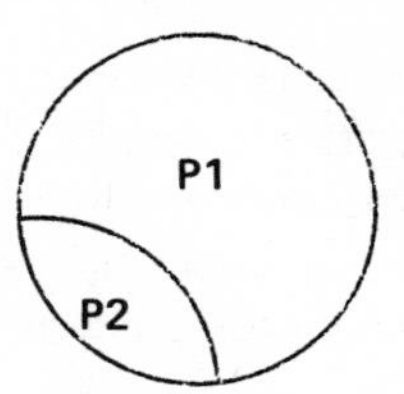

t = t_2 : A second product P2 is substituting the older product P1. (one-to-one substitution)

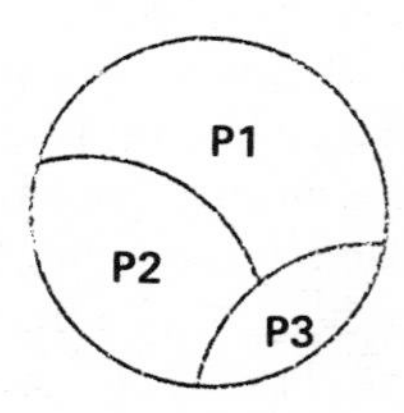

t = t_3 : A third product P3 is substituting older products P1 and P2; P2 is substituting older product P1 but also being substituted by newer product P3; P1 is being substituted by later products P2 and P3.

(Multilevel substitution)

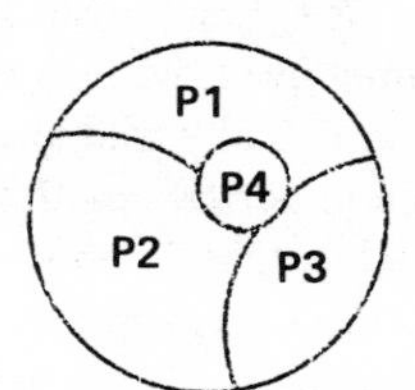

t = t_4: A fourth product P4 is substituting P1, P2 and P3; P2 is substituting P1 but being substituted by P3 and P4; P3 is substituting P1 and P2 but being substituted by P4; P1 is being substituted by P2, P3 and P4.

(Multilevel substitution)

Fig. 1. Schematic diagrams showing the process of multilevel substitution.

where

f = market share of a product at time t,
σ = delay coefficient, $0 \leqslant \sigma \leqslant 1$,
t = time,
C_1, C_2 = constants.

Notes to Eq. (1):

(a) In the case of P1, $f_{\text{P1}} = 1 - f_{\text{OT}}$, where f_{OT} = sum total of the fraction of market share belonging to all newer products (P2 to PN); and in the case of PN, $f_{\text{P}N}$ = market share of the newest product (PN).

(b) The delay coefficient σ is dependent upon the nature of data, expected life cycle and other exogeneous factors [29]. If a stepwise procedure is followed, $\sigma = 0$ may give reasonably accurate forecast.

In the case of intermediate products (PX) the following analytical procedure can be used:

(1) Group all products P1 to PX–1 together, calling it PG1. Forecast the market share of PG1 vs Others using Eq. (1) in the same way as in the case of P1.

(2) Group all products PX+1 to PN together, calling it PGN. Forecast the market share of Others vs PGN using Eq. (1) in the same way as in the case of PN.

(3) Forecast the market share of PX using the following relationship

$$f_{\mathrm{PX}} = 1 - f_{\mathrm{PG1}} - f_{\mathrm{PG}N}$$

The procedure outlined above gives a fairly accurate overall estimation of the trend as long as the principles of substitution are valid. If one is interested in finding out the different rates of substitution and the characteristics of the total market of individual products in addition to its market share, the procedure needs to be modified. A procedure using the computer-based simulation technique of system dynamics developed by Forrester [17, 18] is given next for this purpose. The procedure has been developed for three products only as any higher order problem can be very easily reduced to a number of three product problems.

Figure 2 shows a very simplified system dynamics diagram for a three product multilevel substitution situation. (For a complete list of system dynamics notations, symbols, equations, flow charting, etc., see reference [28]). The three products P1, P2 and P3 (in the order of their appearance in the market) are shown as "levels". $R_{a \cdot b}$ indicates the "rate" at which the older product a is being substituted by the newer product b, and the arrows indicate the direction of gain in market share.

The problem is to find the rate equations corresponding to $R_{1 \cdot 2}$, $R_{1 \cdot 3}$ and $R_{2 \cdot 3}$. The rate of change in the market share in the case of one-to-one substitution can be obtained from Eq. (1) as

$$\frac{df}{dt} = \frac{C_2 f (1-f)^2}{\sigma f + (1-f)} \tag{2}$$

When there are only two products in the market such that product b is substituting product a, then the corresponding rate equation can be written as

$$R_{a \cdot b} = \frac{C_{2a \cdot b}\, f_b\, (1-f_b)^2}{\sigma_{a \cdot b}\; f_b + (1-f_b)} \tag{3}$$

In order to use Eq. (3) for computing the rates $R_{1 \cdot 2}$, $R_{1 \cdot 3}$ and $R_{2 \cdot 3}$, the system given in Fig. 2 is first of all partitioned and grouped as shown in Fig. 3. Note that products P2 and P3 can be grouped together, calling it P23, without violating the principles of substitution as both P2 and P3 are newer products relative to P1 and they together (P23) account for the loss of market share of P1. Similary, P1 and P2 can be grouped together, calling it P12 as both P1 and P2 together account for the gain in market share of P3. Eq. (3) can then be used to find the rates $R_{1 \cdot 23}$ and $R_{12 \cdot 3}$. Now the problem is to find a way for dividing the group contributions to its components.

Let us define two new variables (See Fig. 3 for definitions of $R_{a \cdot b}$ and $R_{a \cdot b(c)}$)

$$W1 = \frac{R_{1 \cdot 3(2)}}{R_{1 \cdot 3(2)} + R_{1 \cdot 2(3)}} \tag{4}$$

$$W2 = \frac{R_{1 \cdot 3(2)}}{R_{2 \cdot 3(1)} + R_{1 \cdot 3(2)}} \tag{5}$$

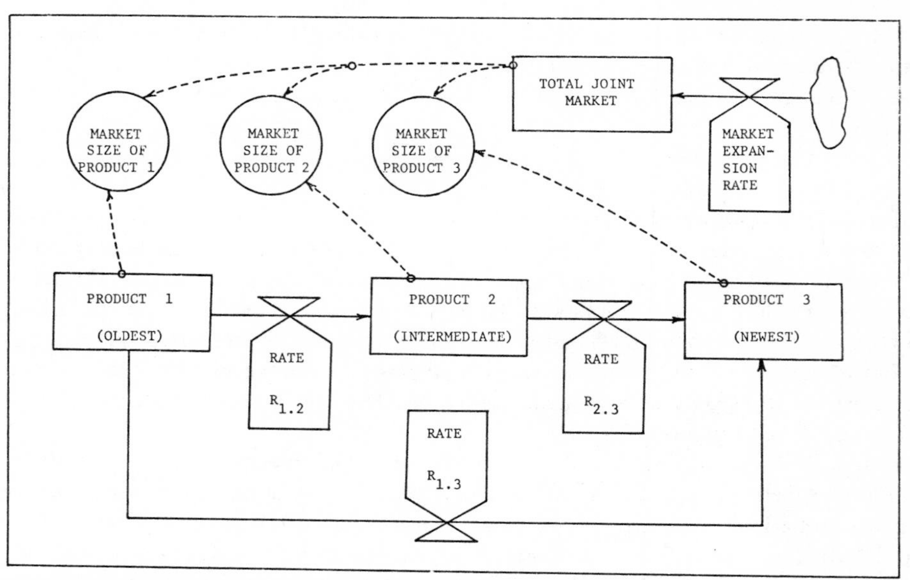

Fig. 2. Simplified system dynamics diagram showing the basic structure of the multilevel substitution model.

which are weightage factors for determining $R_{1\cdot3}$ as follows

$$R'_{1\cdot3} = R_{1\cdot23} \frac{R_{1\cdot3(2)}}{R_{1\cdot3(2)} + R_{1\cdot2(3)}} \tag{6}$$

$$R''_{1\cdot3} = R_{12\cdot3} \frac{R_{1\cdot3(2)}}{R_{2\cdot3(1)} + R_{1\cdot3(2)}} \tag{7}$$

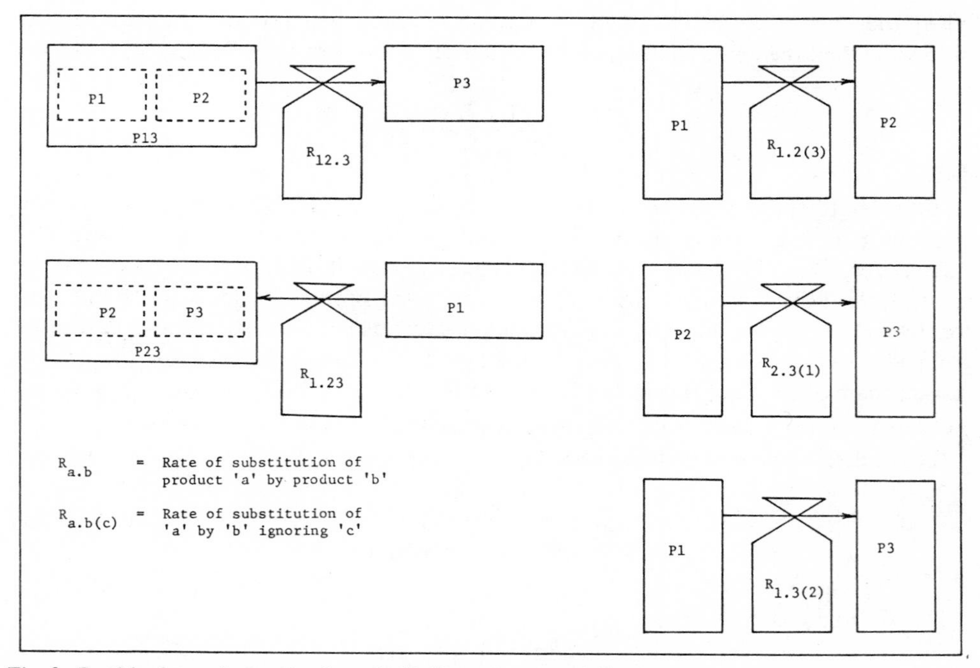

Fig. 3. Partitioning and grouping for substitution rate computation.

Ideally, $R'_{1 \cdot 3}$ should be equal to $R''_{1 \cdot 3}$. However, some discrepancy will occur during the starting and completion stages of substitution. Somewhere in the middle of the substitution process $R'_{1 \cdot 3} = R''_{1 \cdot 3} = R_{1 \cdot 3}$. When there is a discrepancy, as a rule of thumb use the smaller value as this will ensure the nonreversibility of the direction of gain in market share.

Once $R_{1 \cdot 3}$ has been determined, the other rate values can be obtained very easily from the following relationships

$$R_{1 \cdot 23} = R_{1 \cdot 2} + R_{1 \cdot 3} \tag{8}$$

$$R_{12 \cdot 3} = R_{1 \cdot 3} + R_{2 \cdot 3} \tag{9}$$

Regarding the growth of the total market, it has been observed that in many cases the total market grows exponentially. This growth can be represented as

$$M(t) = M_0 e^{Kt} \tag{10}$$

where

$M(t)$ = total market at time t,

M_0 = initial value of $M(t)$,

t = time,

K = constant.

If the time series data of total joint market reveals a growth pattern other than exponential, it can be very easily incorporated using the system dynamics technique. Hence the form of Eq. (10) may vary from case to case. Once a forecast of the total market is made, the market of individual products are obtained by dividing the total market according to the forecasted market share of each product. The complete system dynamics flow diagram is shown in Fig. 4, and the computer program using DYNAMO compiler [22] is given in Appendix 1. Appendix 1 also describes the variables and their interrelationships corresponding to Fig. 4.

ILLUSTRATIVE EXAMPLE OF MULTILEVEL SUBSTITUTION

Members of the International Air Transport Association (IATA) are using a wide variety of aircraft for their domestic and international flights. On the basis of the propulsion system, these aircraft can be classified into three groups—(i) piston, (ii) turbo-prop, and (iii) turbo-jet—in the order of technological development that has successfully entered the commercial aero-industry. The three way substitution problem among piston, turbo-prop and turbo-jet aircraft is used here to illustrate the multilevel forecasting procedure outlined in this paper.

Historical data has been taken from reference [31]. The time of forecast is 1970. Considering the suggestions put forward in reference [29] regarding corrective measure for improving the accuracy of the forecasts, the value of σ in Eq. (1) has been estimated to be zero for this analysis. Equation (1) was used to compute the initial values and TABLE functions for C_2 of the DYNAMO program as indicated in Fig. 4 and Appendix 1 by regression analysis on the historical data using the partitioning and grouping arrangements shown in Fig. 3. The simulation procedure is carried out using Eqs. (3)–(10) and following the sequence as shown in Fig. 4.

The computer program given in Appendix 1 was run on CDC 3600 using a special

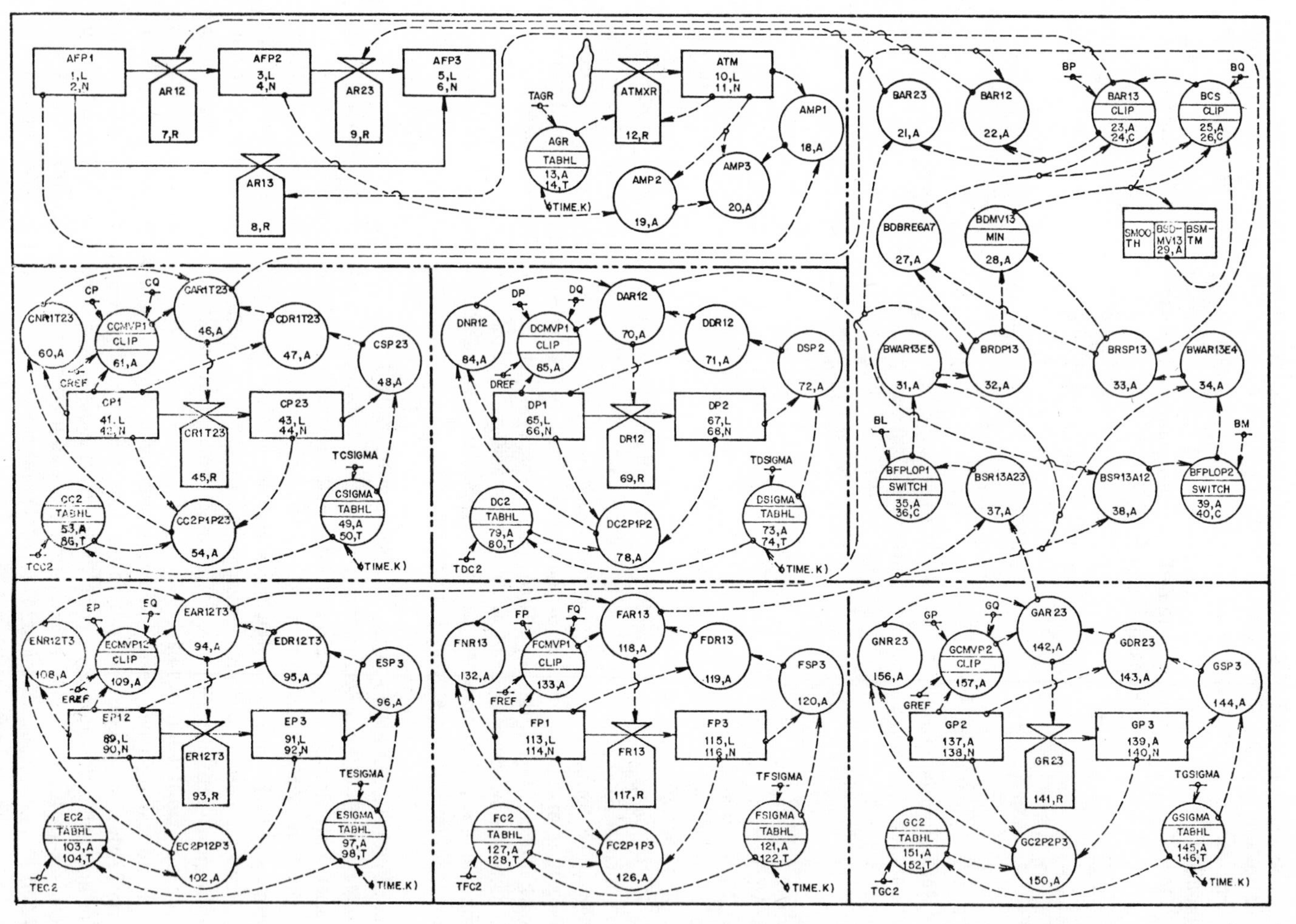

Fig. 4. Complete system dynamics flow chart.

purpose DYNAMO compiler [22]. The outputs are given in Figs. 5(a), 5(b) and 5(c). From these figures the forecasted behavior of a multilevel substitution process can be observed.

2. Dynamic Modelling

Let us go back to Eq. (2) which gives the rate of substitution of the old product (called the defender) by the new one (called the challenger). For any forecast, the delay coefficient σ is a constant, the value of which is selected by the analyst on the basis of the period of forecast, effective life span, data extent and scatteredness, and the point of forecast (see reference [29] for detailed discussions on σ). Therefore, with C_2 also being constant in Eq. (2), we can observe that the rate of substitution at any time t will lie between the following limits

$$C_2 f(1-f)^2 \leqslant \frac{df}{dt} \leqslant C_2 f(1-f) \tag{11}$$

Let us now consider the parameter C_2 to be a time dependent function designated by $\beta(t)$. Equation (2) is thus changed to

$$\frac{d}{dt} f(t) = \frac{\beta(t)\, f(t)\, [1-f(t)]^2}{\sigma f(t) + [1-f(t)]} \tag{12}$$

where

$f(t)$ = market share of the challenger at time t,

$\beta(t)$ = instantaneous rate of change at time t,

σ = a constant, and t = time.

The term $\beta(t)$ is fixed in market structure, but changing in value at a certain rate depending upon its functional form. The proper functional form of $\beta(t)$ can be determined mathematically by identifying the trend of C_2 using historical data up to different points in time with selected values of σ substituted in Eq. (1).

Once the functional form of $\beta(t)$ has been determined, Eq. (12) can be integrated to obtain the forecasting model. However, with even a very simple mathematical expression for $\beta(t)$ the analysis becomes cumbersome. Therefore, the simulation procedure using the system dynamics technique is suggested whereby even irregular functions can be handled very easily.

Before going into the simulation procedure, let us see how one can make the model adaptive to exogenous influences. So far, the model has been formulated using an exploratory approach only. Now, suppose there occurs a significant change in the environment that is likely to affect the substitution pattern forecasted by the trend extrapolation method. Normative approach with subjective judgments can be used to make the forecast adaptive to such circumstances.

As can be observed from Eqs. (2) and (12), the coefficient σ provides an effective mechanism to vary the rate of substitution in either direction—acceleration or retardation. Therefore, instead of assuming σ to be constant for the forecasted period, let us consider it to be a time dependent parameter designated at $\lambda(t)$. Hence, Eq. (12) further changes to

$$\frac{d}{dt} f(t) = \frac{\beta(t) f(t) [1-f(t)]^2}{\lambda(t) f(t) + [1-f(t)]} \tag{13}$$

where $\lambda(t)$ = dynamic response coefficient at time t, and $\beta(t)$, $f(t)$, t as defined earlier.

The dynamic response coefficient $\lambda(t)$ will determine the probably path of the substitution curve under the influence of some exogeneous factor. The prediction of the value of this parameter is a matter of subjective judgment. However, considering the viewpoints advanced in [29] for stepwise forecasting, and also due to the presence of "inertia of trend curves" [3] it may be reasonable to assume that the value of $\lambda(t)$ will be between zero and one. Moreover, given the complexity of market dynamics, the changes in $\lambda(t)$ are more likely to be gradual and continuous. Given these assumptions, $\lambda(t)$ can be explicitly determined by some form of an intuitive method, such as: Delphi, Cross-impact analysis, Scenario, etc. (See references Abt et al. [1], Amara et al. [2], Bright [9], Bright-Schoeman [10], Cetron-Ralph [12], Dalkey [13], Gerardin [19].) A preferable method will be to generate a number of scenarios and develop forecasts corresponding to

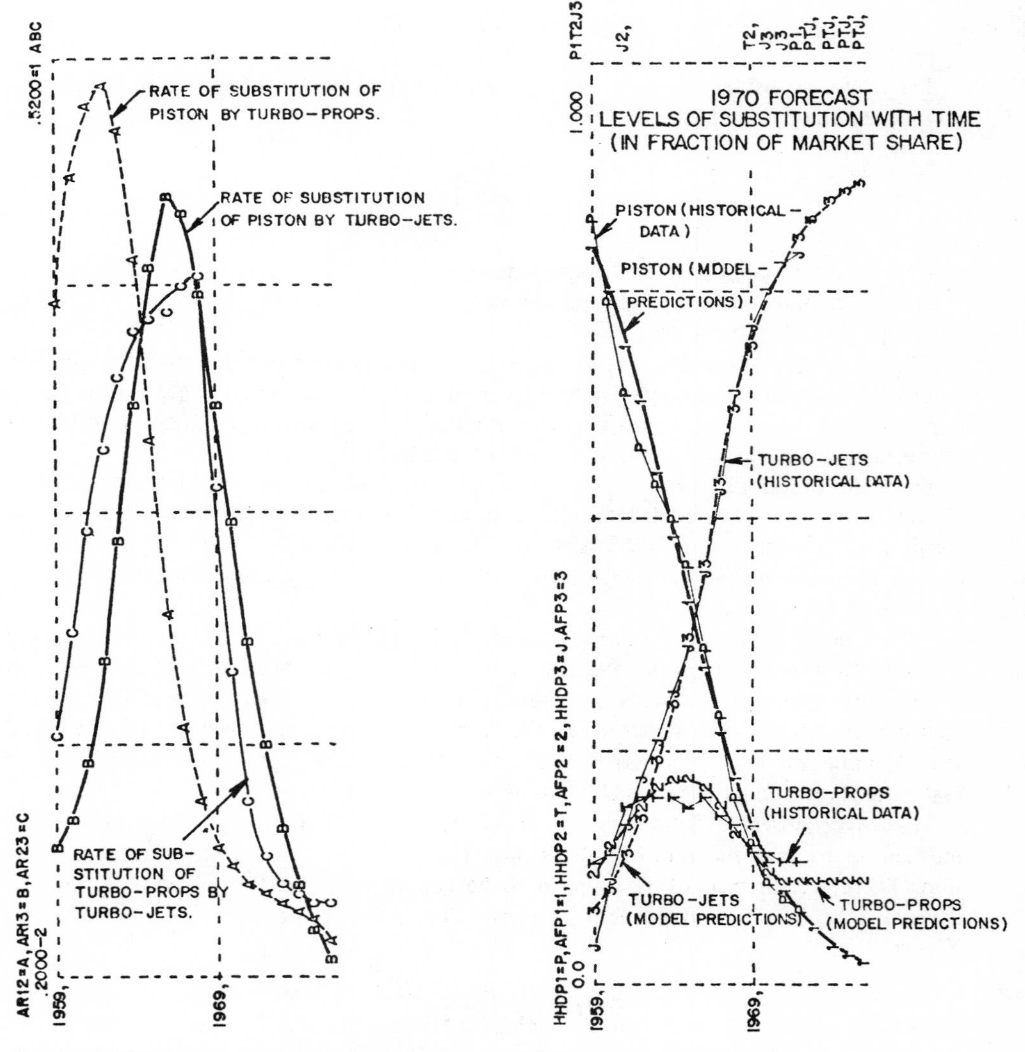

Fig. 5. Multilevel forecasts of rate of substitution, market share and market size.

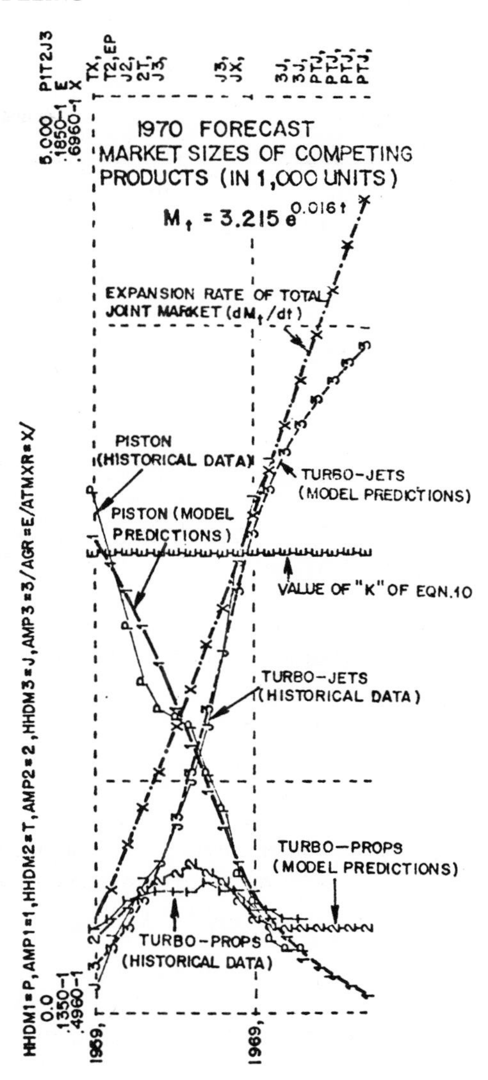

Fig. 5 (cont.). Part (c).

each of these scenarios so as to give the decisionmaker a guide to evaluate different possible consequences.

DYNAMIC SIMULATION PROCEDURE

Using the system dynamics technique, the substitution process can be depicted by a simplified diagram as shown in Fig. 6. Although Fig. 6 is a highly simplified diagram, it shows the interrelationships among the variables contributing to the process of technological substitution.

The arrows (solid line) indicate the direction of market gain, the dotted lines indicate interaction and the direction of information flow for simulation. The market share of the defending product gradually decreases from 100% to 0% as the market share of the

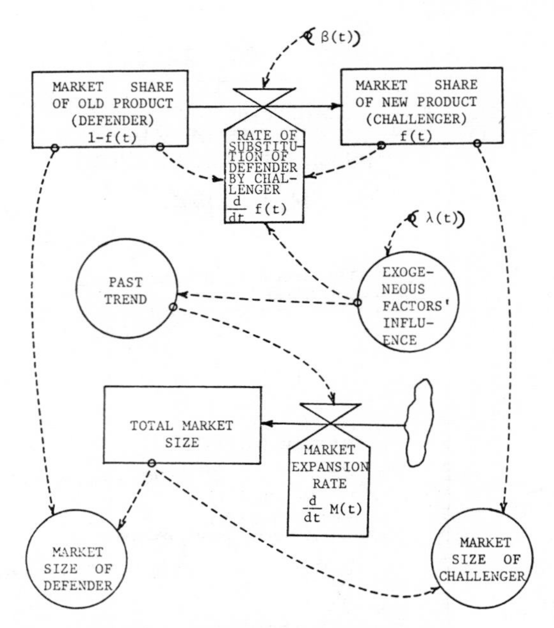

Fig. 6. A simplified system dynamics diagram of technological substitution.

challenging product gradually increases from 0% to 100%, the rate of substitution being given by Eq. (13). The substitution rate is dependent upon the fraction of the market already substituted, the fraction of the market share remaining to be substituted, the instantaneous rate of change, and the dynamic response coefficient. On the other hand the joint total market is determined by the market expansion rate which is influenced by the past trend as well as exogeneous factors. The growth of the total market can be expressed as Eq. (10).

Figure 7 gives the complete system dynamics flow chart for the simulation procedure and the computer program is included in Appendix 2. For $\beta(t)$ and $\lambda(t)$ the procedure can use TABLE functions derived from actual functional relationships. Figures 8 and 9 represent typical examples of $\beta(t)$ and $\lambda(t)$ curves and their conversion into TABLE functions.

ILLUSTRATIVE EXAMPLE

The example selected for illustration is the substitution of Jute by Polypropylene (PP). PP is an oil-based synthetic fiber that competes directly with jute products, such as, carpet backing and jute bags. The unexpected quadruple increase of the crude-oil price in late 1973 has put PP in a disadvantageous position vis-a-vis jute price. However, the increased cost of transportation and world-wide unprecedented inflation is definitely affecting the jute market as well. Under these conditions Eq. (13) is to be used to simulate a reasonable forecast.

In order to use Eq. (13), the instantaneous rate of change $\beta(t)$ and the dynamic response coefficient $\lambda(t)$ have to be specified explicitly. The functional equation of $\beta(t)$ is derived by fitting a curve through the points obtained for C_2 at different time t from the historical data using Eq. (1). The expression obtained for the Jute vs PP case is

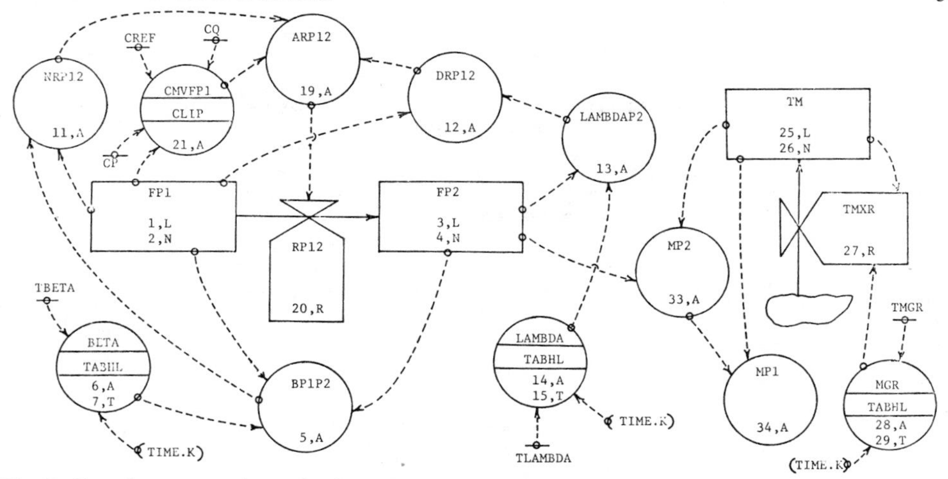

Fig. 7. Complete system dynamics flow diagram for dynamic modelling.

$$\beta(t) = 0.349\, e^{-0.014\, t} \tag{14}$$

In the case of $\lambda(t)$, it should be noted that for the past data, $\lambda(t)$ is nothing but the values of σ at different time t. For the future period, $\lambda(t)$ is to be determined by value judgment on the basis of the changing environmental conditions. For the purpose of illustrating the working principle of the model, three scenarios are considered here.

Scenario 1. It is predicted that the present oil crisis will put PP in a difficult position for a considerable period and jute producing countries will fully utilize this opportunity to develop an effective strategy to decrease the rate of penetration of PP in the jute market.

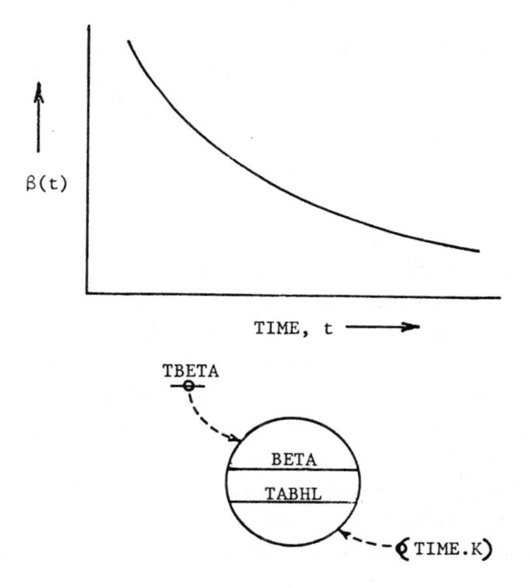

BETA.K = TABHL (TBETA, TIME.K, BN1, BN2, BN3)

Fig. 8. Functional relationship of $\beta(t)$.

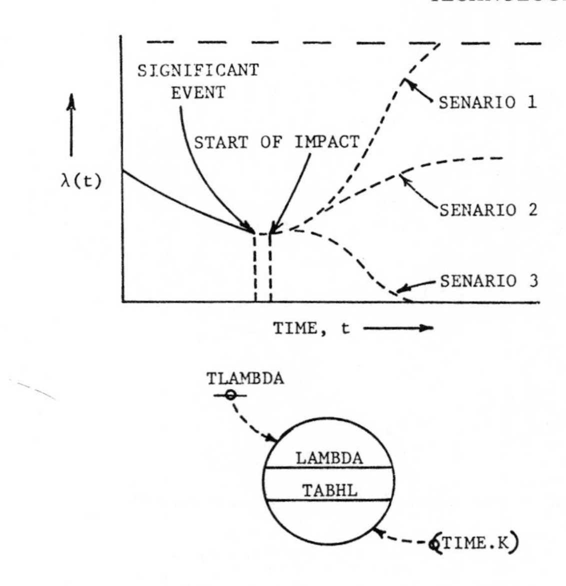

Fig. 9. Functional relationship of $\lambda(t)$.

Scenario 2. It may also happen that the present leverage of jute over PP gained through the oil crisis will be completely over-turned by the discovery of an alternative energy source that may cause a reduction in oil price.

Scenario 3. It is possible that the present food crisis in the principal jute growing countries, like Bangladesh and India, may force them to reduce the jute acreage in favor of rice such that it counterbalances the effect of the oil crisis.

The model forecasts corresponding to these scenarios are shown in Fig. 10, 11 and 12. The historical data are taken from references [20, 21, 27]. The forecast for the years 1980, 1985 and 1990 are circled in the diagrams to show their relative positions corresponding to the three scenarios.

Conclusion

The multilevel forecasting methodology presented in this paper is based upon the commonly agreed "principle of substitution" which may be stated "as a new-technology product enters the market to compete with existing product(s), it can gain consumer acceptance only if it is in a relatively advantageous position in terms of technological and economical aspects, and once the substitution has begun it is highly probably that it will eventually take over the available market". It has been pointed out that whenever there are more than two products in the market to satisfy a particular consumer need, except for the oldest and the newest products, all intermediate products undergo a process of multilevel substitution—they gain market from the older product(s) while at the same time lose market to the new product(s)—following the above principle.

Irrespective of the number of products that are competing in the market simultaneously, the oldest product will lose its market share to all other newer products and the newest product will gain market share from all other older products, and their substitution process will follow the familiar S-shaped pattern. However, to forecast the market

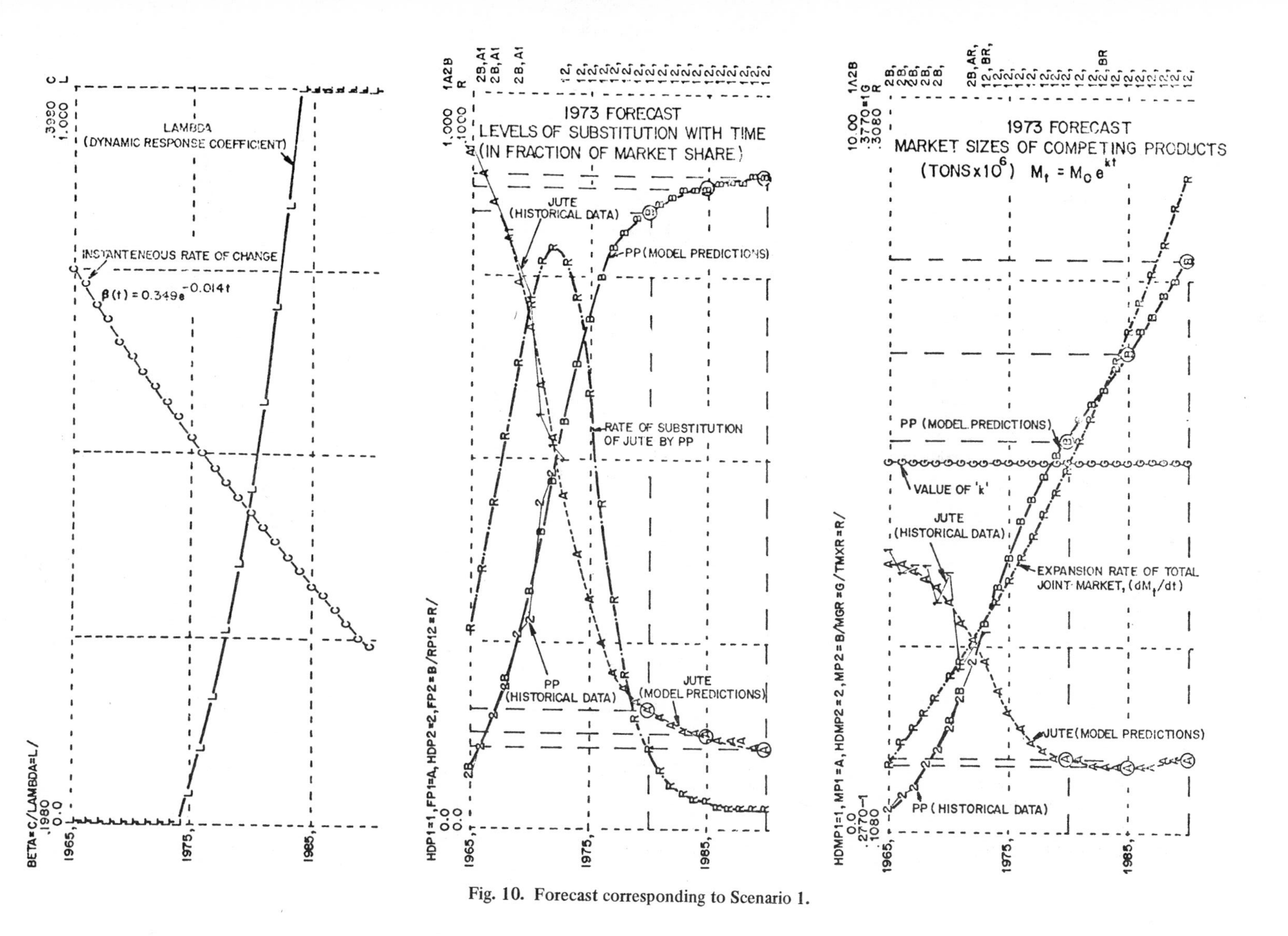

Fig. 10. Forecast corresponding to Scenario 1.

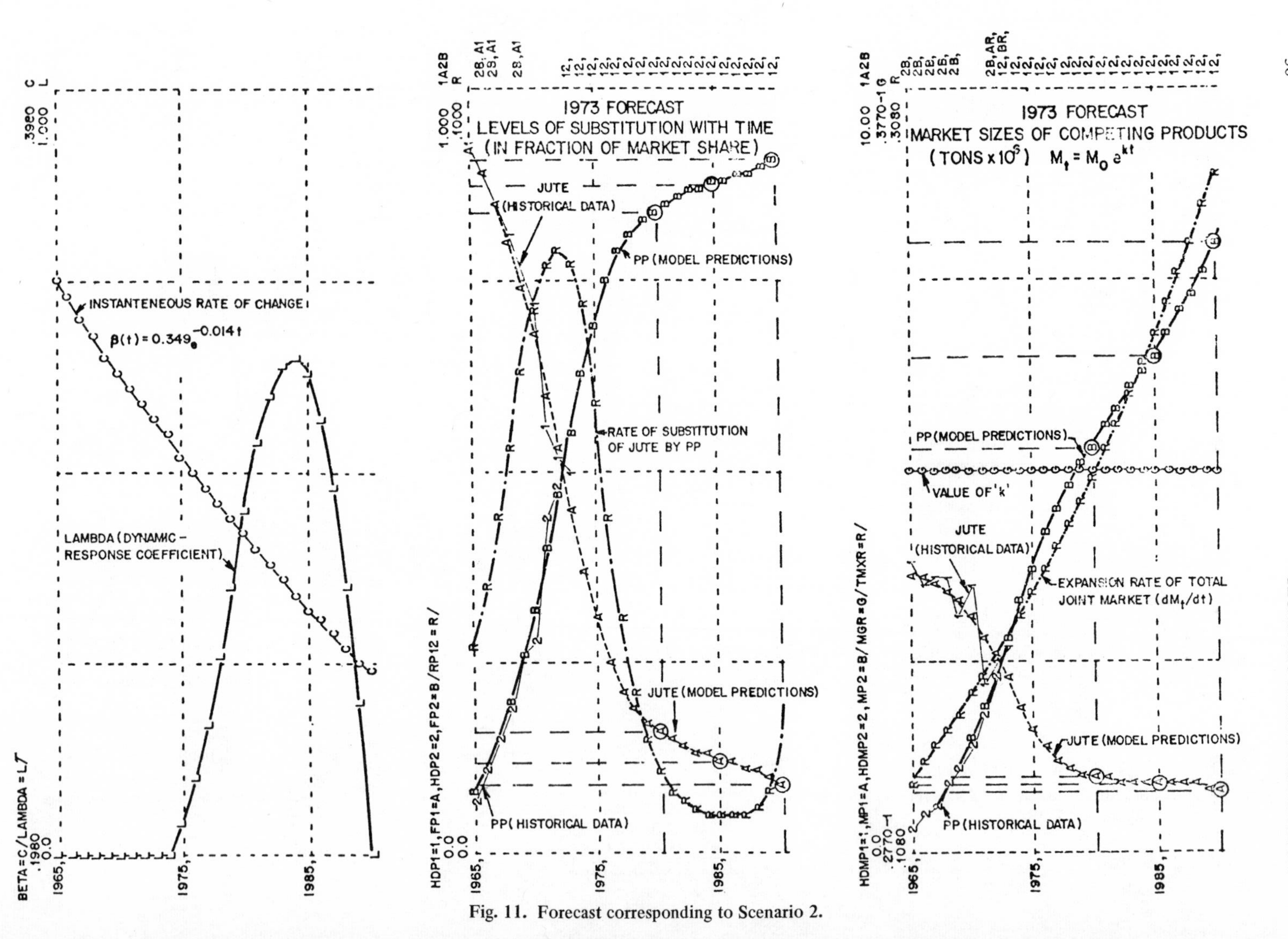

Fig. 11. Forecast corresponding to Scenario 2.

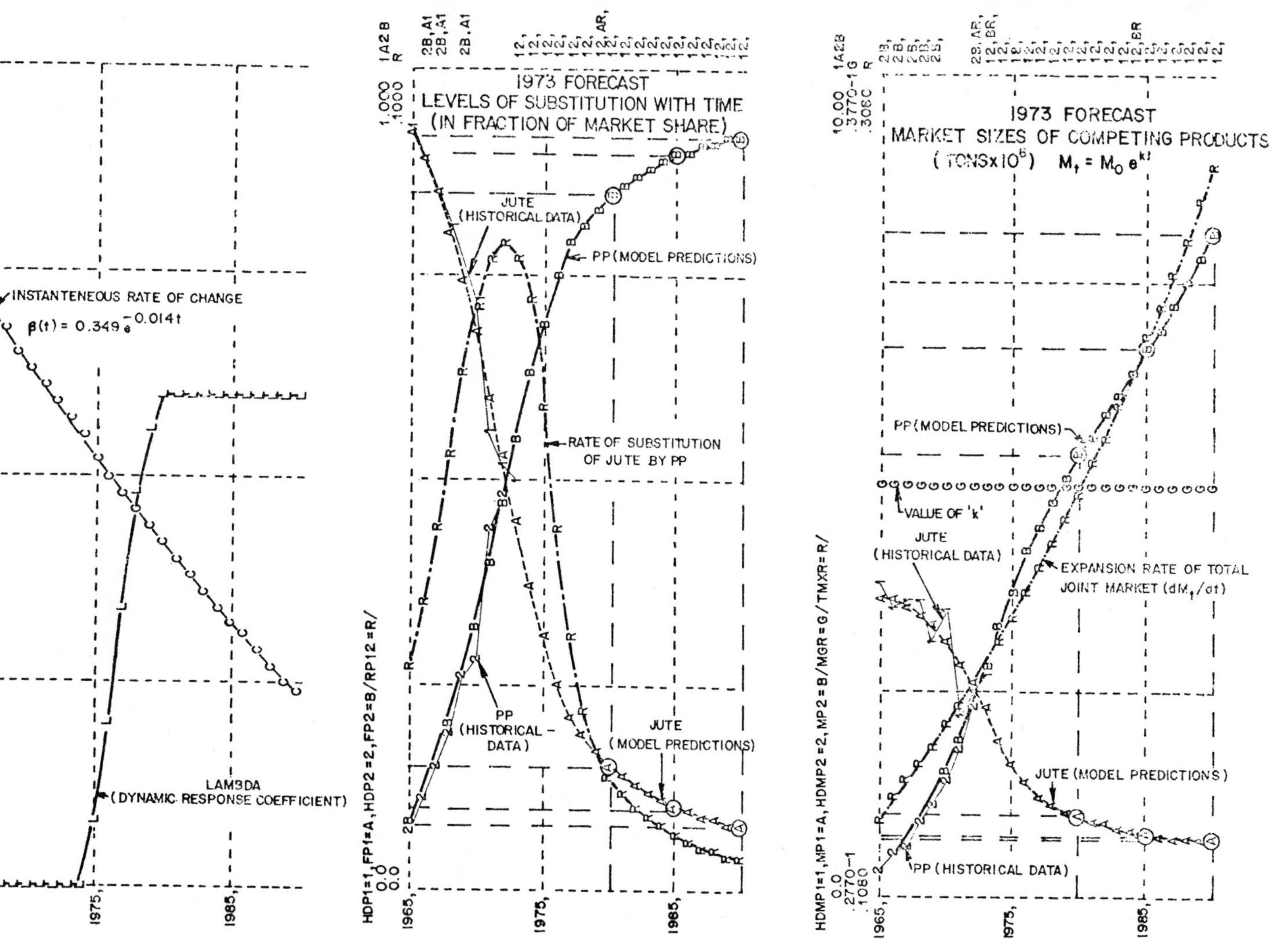

Fig. 12. Forecast corresponding to Scenario 3.

share of any of the intermediate products over time, one needs to convert the problem into an equivalent three-product system and follow a procedure similar to the one presented in this paper.

The second half of this presentation showed how an existing model can be made to incorporate the effect of some significant change over time in the exogeneous factors that influence the course of substitution.

The procedures developed in this paper are conceptually very simple and require a minimum amount of data. However, the use of the system dynamics technique provides a lot of other information as well as a mechanism for manipulation and sensitivity analysis for better decisionmaking. Moreover, when economic data are available the system dynamics modelling can easily take into account the variations due to economic factors.

References

1. Abt, C. C., Foster, R. N., and Rea, R. H., A scenario generating methodology in *A Guide to Practical Technological Forecasting* (Bright and Schoeman, Eds.), Prentice-Hall, Englewood Cliffs, New Jersey, 1973, pp. 199–214.
2. Amara, R. C., and Lipnski, A. J., Some views on the use of judgement, *Technol. Forecast. Soc. Change* **5**, 233–242 (1973).
3. Ayres, R. U., *Technological Forecasting and Long-Range Planning*, McGraw-Hill, New York, 1969.
4. Ayres, R. U., Noble, S., and Overly, D., Technological change as explicit factor of economic growth, in *Effects of Technological Change on and Environmental Implications of an Input-Output Analysis of the United States*, Part I, 1967–2020. (As cited in ref. [30]).
5. Blackman, A. W., The use of dynamic modelling for conditional forecasts of resourse allocation policies, *Technol. Forecast. Soc. Change* **6**, 191–208 (1974).
6. Blackman, A. W., Forecasting through dynamic modelling, *Technol. Forecast. Soc. Change* **3**, 291–307 (1972).
7. Blackman, A. W., A mathematical model for trend forecasts, *Technol. Forecast. Soc. Change* **3**, 441–452 (1972).
8. Blackman, A. W., The market dynamics of technological substitutions, *Technol. Forecast. Soc. Change* **6**, 41–63 (1974).
9. Bright, J. R. (Ed.), *Technological Forecasting for Industry and Government: Methods and Applications*, Prentice-Hall, Englewood Cliffs, New Jersey, 1968.
10. Bright, J. R., and Schoeman, M. E. F. (Eds.), *A Guide to Practical Technological Forecasting*, Prentice-Hall, Englewood Cliffs, New Jersey, 1973.
11. Cetron, M. J., and Monahan, T. I., An evaluation and appraisal of various approaches to technological forecasting in *Technological Forecasting for Industry and Government: Methods and Applications* (Bright, J. R., Ed.), Prentice-Hall, Englewood Cliffs, New Jersey, 1968, pp. 144–182.
12. Cetron, M. J., and Ralph, C. A.; *Industrial Applications of Technological Forecasting: Its Utilization in R&D Management*, Wiley-Interscience, New York, 1971.
13. Dalkey, N. C., An elementary cross-impact model; *Technol. Forecast. Soc. Change* **3**, 341–351 (1972).
14. Fisher, J. C., and Pry, R. H.; A simple substitution model of technological change, *Technol. Forecast. Soc. Change* **3**, 75–88 (1971).
15. Floyd, A., Trend forecasting: A methodology for figure of merit in *Technological Forecasting for Industry and Government: Methods and Applications* (Bright, J. R., Ed.), Prentice-Hall, Englewood Cliffs, New Jersey, 1968, pp. 95–109.
16. This form of the Floyd model is given in ref. [7].
17. Forrester, J. W., *Industrial Dynamics*, The M.I.T. Press, Cambridge, Mass., 1961.
18. Forrester, J. W., *Principles of Systems*, Wright Allen Press, Mass., 1968.
19. Gerandin, L., Study of alternative futures: A scenario writing method in *A Guide to Practical Technological Forecasting* (Bright and Schoeman, Ed.), Prentice-Hall, Englewood Cliffs, New Jersey, 1973, pp. 276–288.
20. Ghazi, A. U., An Investigation into the Prospect of Jute Manufacturing Industry of Bangladesh, Master of Engineering Thesis, Industrial Development and Management Division, Asian Institute of Technology, 1974.

21. Impact of Synthetics on Jute and Allied Fibres, *Commodity Bulletin Series,* No. 46, F.A.O., U.N., Rome, 1969.
22. Instructions for the use of DYNAMO, Manual prepared for the A.I.T. DYNAMO processor, Asian Institute of Technology, Bangkok, Thailand, 1974.
23. Jantsch, E., *Technological Forecasting in Perspective*, OECD, Paris, 1967.
24. Mansfield, E., Technical change and the rate of immitation, *Econometrica* **29** (No. 4), 741–765 (Oct. 1961).
25. Martino, J. P., The effect of errors in estimating the upper limit of a growth curve; *Technol. Forecast. Soc. Change* **4**, 77–84 (1972).
26. Martino, J. P., *Technological Forecasting for Decision Making*, Elsevier, New York, 1972.
27. *Production Yearbook*, FAO, Rome, Vol. 27, 1973 and earlier volumes.
28. Sharif, M. N. and Ghazi, A. U., A procedure for adapting technological forecasting models, *Technol. Forecast. Soc. Change* **7**, 99–106 (1975).
29. Sharif, M. N., and Kabir, C., A generalized model for forecasting technological substitution, *Technol. Forecast. Soc. Change* **8**, 353–364 (1976).
30. Stern, M. O., Ayres, R. U., and Shapanka, A., A model for forecasting the substitution of one technology for another, *Technol. Forecast. Soc. Change* **7**, 57–79 (1975).
31. *World Air Transport Statistics*, No. 17 (1972), International Air Transport Association, Switzerland, and earlier issues.

APPENDIX 1
Computer Program for Multilevel Substitution Model

```
NOTE
1) IN ORDER TO FOLLOW FIG.4 EASILY, THE ENTIRE FLOW
   DIAGRAM HAS BEEN DIVIDED INTO SEVEN SECTIONS. THE
   NAMING OF EACH SECTION HAS BEEN DONE BY ASSIGNING
   AN ALPHABET TO EACH ONE OF THEM.THE FIRST ALPHABET
   OF EACH VARIABLE OR CONSTANT REPRESENTS THE SECTION
   TO WHICH IT BELONGS. IT SHOULD BE NOTED THAT THE
   DIVIDING LINES OF THE SECTIONS HAVE NOTHING TO DO
   NEITHER WITH THE FORMULATION NOR WITH THE COMPUTATIONAL
   PROCEDURE OF THE MODEL.

2) THE NAMES AND ACRONYMS OF VARIABLES OF FIG. 4
   FOLLOW THE DYNAMO EQUATIONS.

3) IN CASE OF THREE-WAY SUBSTITUTION PROBLEM -
   PRODUCT 1 IS THE OLDEST,
   PRODUCT 2 IS THE INTERMADIATE AND
   PRODUCT 3 IS THE MOST RECENT.

4) DYNAMO EQUATIONS FOR PLOTING HISTORICAL DATA (AS
   SHOWN IN FIG. 5 ) ARE NOT INCLUDED IN THIS PROGRAM.

5) TO DETERMINE VALUES FOR SIGMA( DELAY COEFFICIENT )
   IN SECTIONS C,D,E,F AND G, IT IS SUGGESTED TO USE THE
   EMPIRICAL CURVES GIVEN IN REF. 29 UNLESS THE DECISION
   MAKER USES HIS OWN JUDGEMENT TO SELECT VALUES
   CONSIDERING VARIOUS EXOGENEOUS FACTORS.

6) NUMERICAL VALUES ASSIGNED TO DIFFERENT DYNAMO
   EQUATIONS PERTAIN TO THE ILLUSTRATIVE EXAMPLE IN
   PART ONE.


SECTION A

L    AFP1.K=AFP1.J-(DT)(AR12.JK+AR13.JK)                          1
        AFP1 - PRODUCT 1 (IN FRACTION OF TOTAL JOINT MARKET).
N    AFP1=.801                                                    2
L    AFP2.K=AFP2.J+(DT)(AR12.JK-AR23.JK)                          3
        AFP2 - PRODUCT 2 (IN FRACTION OF TOTAL JOINT MARKET).
```

```
N       AFP2=,122                                                         4
L       AFP3,K=AFP3.J+(DT)(AR23.JK+AR13.JK)                               5
           AFP3 - PRODUCT 3 (IN FRACTION OF TOTAL JOINT MARKET),
N       AFP3=,077                                                         6
R       AR12,KL=BAR12.K                                                   7
           AR12 - RATE OF SUBSTITUTION OF PRODUCT 1 BY
                  PRODUCT 2,
R       AR13,KL=BAR13.K                                                   8
           AR13 - RATE OF SUBSTITUTION OF PRODUCT 1 BY
                  PRODUCT 3 ,
R       AR23,KL=BAR23.K                                                   9
           AR23 - RATE OF SUBSTITUTION OF PRODUCT 2 BY
                  PRODUCT 3 ,
L       ATM,K=ATM,J+DT*ATMXR,JK                                          10
           ATM - TOTAL JOINT MARKET OF PRODUCTS 1,2,3 (QUANTITY),
N       ATM=3,215                                                        11
R       ATMXR,KL=ATM,K*AGR,K                                             12
           ATMXR - MARKET EXPANSION RATE OF ATM (QTY. / UNIT TIME),
A       AGR,K=TABHL(TAGR,TIME,K,AN1,AN2,AN3)                             13
           AGR - GROWTH RATE (PERCENT / UNIT TIME),
T       TAGR=,016/,016                                                   14
C       AN1=1959                                                         15
C       AN2=1976                                                         16
C       AN3=17                                                           17
A       AMP1,K=AFP1.K*ATM.K                                              18
           AMP1 - MARKET SIZE OF PRODUCT 1 (QUANTITY),
A       AMP2,K=AFP2.K*ATM.K                                              19
           AMP2 - MARKET SIZE OF PRODUCT 2 (QUANTITY),
A       AMP3,K=ATM,K-AMP1.K-AMP2,K                                       20
           AMP3 - MARKET SIZE OF PRODUCT 3 (QUANTITY),

           SECTION B

A       BAR23,K=EAR12T3.K-BAR13.K                                        21
           BAR23 - DETERMINES RATE FOR AR23.
A       BAR12,K=CAR1T23,K-BAR13,K                                        22
           BAR12 - DETERMINES RATE FOR AR12.
A       BAR13,K=CLIP(BCS,K,BDMV13.K,BDBRE6A7.K,BP)                       23
           BAR13 - DETERMINES RATE FOR AR13.
C       BP=-,03                                                          24
A       BCS,K=CLIP(BDMV13.K,BSDMV13.K,BDBRE6A7,K,BQ)                     25
           BCS - CONTROLS SMOOTHING.
C       BQ=,03                                                           26
A       BDBRE6A7,K=BRSP13.K-BRDP13.K                                     27
           BDBRE6A7 - DIFFERENCE BETWEEN RATES OF EQS. 6 AND 7,
A       BDMV13,K=MIN(BRDP13.K,BRSP13.K)                                  28
           BDMV13 - DETERMINES VALUE FOR RATE AR13,
A       BSDMV13,K=SMOOTH(BDMV13,JK,BSMTM)                                29
           BSDMV13 - SMOOTHS BDMV13.
C       BSMTM=1,5                                                        30
A       BWAR13E5,K=FAR13,K/BFPLOP1,K                                     31
           BWAR13E5 - WEIGHTAGE FOR RATE AR13 (AS PER EQ.(5)).
A       BRDP13,K=EAR12T3,K*BWAR13E5.K                                    32
           BRDP13 - RATE DOUBLE PRIME AR13 (AS PER EQ.(7)).
A       BRSP13,K=CAR1T23,K*BWAR13E4.K                                    33
           BRSP13 - RATE SINGLE PRIME AR13 (AS PER EQ.(6)).
A       BWAR13E4,K=FAR13,K/BFPLOP2.K                                     34
           BWAR13E4 - WEIGHTAGE FOR RATE AR13 (AS PER EQ.(4)).
A       BFPLOP1.K=SWITCH(BL,BSR13A23.K,BSR13A23.K)                       35
           BFPLOP1 - FUNCTION PERFORMING LOGICAL OPERATION 1.
C       BL=1,0                                                           36
A       BSR13A23,K=FAR13,K+GAR23.K                                       37
           BSR13A23 - SUMMATION OF RATES FAR13 AND GAR23.
A       BSR13A12,K=FAR13,K+DAR12.K                                       38
           BSR13A12 - SUMMATION OF RATES FAR13 AND DAR12.
A       BFPLOP2,K=SWITCH(BM,BSR13A12.K,BSR13A12.K)                       39
           BFPLOP2 - FUNCTION PERFORMING LOGICAL OPERATION 2.
C       BM=1,0                                                           40
```

```
        SECTION C

L    CP1.K=CP1.J-DT*CR1T23.JK                                        41
        CP1 - PRODUCT 1 (IN FRACTION OF TOTAL JOINT MARKET).
N    CP1=.801                                                        42
L    CP23.K=CP23.J+DT*CR1T23.JK                                      43
        CP23 - PRODUCTS 2 AND 3 COMBINED (IN FRACTION OF
                TOTAL JOINT MARKET).
N    CP23=.199                                                       44
R    CR1T23.KL=CAR1T23.K                                             45
        CR1T23 - RATE OF SUBSTITUTION OF CP1 BY CP23
A    CAR1T23.K=(CCMVP1.K)(CNR1T23.K/CDR1T23.K)                       46
        CAR1T23 - DETERMINES RATE FOR CR1T23.
A    CDR1T23.K=CP1.K+CSP23.K                                         47
A    CSP23.K=CSIGMA.K*CP23.K                                         48
A    CSIGMA.K=TABHL(TCSIGMA,TIME.K,CN1,CN2,CN3)                      49
        CSIGMA - DELAY COEFFICIENT (DIMENSIONLESS).
T    TCSIGMA=0/0                                                     50
C    CN1=1959                                                        51
C    CN2=1976                                                        52
C    CN3=17                                                          53
A    CC2P1P23.K=CC2.K*CP1.K*CP23.K                                   54
A    CC2.K=TABHL(TCC2,CSIGMA.K,CM1,CM2,CM3)                          55
        CC2 - COEFFICIENT OF TIME (EQ.(1)).
T    TCC2=.2969/.6597                                                56
C    CM1=0                                                           57
C    CM2=1                                                           58
C    CM3=1                                                           59
A    CNR1T23.K=CP1.K*CC2P1P23.K                                      60
A    CCMVP1.K=CLIP(CP,CQ,CP1.K,CREF)                                 61
        CCMVP1 - CONTROLS MINIMUM VALUE OF CP1.
C    CP=1.0                                                          62
C    CQ=0                                                            63
C    CREF=0                                                          64

        SECTION D

L    DP1.K=DP1.J-DT*DR12.JK                                          65
        DP1 - PRODUCT 1 (FRACTION OF TOTAL MARKET CONSISTING
                OF PRODUCTS 1 AND 2).
N    DP1=.852                                                        66
L    DP2.K=DP2.J+DT*DR12.JK                                          67
        DP2 - PRODUCT 2 (FRACTION OF TOTAL MARKET CONSISTING
                OF PRODUCTS 1 AND 2).
N    DP2=.148                                                        68
R    DR12.KL=DAR12.K                                                 69
        DR12 - RATE OF SUBSTITUTION OF PRODUCT 1 BY PRODUCT 2
                IGNORING THE PRESENCE OF PRODUCT 3.
A    DAR12.K=(DCMVP1.K)(DNR12.K/DDR12.K)                             70
        DAR12 - DETERMINES RATE FOR DR12.
A    DDR12.K=DP1.K+DSP2.K                                            71
A    DSP2.K=DSIGMA.K*DP2.K                                           72
A    DSIGMA.K=TABHL(TDSIGMA,TIME.K,DN1,DN2,DN3)                      73
        DSIGMA - DELAY COEFFICIENT (DIMENSIONLESS).
T    TDSIGMA=0/0                                                     74
C    DN1=1959                                                        75
C    DN2=1976                                                        76
C    DN3=17                                                          77
A    DC2P1P2.K=DC2.K*DP1.K*DP2.K                                     78
A    DC2.K=TABHL(TDC2,DSIGMA.K,DM1,DM2,DM3)                          79
T    TDC2=.1699/.2448                                                80
C    DM1=0                                                           81
C    DM2=1                                                           82
C    DM3=1                                                           83
A    DNR12.K=DP1.K*DC2P1P2.K                                         84
A    DCMVP1.K=CLIP(DP,DQ,DP1.K,DREF)                                 85
        DCMVP1 - CONTROLS MINIMUM VALUE OF DP1.
C    DP=1.0                                                          86
```

```
C       DQ=0                                                          87
C       DREF=0                                                        88

           SECTION E

L       EP12.K=EP12.J-DT*ER12T3.JK                                    89
           EP12 - PRODUCTS 1 AND 2 COMBINED (IN FRACTION OF
                  TOTAL JOINT MARKET).
N       EP12=.923                                                     90
L       EP3.K=EP3.J+DT*ER12T3.JK                                      91
           EP3 - PRODUCT 3 (IN FRACTION OF TOTAL JOINT MARKET).
N       EP3=.077                                                      92
R       ER12T3.KL=EAR12T3.K                                           93
           ER12T3 - RATE OF SUBSTITUTION OF EP12 BY EP3.
A       EAR12T3.K=(ECMVP12.K)(ENR12T3.K/EDR12T3.K)                    94
           EAR12T3 - DETERMINES RATE FOR ER12T3.
A       EDR12T3.K=EP12.K+ESP3.K                                       95
A       ESP3.K=ESIGMA.K*EP3.K                                         96
A       ESIGMA.K=TABHL(TESIGMA,TIME.K,EN1,EN2,EN3)                    97
           ESIGMA - DELAY COEFFICIENT (DIMENSIONLESS).
T       TESIGMA=0/.8                                                  98
C       EN1=1968                                                      99
C       EN2=1976                                                     100
C       EN3=8                                                        101
A       EC2P12P3.K=EC2.K*EP12.K*EP3.K                                102
A       EC2.K=TABHL(TEC2,ESIGMA.K,EM1,EM2,EM3)                       103
           EC2 - COEFFICIENT OF TIME (EQ.(1)).
T       TEC2=.3341/.4861                                             104
C       EM1=0                                                        105
C       EM2=.8                                                       106
C       EM3=.8                                                       107
A       ENR12T3.K=EP12.K*EC2P12P3.K                                  108
A       ECMVP12.K=CLIP(EP,EQ,EP12.K,EREF)                            109
           ECMVP12 - CONTROLS MINIMUM VALUE OF EP12.
C       EP=1.0                                                       110
C       EQ=0                                                         111


C       EREF=0                                                       112


           SECTION F

L       FP1.K=FP1.J-DT*FR13.JK                                       113
           FP1 - PRODUCT 1 (IN FRACTION OF TOTAL MARKET CONSISTING
                 OF PRODUCTS 1 AND 3).
N       FP1=.912                                                     114
L       FP3.K=FP3.J+DT*FR13.JK                                       115
           FP3 - PRODUCT 3 (IN FRACTION OF TOTAL MARKET CONSISTING
                 OF PRODUCTS 1 AND 3).
N       FP3=.088                                                     116
R       FR13.KL=FAR13.K                                              117
           FR13 - RATE OF SUBSTITUTION OF PRODUCT 1 BY PRODUCT 3
                  IGNORING THE PRESENCE OF PRODUCT 2.
A       FAR13.K=(FCMVP1.K)(FNR13.K/FDR13.K)                          118
           FAR13 - DETERMINES RATE FOR FR13.
A       FDR13.K=FP1.K+FSP3.K                                         119
A       FSP3.K=FSIGMA.K*FP3.K                                        120
A       FSIGMA.K=TABHL(TFSIGMA,TIME.K,FN1,FN2,FN3)                   121
           FSIGMA - DELAY COEFFICIENT (DIMENSIONLESS).
T       TFSIGMA=0/0                                                  122
C       FN1=1959                                                     123
C       FN2=1976                                                     124
C       FN3=17                                                       125
A       FC2P1P3.K=FP1.K*FP3.K*FC2.K                                  126
```

```
A       FC2.K=TABHL(TFC2,FSIGMA.K,FM1,FM2,FM3)                          127
           FC2 - COEFFICIENT OF TIME (EQ.(1)).
T       TFC2=.3797/.7469                                                128
C       FM1=0                                                           129
C       FM2=1                                                           130
C       FM3=1                                                           131
A       FNR13.K=FP1.K*FC2P1P3.K                                         132
A       FCMVP1.K=CLIP(FP,FQ,FP1.K,FREF)                                 133
           FCMVP1 - CONTROLS MINIMUM VALUE OF FP1.
C       FP=1.0                                                          134
C       FQ=0                                                            135
C       FREF=0                                                          136



           SECTION G

L       GP2.K=GP2.J-DT*GR23.JK                                          137
           GP2 - PRODUCT 2 (IN FRACTION OF TOTAL MARKET CONSISTING
                 OF PRODUCTS 2 AND 3).
N       GP2=.679                                                        138
L       GP3.K=GP3.J+DT*GR23.JK                                          139
           GP3 - PRODUCT 3 (IN FRACTION OF TOTAL MARKET CONSISTING
                 OF PRODUCTS 2 AND 3).
N       GP3=.321                                                        140
R       GR23.KL=GAR23.K                                                 141
           GR23 - RATE OF SUBSTITUTION OF PRODUCT 2 BY PRODUCT 3
                  IGNORING THE PRESENCE OF PRODUCT 1.
A       GAR23.K=(GCMVP2.K)(GNR23.K/GDR23.K)                             142
           GAR23 - DETERMINES RATE FOR GR23.
A       GDR23.K=GP2.K+GSP3.K                                            143
A       GSP3.K=GSIGMA.K*GP3.K                                           144
A       GSIGMA.K=TABHL(TGSIGMA,TIME.K,GN1,GN2,GN3)                      145
           GSIGMA - DELAY COEFFICIENT ( DIMENSIONLESS).
T       TGSIGMA=0/0                                                     146
C       GN1=1959                                                        147
C       GN2=1976                                                        148
C       GN3=17                                                          149
A       GC2P2P3.K=GC2.K*GP2.K*GP3.K                                     150
A       GC2.K=TABHL(TGC2,GSIGMA.K,GM1,GM2,GM3)                          151
           GC2 - COEFFICIENT OF TIME (EQ.(1)).
T       TGC2=.2332/.6252                                                152
C       GM1=0                                                           153
C       GM2=1                                                           154
C       GM3=1                                                           155
A       GNR23.K=GP2.K*GC2P2P3.K                                         156
A       GCMVP2.K=CLIP(GP,GQ,GP2.K,GREF)                                 157
           GCMVP2 - CONTROLS MINIMUM VALUE OF GP2.
C       GP=1.0                                                          158
C       GQ=0                                                            159
C       GREF=0                                                          160
```

APPENDIX 2
Computer Program for Dynamic Modeling

```
NOTE

1) THE NAMES AND ACRONYMS OF VARIABLES OF
   FIG. 9 FOLLOW THE DYNAMO EQUATIONS.
2) DYNAMO EQUATIONS FOR PLOTING HISTORICAL
   DATA (AS SHOWN IN FIGS. 10,11 AND 12) ARE
   NOT INCLUDED IN THIS PROGRAM.
3) NUMERICAL VALUES ASSIGNED TO DIFFERENT
   DYNAMO EQUATIONS PERTAIN TO THE ILLUSTRATIVE
   EXAMPLE IN PART TWO.
```

```
L      FP1.K=FP1.J-DT*RP12.JK                                   1
          FP1 = MARKET SHARE OF PRODUCT 1 IN FRACTION
N      FP1=.916                                                 2
L      FP2.K=FP2.J+DT*RP12.JK                                   3
          FP2 = MARKET SHARE OF PRODUCT 2 IN FRACTION
N      FP2=.084                                                 4
A      BP1P2.K=BETA.K*FP1.K*FP2.K                               5
A      BETA.K=TABHL(TBETA,TIME.K,BN1,BN2,BN3)                   6
          BETA = INSTANTENEOUS RATE OF CHANGE AT TIME T.
T      TBETA=.349/.339/.329/.320/.312/.303/.295/.286/           7
X      .278/.271/.263/.256/.249/.245
C      BN1=1965                                                 8
C      BN2=1991                                                 9
C      BN3=2                                                   10
A      NRP12.K=BP1P2.K*FP1.K                                   11
A      DRP12.K=FP1.K+LAMBDAP2.K                                12
A      LAMBDAP2.K=FP2.K*LAMBDA.K                               13
A      LAMBDA.K=TABHL(TLAMBDA,TIME.K,LN1,LN2,LN3)              14
          LAMBDA = DYNAMIC RESPONSE COEFFICIENT AT TIME T.
T      TLAMBDA=0/.04/.1/.17/.24/.28/.3                         15
C      LN1=1974                                                16
C      LN2=1980                                                17
C      LN3=1.0                                                 18
A      ARP12.K=(NRP12.K/DRP12.K)(CMVFP1.K)                     19
          ARP12 = DETERMINES RATE FOR RP12
R      RP12.KL=ARP12.K                                         20
          RP12 = RATE OF SUBSTITUTION OF PRODUCT 1
                 BY PRODUCT 2
A      CMVFP1.K=CLIP(CP,CQ,FP1.K,CREF)                         21
          CMVFP1 = CONTROLS MINIMUM VALUE OF LEVEL FP1.K
C      CP=1.0                                                  22
C      CQ=0                                                    23
C      CREF=0                                                  24
L      TM.K=TM.J+DT*TMXR.JK                                    25
          TM = TOTAL JOINT MARKET
N      TM=3.88                                                 26
```

```
R       TMXR.KL=TM.K*MGR.K                                          27
           TMXR = EXPANSION RATE OF TOTAL JOINT MARKET
A       MGR.K=TABHL(TMGR,TIME.K,TN1,TN2,TN3)                        28
           MGR = ANNUAL MARKET GROWTH IN FRACTION
T       TMGR=.0327/.0327                                            29
C       TN1=1974                                                    30
C       TN2=1990                                                    31
C       TN3=16                                                      32
A       MP2.K=TM.K*FP2.K                                            33
           MP2 = MARKET SIZE OF PRODUCT 2
A       MP1.K=TM.K-MP2.K                                            34
           MP1 = MARKET SIZE OF PRODUCT 1
```

The Normal Distribution as a Model of Technological Substitution

EARL STAPLETON

ABSTRACT

A model for forecasting technological substitution, based on the use of the normal distribution, is presented. Both mathematical and graphical techniques are discussed. Examples from the literature are used to illustrate the model.

Introduction

Experience has shown that many technologies tend to follow a predictable pattern of growth, which can be described as a symmetrical S-shaped curve. Originally developed to describe the growth of organisms in a fixed environment, the Pearl Curve [1] has been found to describe this behavior for technological change as well. More recently, Fisher and Pry [2] have developed a model which is based on a special form of the Pearl Curve.

Regardless of which model is used, the same three assumptions concerning the behavior of technological changes are involved. These are:

(1) The growth follows an S-shaped pattern with time, and change exhibits an exponential increase in the early periods.

(2) There is an upper limit beyond which the change will not proceed (the limit is implicit in the Fisher-Pry model and will be discussed in more detail in the next section).

(3) The curve is symmetrical in shape, with a transition point equidistant between the beginning and end of the technological change.

FISHER-PRY MODEL–UPPER LIMIT ASSUMPTION

The Pearl model uses an explicit upper limit, which must be estimated or otherwise obtained. The Fisher-Pry model uses an implicit upper limit which is not at first obvious. In the case of technological substitutions, which is the topic of the Fisher-Pry work, the upper limit is always 100% substitution. However, any change could be scaled in a similar manner as a percentage of some limit, and the model would then work equally well. Additionally, in the case of substitutions, less than 100% takeover could be estimated and the fraction of effective substitution computed relative to this limit.

Thus, the Fisher-Pry model includes all the parameters found in the Pearl formulation, but uses simpler computations. However, it still requires transformations and conversions to obtain forecasts. While this poses no insurmountable problem, the alternative proposed in this paper represents an improvement in computational simplicity.

This article appeared in *Technological Forecasting and Social Change*, Vol. 8, No. 3, 1976.

TABLE 1[a]

Areas Under the Normal Curve
Proportion of Total Area Under the Curve that is Under the Portion of the Curve
from $-\infty$ to $\frac{X_i - X'}{\sigma'}$. (X_i Represents any Desired Value of the Variable X)

$\frac{X_i - X'}{\sigma'}$	0.09	0.08	0.07	0.06	0.05	0.04	0.03	0.02	0.01	0.00
–3.5	0.00017	0.00017	0.00018	0.00019	0.00019	0.00020	0.00021	0.00022	0.00022	0.00023
–3.4	0.00024	0.00025	0.00026	0.00027	0.00028	0.00029	0.00030	0.00031	0.00033	0.00034
–3.3	0.00035	0.00036	0.00038	0.00039	0.00040	0.00042	0.00043	0.00045	0.00047	0.00048
–3.2	0.00050	0.00052	0.00054	0.00056	0.00058	0.00060	0.00062	0.00064	0.00066	0.00069
–3.1	0.00071	0.00074	0.00076	0.00079	0.00082	0.00085	0.00087	0.00090	0.00094	0.00097
–3.0	0.00100	0.00104	1.00107	0.00111	0.00114	0.00118	0.00122	0.00126	0.00131	0.00135
–2.9	0.0014	0.0014	0.0015	0.0015	0.0016	0.0016	0.0017	0.0017	0.0018	0.0019
–2.8	0.0019	0.0020	0.0021	0.0021	0.0022	0.0023	0.0023	0.0024	0.0025	0.0026
–2.7	0.0026	0.0027	0.0028	0.0029	0.0030	0.0031	0.0032	0.0033	0.0034	0.0035
–2.6	0.0036	0.0037	0.0038	0.0039	0.0040	0.0041	0.0043	0.0044	0.0045	0.0047
–2.5	0.0048	0.0049	0.0051	0.0052	0.0054	0.0055	0.0057	0.0059	0.0060	0.0062
–2.4	0.0064	0.0066	0.0068	0.0069	0.0071	0.0073	0.0075	0.0078	0.0080	0.0082
–2.3	0.0084	0.0087	0.0089	0.0091	0.0094	0.0096	0.0099	0.0102	0.0104	0.0107
–2.2	0.0110	0.0113	0.0116	0.0119	0.0122	0.0125	0.0129	0.0132	0.0136	0.0139
–2.1	0.0143	0.0146	0.0150	0.0154	0.0158	0.0162	0.0166	0.0170	0.0174	0.0179
–2.0	0.0183	0.0188	0.0192	0.0197	0.0202	0.0207	0.0212	0.0217	0.0222	0.0228
–1.9	0.0233	0.0239	0.0244	0.0250	0.0256	0.0262	0.0268	0.0274	0.0281	0.0287
–1.8	0.0294	0.0301	0.0307	0.0314	0.0322	0.0329	0.0336	0.0344	0.0351	0.0359
–1.7	0.0367	0.0375	0.0384	0.0392	0.0401	0.0409	0.0418	0.0427	0.0436	0.0446
–1.6	0.0455	0.0465	0.0475	0.0485	0.0495	0.0505	0.0516	0.0526	0.0537	0.0548

(Table 1 Continued.)

–1.5	0.0559	0.0571	0.0582	0.0594	0.0606	0.0618	0.0630	0.0643	0.0652	0.0668
–1.4	0.0681	0.0694	0.0708	0.0721	0.0735	0.0749	0.0764	0.0778	0.0793	0.0808
–1.3	0.0823	0.0838	0.0853	0.0869	0.0885	0.0901	0.0918	0.0934	0.0951	0.0968
–1.2	0.0985	0.1003	0.1020	0.1038	0.1057	0.1075	0.1093	0.1112	0.1131	0.1151
–1.1	0.1170	0.1190	0.1210	0.1230	0.1251	0.1271	0.1292	0.1314	0.1335	0.1357
–1.0	0.1379	0.1401	0.1423	0.1446	0.1469	0.1492	0.1515	0.1539	0.1562	0.1587
–0.9	0.1611	0.1635	0.1660	0.1685	0.1711	0.1736	0.1762	0.1788	0.1814	0.1841
–0.8	0.1867	0.1894	0.1922	0.1949	0.1977	0.2005	0.2033	0.2061	0.2090	0.2119
–0.7	0.2148	0.2177	0.2207	0.2236	0.2266	0.2297	0.2327	0.2358	0.2389	0.2420
–0.6	0.2451	0.2483	0.2514	0.2546	0.2578	0.2611	0.2643	0.2676	0.2709	0.2743
–0.5	0.2776	0.2810	0.2843	0.2877	0.2912	0.2946	0.2981	0.3015	0.3050	0.3085
–0.4	0.3121	0.3156	0.3192	0.3228	0.3264	0.3300	0.3336	0.3372	0.3409	0.3446
–0.3	0.3483	0.3520	0.3557	0.3594	0.3632	0.3669	0.3707	0.3745	0.3783	0.3821
–0.2	0.3859	0.3897	0.3936	0.3974	0.4013	0.4052	0.4090	0.4129	0.4168	0.4207
–0.1	0.4247	0.4286	0.4325	0.4364	0.4404	0.4443	0.4483	0.4522	0.4562	0.4602
–0.0	0.4641	0.4681	0.4721	0.4761	0.4801	0.4840	0.4880	0.4920	0.4960	0.5000
+0.0	0.5000	0.5040	0.5080	0.5120	0.5160	0.5199	0.5239	0.5279	0.5319	0.5359
+0.1	0.5398	0.5438	0.5478	0.5517	0.5557	0.5596	0.5636	0.5675	0.5714	0.5753
+0.2	0.5793	0.5832	0.5871	0.5910	0.5948	0.5987	0.6026	0.6064	0.6103	0.6141
+0.3	0.6179	0.6217	0.6255	0.6293	0.6331	0.6368	0.6406	0.6443	0.6480	0.6517
+0.4	0.6554	0.6591	0.6628	0.6664	0.6700	0.6736	0.6772	0.6808	0.6844	0.6879
+0.5	0.6915	0.6950	0.6985	0.7019	0.7054	0.7088	0.7123	0.7157	0.7190	0.7224
+0.6	0.7257	0.7291	0.7324	0.7357	0.7389	0.7422	0.7454	0.7486	0.7517	0.7549
+0.7	0.7580	0.7611	0.7642	0.7673	0.7704	0.7734	0.7764	0.7794	0.7823	0.7852
+0.8	0.7881	0.7910	0.7939	0.7967	0.7995	0.8023	0.8051	0.8079	0.8106	0.8133
+0.9	0.8159	0.8186	0.8212	0.8238	0.8264	0.8289	0.8315	0.8340	0.8365	0.8389
+1.0	0.8413	0.8438	0.8461	0.8485	0.8508	0.8531	0.8554	0.8577	0.8599	0.8621

(Table 1 continued on following page)

TABLE 1 *(continued)*

$\frac{X_i - X'}{\sigma'}$	0.00	0.01	0.02	0.03	0.04	0.05	0.06	0.07	0.08	0.09
+1.1	0.8643	0.8665	0.8686	0.8708	0.8729	0.8749	0.8770	0.8790	0.8810	0.8830
+1.2	0.8849	0.8869	0.8888	0.8907	0.8925	0.8944	0.8962	0.8980	0.8997	0.9015
+1.3	0.9032	0.9049	0.9066	0.9082	0.9099	0.9115	0.9131	0.9147	0.9162	0.9177
+1.4	0.9192	0.9207	0.9222	0.9236	0.9251	0.9265	0.9279	0.9292	0.9306	0.9319
+1.5	0.9332	0.9345	0.9357	0.9370	0.9382	0.9394	0.9406	0.9418	0.9429	0.9441
+1.6	0.9452	0.9463	0.9474	0.9484	0.9495	0.9505	0.9515	0.9525	0.9535	0.9545
+1.7	0.9554	0.9564	0.9573	0.9582	0.9591	0.9599	0.9608	0.9616	0.9625	0.9633
+1.8	0.9641	0.9649	0.9656	0.9664	0.9671	0.9678	0.9686	0.9693	0.9699	0.9706
+1.9	0.9713	0.9719	0.9726	0.9732	0.9738	0.9744	0.9750	0.9756	0.9761	0.9767
+2.0	0.9773	0.9778	0.9783	0.9788	0.9793	0.9798	0.9803	0.9808	0.9812	0.9817
+2.1	0.9821	0.9826	0.9830	0.9834	0.9838	0.9842	0.9846	0.9850	0.9854	0.9857
+2.2	0.9861	0.9864	0.9868	0.9871	0.9875	0.9878	0.9881	0.9884	0.9887	0.9890
+2.3	0.9893	0.9896	0.9898	0.9901	0.9904	0.9906	0.9909	0.9911	0.9913	0.9916
+2.4	0.9918	0.9920	0.9922	0.9925	0.9927	0.9929	0.9931	0.9932	0.9934	0.9936
+2.5	0.9938	0.9940	0.9941	0.9943	0.9945	0.9946	0.9948	0.9949	0.9951	0.9952
+2.6	0.9953	0.9955	0.9956	0.9957	0.9959	0.9960	0.9961	0.9962	0.9963	0.9964
+2.7	0.9965	0.9966	0.9967	0.9968	0.9969	0.9970	0.9971	0.9972	0.9973	0.9974
+2.8	0.9974	0.9975	0.9976	0.9977	0.9977	0.9978	0.9979	0.9979	0.9980	0.9981
+2.9	0.9981	0.9982	0.9983	0.9983	0.9984	0.9984	0.9985	0.9985	0.9986	0.9986
+3.0	0.99865	0.99869	0.99874	0.99878	0.99882	0.99886	0.99889	0.99893	0.99896	0.99900
+3.1	0.99903	0.99906	0.99910	0.99913	0.99915	0.99918	0.99921	0.99924	0.99926	0.99929
+3.2	0.99931	0.99934	0.99936	0.99938	0.99940	0.99942	0.99944	0.99946	0.99948	0.99950
+3.3	0.99952	0.99953	0.99955	0.99957	0.99958	0.99960	0.99961	0.99962	0.99964	0.99965
+3.4	0.99966	0.99967	0.99969	0.99970	0.99971	0.99972	0.99973	0.99974	0.99975	0.99976
+3.5	0.99977	0.99978	0.99978	0.99979	0.99980	0.99981	0.99981	0.99982	0.99983	0.99983

[a]Reproduced from Grant, Eugene L., *Statistical Quality Control,* McGraw-Hill, New York, 1946, pp. 534–535.

Alternative Approach

The alternative function which conforms to the three conditions noted earlier is a cumulative normal distribution. While the mathematics related to this function is more complex than that of the Fisher-Pry model, it has the advantage of being more completely explored. Thus, one has access to a standard table, and can avoid all computations except those related to standard regression analysis.[1]

In particular, an "area under the normal curve" table (Table 1) relates the two basic measures required for forecasting. The body of the table may be interpreted as fraction of completion, or substitution, values. The proportion-of-total-area column is this value normalized to a linear variate by using the standard deviation also expressed as a fraction. In this manner, a set of values can be obtained which, when plotted on rectangular coordinate paper, will fall on a straight line *if the three previously stated assumptions are valid.*

APPLICATIONS

The use of the "normal curve model" is straightforward. The principal steps are listed below:

(1) Determine the fraction of completion of the technological change for each data point relative to the assumed limit.

(2) Transform the data to linear variates using the area-under-the-normal-curve table as discussed earlier.

(3) Using the values obtained from the table, fit the data, along with the associated dates, directly to the linear form $y = a + bx$.

To illustrate the technique, the first example from the Fisher-Pry work has been reworked using these steps. The data are shown in Table 2. The resulting curve is

TABLE 2

Synthetic vs. Natural Fiber Substitution Data

Date	Fraction synthetic	Transformed value
1930	0.044	−1.71
1935	0.079	−1.42
1940	0.10	−1.28
1945	0.14	−1.08
1950	0.22	−0.77
1955	0.28	−0.58
1960	0.29	−0.55
1965	0.43	−0.17
1967	0.47	−0.08

[1] An Approach not very different from the present one can be found in *KTBL-Berichte über Landtechnik 129, Methoden der Produktforschung am Beispiel des Mähdreschers im OECD Gebiet Europas 1969,* pp. 36–48. This study assumes a normal distribution of the values of yearly *increase* in the diffusion of technology.

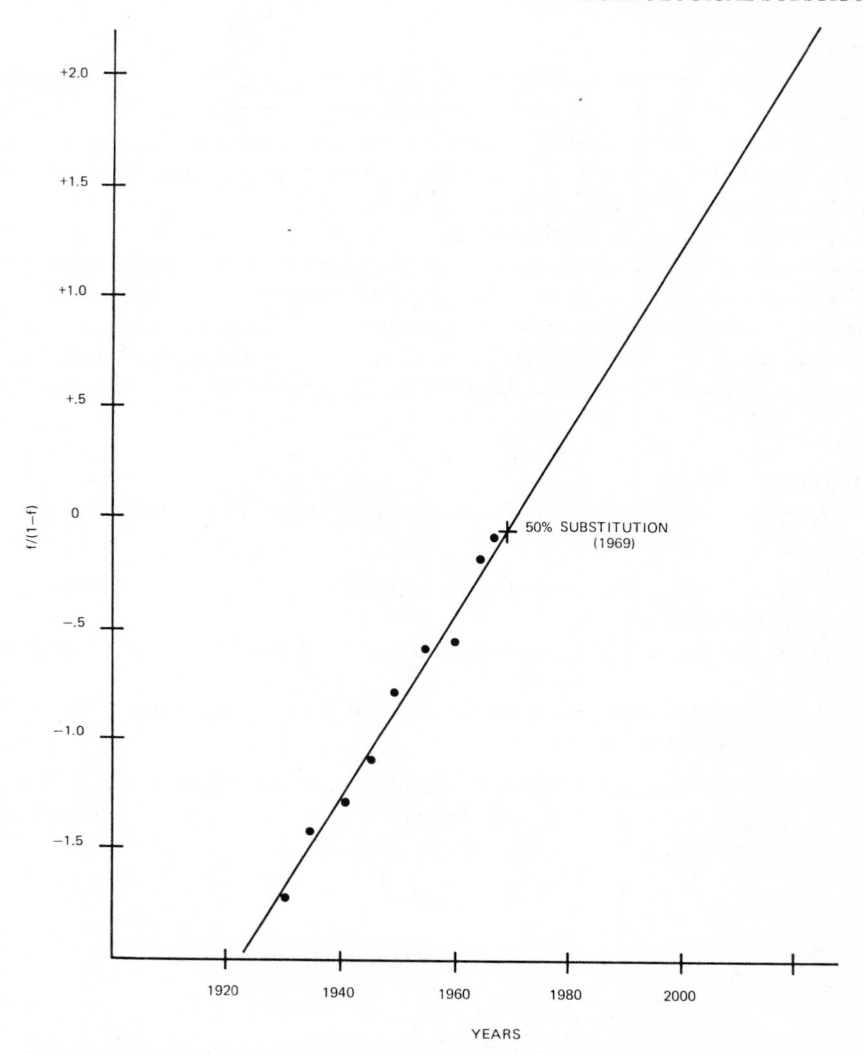

FIG. 1a. Synthetic for natural fiber substitution vs. years; Fisher-Pry model fit to data.

illustrated in Fig. 1(a), with the Fisher-Pry equivalent in Fig. 1(b). The equation obtained is $y = -2.95 + 0.042x$. The 50% completion and 90% completion dates obtained are identical to those in the Fisher-Pry work.

To demonstrate the ability of the approach in non-substitution circumstances, an example [3] from the Pearl model has also been reworked. The basic data may be found in Table 3. The transformed values are plotted in Fig. 2(a), with the Pearl equivalent shown in Fig. 2(b). The equation obtained is $y = -1.99 + 0.038x$.

In this instance, the results are also comparable. Using the standard Pearl Formulation, a forecast of 688.3 was obtained for the year 2000, while the normal curve model gives 672.5 for the year 2000. It should be recalled, however, that in the Pearl model it is not possible to minimize the sum of the squares of the differences between the original data

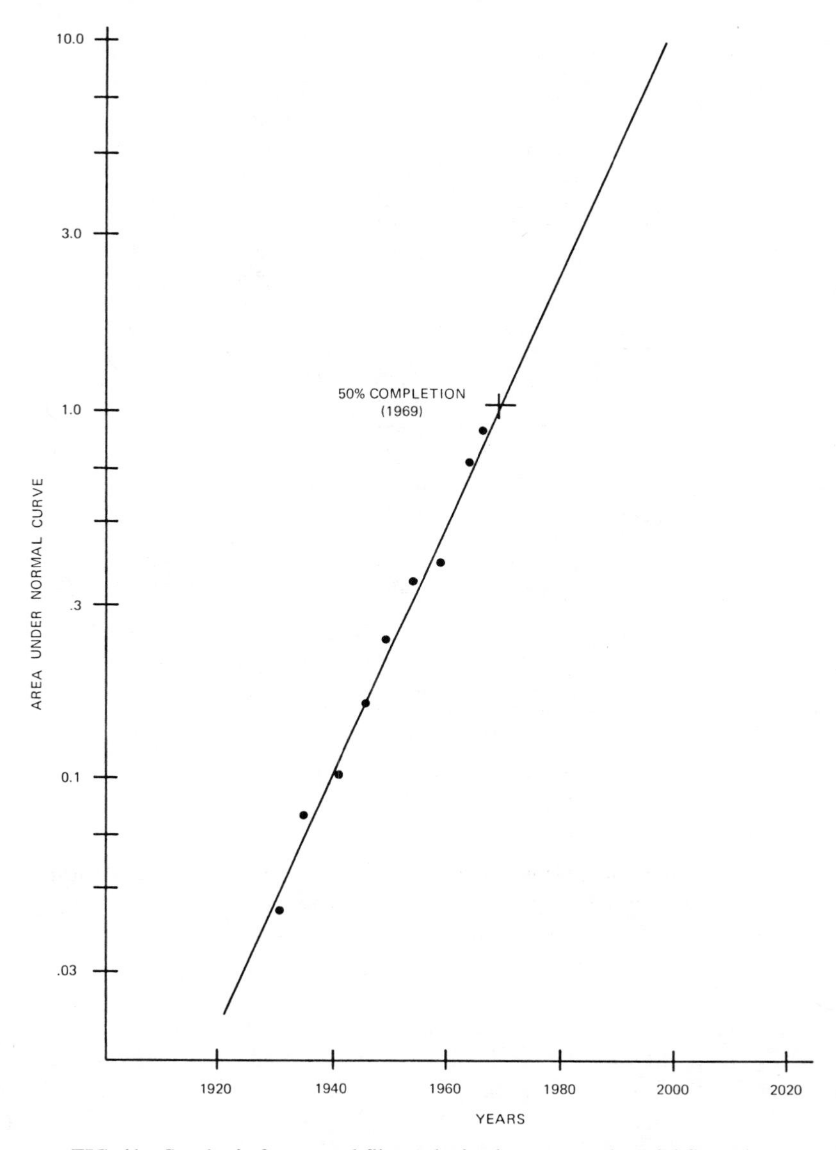

FIG. 1b. Synthetic for natural fiber substitution vs. years; model fit to data.

and the fitted curve. The procedure described in this paper does fit directly, and accounts for the variation in results.

GRAPHICAL APPROACH

In those instances where a timely, but less precise, forecast would be more relevant than a mathematical forecast, a graphical approach can be employed. Fractional completion (substitution) figures computed as in step one above may be plotted directly onto

TABLE 3

Number of Telephones per Capita Data

Date	Number of Telephones	Fraction of Limit[a]	Transformed Value
1880	1.1	0.0016	–2.95
'85	2.7	0.0028	–2.77
90	3.6	0.005	–2.58
95	4.8	0.007	–2.46
1900	17.6	0.025	–1.96
'05	48.8	0.070	–1.48
10	82.0	0.118	–1.19
15	103.9	0.149	–1.04
20	123.4	0.177	–0.93
25	144.6	0.208	–0.81
30	162.6	0.233	–0.73
35	136.4	0.196	–0.86
40	165.1	0.237	–0.72
45	198.1	0.284	–0.57
50	280.8	0.403	–0.25
55	337.2	0.484	–0.04
60	407.8	0.585	+0.22
65	478.2	0.686	+0.49
66	498.7	0.716	+0.57
67	518.3	0.744	+0.66

[a]Limit is 696.9 phones per thousand.

probability paper. The scale of this paper is so designed[2] that the ogive of a normal curve will plot to a straight line. Once plotted, a free-form fit may be made to the points, and the forecast read directly from the graph.

Summary

From these examples, it can be seen that a sample transformation can be used on symmetrical growth curve models to improve computational efficiency, once an estimate of the upper limit of growth has been made. The approach developed in this paper admittedly lacks the intuitive "feel" of the Pearl or Fisher-Pry models. However, this should not be of concern as, to quote Martino:

> The validity of the forecast is dependent much more on the validity of the data, and of the basic assumptions, than on the mathematics. The mathematical fitting technique is primarily an objective means of extracting the significance of the past data and our assumptions. It cannot in any sense add validity to these. [4]

The use of the "normal curve approach," then, provides a technique whereby results can be quickly and easily obtained for a variety of cases where previous techniques did not lend themselves to straightforward computation and analysis.

---

[2] The scale is actually linear when measured in terms of standard deviations.

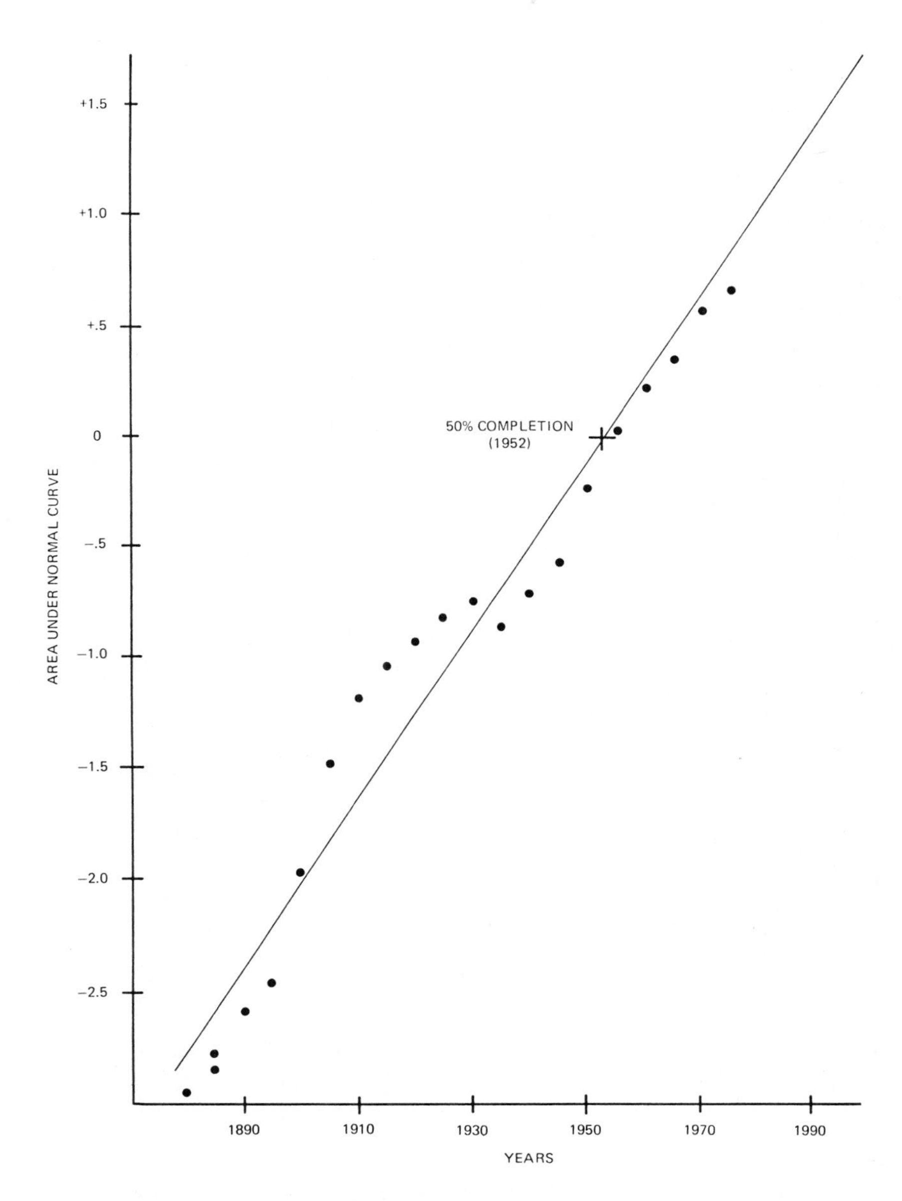

FIG. 2a. Regression line fitted to transformed data on numbers of telephones.

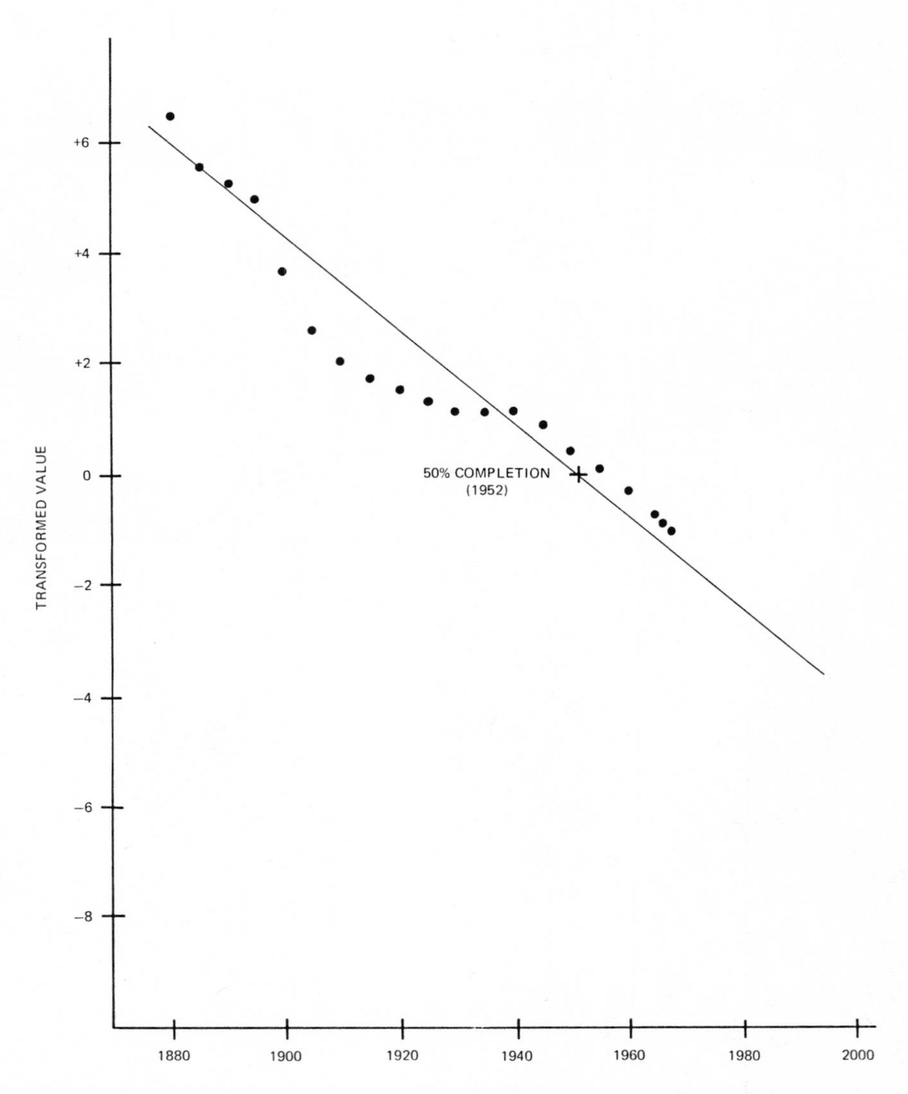

FIG. 2b. Regression line fitted to transformed data on number of telephones using Pearl curve.

References

1. Pearl, R., *The Biology of Population Growth,* Alfred A. Knopf, New York, 1925.
2. Fisher, J.C., and R.H. Pry, A simple substitution model of technological change, *Technol. Forecasting Soc. Change* (1972).
3. Martino, Joseph P., *Technological Forecasting for Decision Making,* American Elsevier, 1968 pp. 119–121.
4. Martino, Joseph P., *op. cit.,* p. 122.

Part 2.

Determinants of Substitution Rate

Part 2. Determinants of Substitution Rate

Introduction

Although the articles in the preceding chapter provide strong support for the hypothesis of an S-shaped technological substitution pattern, they do not explain the determinants of its parameters. Consider, for example, the mathematical expression for the logistic curve,

$$Y = \frac{K}{1 + \exp[-(a + bt)]},$$

where, for the technological substitution studies, Y is the fraction of the market captured by the new product, K is the asymptotic equilibrium value of Y (which, it is usually assumed, will be approximately equal to unity), t is time, a is a constant locating the curve on the time scale, and b is the annual fractional growth of the new product. A parameter of particular interest is b, the rate at which technological substitution proceeds towards possible completion. The parameter b has a similar meaning in the "diffusion of innovation" studies. There it may be regarded as the rate of adoption of the innovation (also sometimes called the rate of imitation). The question that naturally arises is: What determines the rate of technological substitution?

Following the pioneering work of Zvi Griliches in the 1950's on the diffusion of hybrid corn seed,[1] Mansfield hypothesized that, *inter alia*, the rate of adoption of an innovation is a direct function of the profitability of employing the innovation and decreasing function of the size of investment required to use it.[2] He studied the speed with which twelve important innovations were adopted by firms in four industries (bituminous coal, iron and steel, brewing, and railroads) and found wide variation in the speed of adoption. Testing of the model provided strong support for the hypothesis that, *ceteris paribus*, the more profitable the innovation and smaller the required investment, the greater the rate of imitation.

The four articles in this chapter provide us with further evaluations and significant extensions of Mansfield's model. The first article by Blackman corroborates Mansfield's findings in an explanation of the substitution of the turbojet engine in aircraft by the first generation turbofan and the substitution of the latter, in turn, by the second generation turbofan. The rate at which a new product innovation displaces an existing product in a given market is found to be an increasing function of (1) the proportion of the firms already using the new product and (2) the profitability of the new product relative to the old product, and, on the other hand, a decreasing function of the investment required to adopt the new product.

In a significant extension of this work, the second article by Blackman, Seligman, and

---

[1] Griliches, Z., Hybrid corn: An exploration in the economics of technical change, *Econometrica* **25** (1957), pp. 501–522.

[2] Mansfield, E., Technical change and the rate of imitation, *Econometrica* **29** (October 1961), pp. 741–766.

Sogliero presents the development of an innovation index that indicates the relative tendency of the various sectors to innovate. The index is derived from various input variables that measure the extent to which resources are allocated to achieve innovation, and from output variables that measure the extent to which new product and process innovation is achieved. The weights assigned to the individual variables are determined by the factor analysis technique. The innovation index is applied to rank-order various industrial sectors, and changes in innovation characteristics over time are indicated by index values determined for 1960, 1963 and 1970. An important result of the study is that, *inter alia*, the rate of technological substitution is a direct function of the innovation index. Relationships are established between values of the innovation index and inter-industry differences in the dynamics of the development of markets created through technological innovation. A considerably refined forecast can therefore be made of the rates at which markets will develop for the new technological innovation. The methodological structure developed can also be used to establish criteria useful in evaluating and planning new technology-based products.

Further applications of this methodology are described in the third article by Blackman, which also presents procedures useful in the absence of historical data. The performance of these techniques is found to be uniformly good in a variety of cases including substitution of fiberglass for wood, acceptance in the U.S. of electric valve heaters, freezers, electric refrigerators, dishwashers, color TV, etc.

In the last article in this chapter, Bundgaard-Nielsen is concerned with the international diffusion of new technology, a subject in which research has barely begun.[3,4] The rates of diffusion of new techniques in the steel industry (continuous casting of steel and the use of converters) are obtained in the same way as in Mansfield's work. However, the analysis focuses primarily on how a given country's industrial growth rate affects the rate of adoption of the new technique in that country. While the results of this study confirm the commonly held opinion that a fast growing industry is more likely to invest in new techniques of production than a stagnant industry, they also suggest that the impact of the growth rate is more complex and may lead to an increased as well as decreased rate of diffusion, depending upon the particular situation prevailing in the industry. Admittedly, our knowledge of the factors influencing the international diffusion of technology is far from adequate. More such studies are required of this subject that is becoming increasingly important today. The article by Bundgaard-Nielsen might serve as a step toward the development of an international innovation index, on the lines suggested by Blackman et al., whereby the innovation propensity of different countries could be characterized.

---

[3] For an attempt in this direction undertaken jointly by the British National Institute of Economic and Social Research, the Swedish International Institute for Economic and Social Research and other European institutes, see: The diffusion of new technology—A study of ten processes in nine countries, *National Institute Economic Review,* May 1969. See also the March 1974 issue of the *Journal of Economic History* (Vol. 34), which is almost entirely devoted to the interaction diffusion of technology.

[4] Another area requiring considerable further research is diffusion and substitution of innovations in developing countries. See, however, L. A. Brown and B. Lentnek, Innovation diffusion in a developing economy: A mesoscale view, *Economic Development and Cultural Change* **21** (1973), pp. 274–292. See also the special issue of *Technol. Forecast. Soc. Change,* Vol 7, No. 3, 1975, for relevant articles.

The Rate of Innovation in the Commercial Aircraft Jet Engine Market

A. WADE BLACKMAN, JR.

I. Summary

The applicability of a deterministic model which was developed by Mansfield (Reference 1) and which describes the rate at which new product innovations are adopted was applied to the commercial aircraft jet engine market. The model was found to agree well with (1) historical market share data related to the displacement of the turbojet engine by the first generation turbofan, and with (2) forecasts of future market shares related to the displacement of the first generation turbofan by the second generation turbofan. The rate at which a new product innovation displaces an existing product in a given market appears to be an increasing function of: (1) the proportion of firms already using the new product, and (2) the profitability of the new product relative to the old product, and a decreasing function of the size of the investment required to adopt the new product.

II. Introduction

Once a new product innovation has been introduced into the market, the rate at which the new innovation attains acceptance and displaces the products which formally dominated the market is of prime importance because of the influence of the market share attained at a given time by a new product on cash flows and return on investment.

The product displacement rate problem has been investigated extensively in Mansfield (Reference 1) in which a study was made of the rate at which innovations spread from enterprise to enterprise in four industries: bituminous coal, iron and steel, brewing, and railroads. A total of twelve innovations was studied, which consisted of three from each of the four industries. A deterministic model was developed which was built around the hypothesis that the probability that a firm will introduce a new technique is an increasing function of (1) the proportion of firms already using it, and (2) the profitability of the new technique.and a decreasing function of the size of the investment required. Although interindustry differences in the rate of innovation were found to exist, an equation of the form predicted by the model was found to agree well with test data for a given industry. As yet, the application of this model to the aircraft engine market has not been investigated. The objective of the work described herein was to determine the usefulness of the model discussed in Reference 1 for predicting the rate of innovation in the commercial aircraft jet engine market.

III. Analysis

The development of the analytical model given below follows the development in Reference 1 with the exception of changes in boundary-condition assumptions and notation.

This article appeared in *Technological Forecasting and Social Change*, Vol. 2, Nos. 3-4, 1971.

Let $m(t)$ equal the market share captured at time t by a new innovation which is displacing an old product in a given market, L equal to the upper limit on the market share which the new product innovation can capture in the long term, and $\Delta(t)$ equal the change in the market share achieved by the new product innovation between t and $t+1$.

$$\Delta(t)=\frac{m(t+1)-m(t)}{L-m(t)} \tag{1}$$

Following Reference 1, it was assumed that $\Delta(t)$ is a function of (1) the market share obtained at time t divided by the upper limit on the long-term market share which can be captured, $m(t)/L$; (2) the profitability of employing the new product innovation, π; and (3) the size of the investment required to install it, S.

$$\Delta(t)=f\left(\frac{m(t)}{L}, \pi, S\right) \tag{2}$$

As discussed in Reference 1, it would be expected that $\Delta(t)$ would increase as the market share increases, because as more information and experience are accumulated on the new product innovation, its adoption becomes less risky, and a "bandwagon" effect occurs. Also, the more profitable an investment in a new product innovation relative to other investment opportunities, the greater will be the probability that the market share of the new product will increase. For equally profitable new product innovations, $\Delta(t)$ should tend to be smaller for those innovations requiring large investments, because potential buyers will be more cautious about adopting the new product innovation if a large investment is required. Finally, for equally profitable products requiring equal investment, $\Delta(t)$ can be expected to vary among industries because of the different characteristics of risk and rate-of-return in different industries, different investment evaluation criteria, etc.

It was assumed that Equation 2 can be approximated by a Taylor's expansion that drops third- and higher-order terms and that the coefficient of $(m(t)/L)$ is zero based on the data of Reference 1.

$$\Delta(t)=C_1+C_2\frac{m(t)}{L}+C_3\,\pi+C_4\,S+C_5\frac{\pi m(t)}{L}+C_6\,S\frac{m(t)}{L}+C_7\,\pi S \tag{3}$$
$$+C_8\,\pi^2+C_9\,S^2+C_{10}\frac{(m(t))^2}{L^2}+\ldots$$

Substituting Equation 3 into Equation 1

$$m(t+1)-m(t)=\Big[L-m(t)\Big]\Big[C_1+C_2\frac{m(t)}{L}+\ldots C_9\,S^2+\ldots\Big] \tag{4}$$

For small time increments

$$m(t+1)-m(t)=\frac{dm(t)}{dt}=\Big[L-m(t)\Big]\Big[Q+ø\frac{m(t)}{L}\Big] \tag{5}$$

Where Q is the sum of all terms in Equation 3 not containing $m(t)/L$ and ϕ is the sum of all terms involving $m(t)/L$ in Equation 3.

Equation 5 has a solution

$$m(t)=\frac{L\{exp\,[l+(Q+\phi)t]-Q/\phi\}}{1+exp\,[l+(Q+\phi)t]} \tag{6}$$

where l is a constant of integration.

Constraints must be imposed on Equation 6 in order to allow it to represent the innovation process. First, as time goes backward the market share of the innovation must tend toward zero

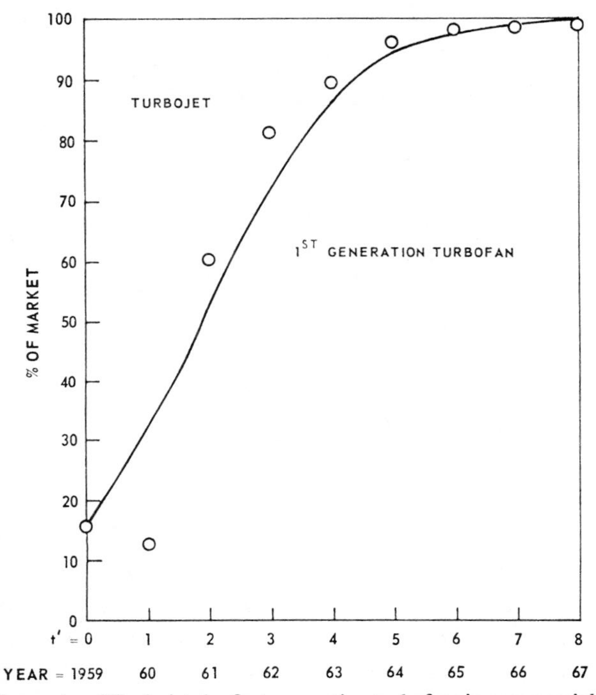

Fig. 1. Displacement dynamics of Turbojets by first generation turbofans in commercial jet engine market.

$$\lim_{t \to -\infty} m(t) = 0 \tag{7}$$

This constraint requires Q to be zero, and Equation 6 becomes

$$m(t) = L\ [1 + exp - (l + \phi t)]^{-1} \tag{8}$$

If t is replaced by t' which is defined as

$$t' \equiv t - t_1 \tag{9}$$

where t_1 is the year in which the first market penetration occurs, an additional constraint may be written as

$$m(t') = N_o \quad \text{when } t = t_1 \tag{10}$$

N_o is defined as the market share obtained by the new product innovation at the end of its first year on the market. Applying these constraints to Equation 8 yields

$$m(t - t_1) = L[1 + \left(\frac{L}{N_o} - 1\right)\ exp - \phi\ (t - t_1)]^{-1} \tag{11}$$

which can be rewritten as

$$\ln \left[\frac{m}{L - m}\right] = -\ln\left(\frac{L}{N_o} - 1\right) + \phi\ (t - t_1) \tag{12}$$

where the functional notation on m has been dropped.

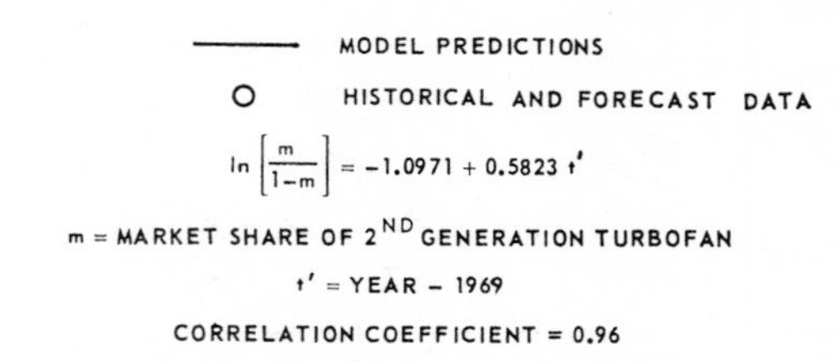

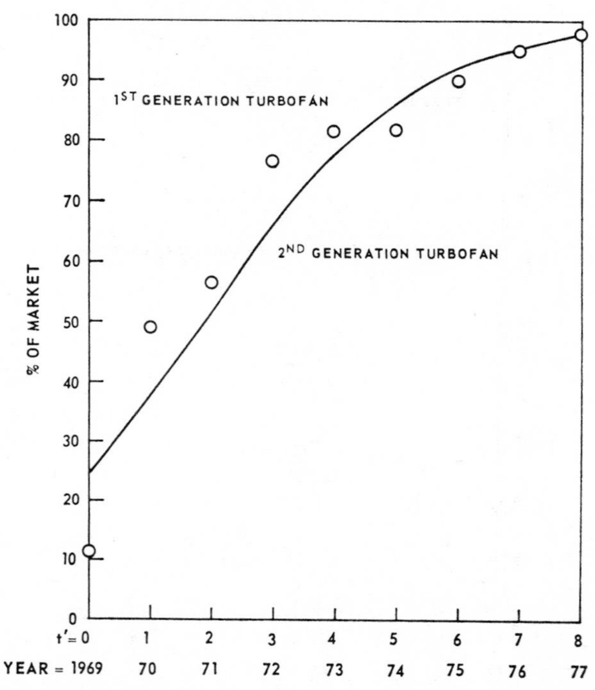

Fig. 2. Displacement dynamics of first generation turbofan by second generation turbofan in commercial jet engine market.

IV. Results and Discussion

To check the applicability of the product innovation rate model expressed by Equation 12 to the commercial aircraft engine market, the rate of displacement of the commercial turbojet engines by the first generation turbofan engines was investigated. Data for the free-world market shares of turbojets and first generation turbofans were used in a linear regression analysis, to examine the extent to which Equation 12 correlated the data. The results are presented in Fig. 1, which show that excellent agreement between the model and the data was obtained, as indicated by a multiple correlation coefficient of 0.97.

The model was next used to investigate the displacement dynamics of the first generation turbofan by the second generation turbofan. Sales forecast data were utilized in a linear regression analysis as before to examine the extent to which Equation 12 correlated the data. The results are shown in Fig. 2. Again, good agreement was obtained between the model and the data. A multiple correlation coefficient of 0.96 resulted.

Figure 3 presents the two sets of data on the same plot.

It is shown in Reference 1 that the ϕ parameter in Equation 12 which governs the rate

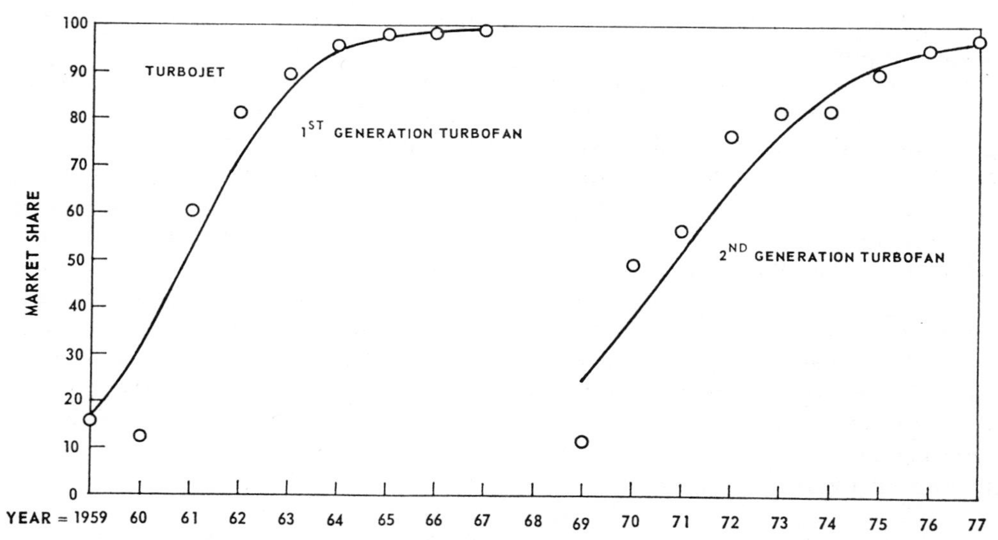

Fig. 3. Relative substitution rates of first and second generation turbofans in commercial jet engine market.

of displacement of an existing product's market share by a competing new product innovation can be expressed as

$$\phi = Z + 0.530\,\pi - 0.027\,S \tag{13}$$

where Z is a constant representative of a given industry; π is a profitability index which can be approximated by the average rate of return from the innovation divided by the cost of capital for the industry; and S is an investment index determined by the average

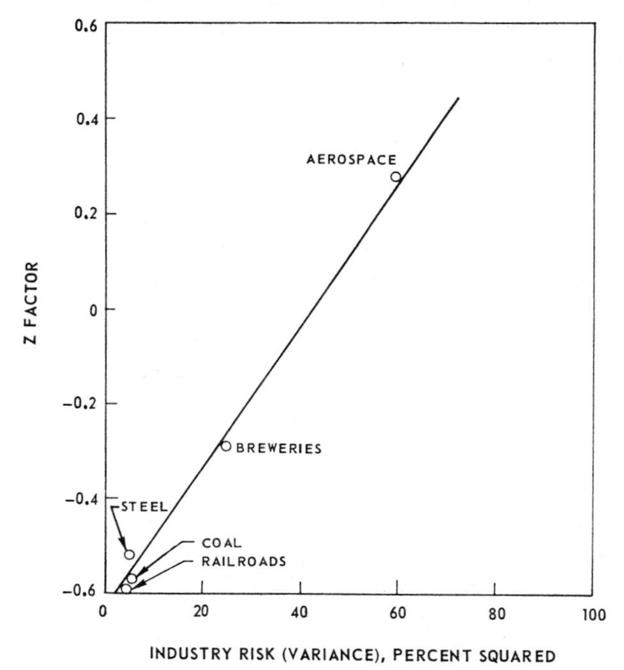

Fig. 4. Variation of Z factor with industry risk.

initial investment in the innovation as a percentage of the total assets of the firms introducing the product innovation.

Using the two values of ϕ obtained from the regression analysis (see equations given in Figs. 1 and 2) the constant, Z, in Equation 13 applicable to the first and second generation turbofan substitutions was evaluated from estimates of π and S. To make these calculations, it was assumed that the rate of substitution of new jet engine innovations was fundamentally governed by the rate of adoption of new types of aircraft employing the new engines. Based on the values of the Federal Aviation Agency (FAA) and the Civil Aeronautics Board (CAB) allowable rates of return, a before-tax return on total investment of 22.8 percent and a before-tax return on flyaway-aircraft price of 30 percent were used in Reference 2 as criteria for evaluating investments in the aircraft industry. Using these same criteria, a value of 1.3 for π results. It was assumed that π was the same value for the first and second generation substitutions. The following relationship was obtained between the two values of S for the first and second generation turbofan by inserting the applicable values for ϕ into Equation 13 for each case and then subtracting the two equations to eliminate π and Z:

$$S_2 = 10.55 + S_1 \tag{14}$$

It was assumed that the initial investment by the airlines in aircraft employing a jet engine innovation would be governed by the purchase of sufficient equipment to serve a constant percentage of the total passenger demand. It follows that

$$\frac{S_1}{S_2} = \left(\frac{D_1}{D_2}\right)\left(\frac{P_1}{P_2}\right)\left(\frac{C_2}{C_1}\right)\left(\frac{A_2}{A_1}\right) \tag{15}$$

where D=total passenger demand; P=aircraft purchase price; C=aircraft passenger capacity; A=average total assets of firms adopting the innovation. The subscripts 1 and 2 refer to the first and second generation turbofan substitutions, respectively.

The total passenger demand increased by a factor of approximately three from 1959, the year of initial first generation turbofan substitution, to 1969, the year of initial second generation turbofan substitution. The price ratio was approximately 5 million divided by 20 million; the capacity ratio, C_2/C_1 was approximately 380 divided by 140 and the asset ratio A_2/A_1, was approximately 0.773 million to 0.539 million. To evaluate the A_2/A_1 ratio, the assets of four leading trunk airlines were obtained from Moody's and an average for the four firms was computed for 1962 and 1968. Equations 14 and 15 were then solved for values of S_1 and S_2 which then allowed a value of Z to be determined from Equation 13. The following equation for the rate of innovation resulted:

$$\phi = 0.28 + 0.530\,\pi - 0.027\,S \tag{16}$$

As discussed in Reference 1, the value of Z in Equation 16 is a characteristic of a given industry and is indicative of the propensity of that industry toward innovation. If the Z value obtained in Equation 16 is compared with the values given in Reference 1 for the brewing, coal, steel, and railroad industries, the innovative propensity in the jet engine industry is relatively high, which probably is a reflection of the greater emphasis on research and development in aerospace vis-à-vis the industries studied in Reference 1.

It can be hypothesized that the propensity of an industry to innovate is related to the risk characteristics of the industry, with those industries which exhibit higher risk also exhibiting a greater innovative propensity. Data are given in Reference 3 which compare the risk characteristics of various industries in terms of the variance of their historical rate-of-return based on book value. When these data were plotted against the values of the Z factors from Reference 1 and the value obtained from Equation 16, the

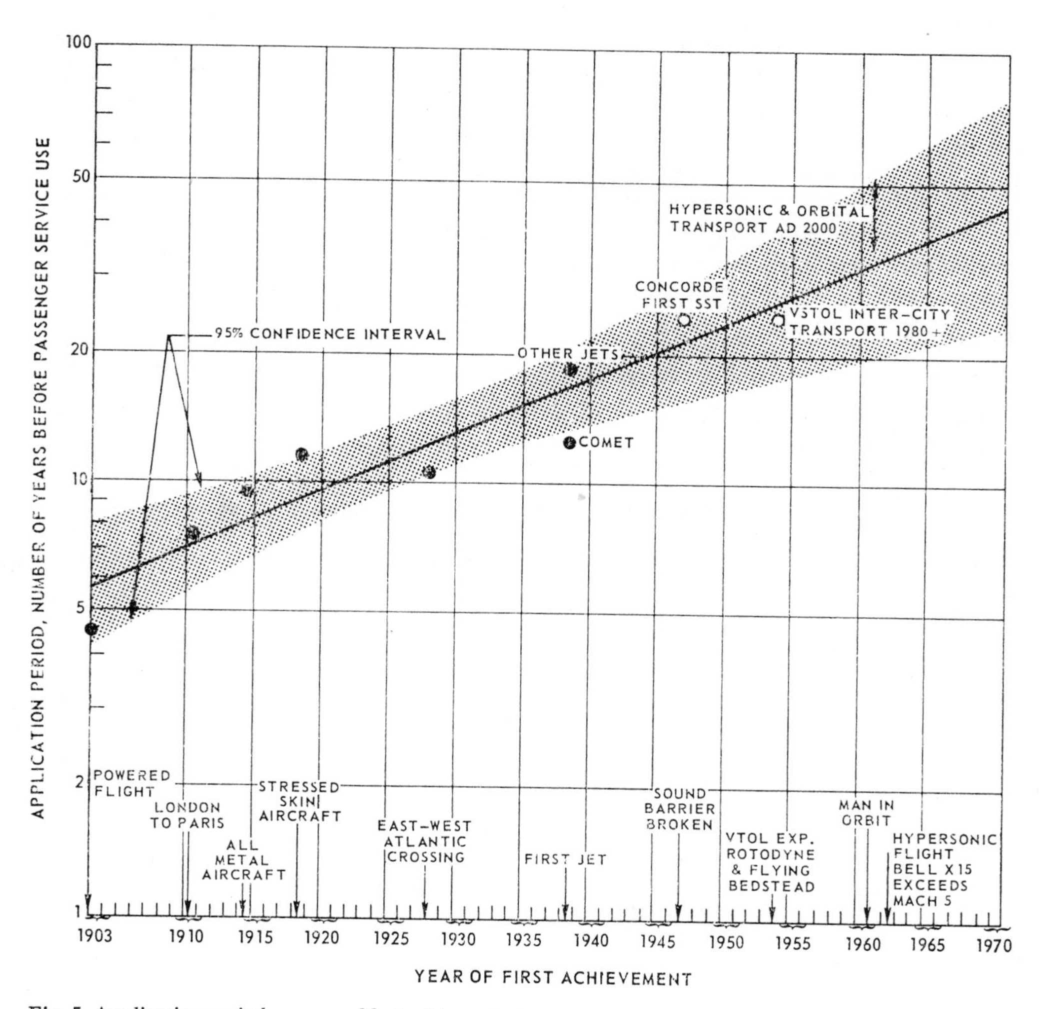

Fig. 5. Application period vs. year of first achievement.

correlation shown in Fig. 4 resulted. Although this correlation appears to support the hypothesis, innovation data for many more industries are needed before much confidence should be placed in the correlation. Because there is a large amount of scatter in the risk characteristics of the various industries given in Reference 3, it is believed that much more scatter would show up in the correlation of Fig. 4 if it were possible to include a greater number of industries.

Comparison of the ϕ values which were obtained for the first and second generation turbofan substitutions indicates a slower rate of innovation in the future as compared with that which has prevailed in the past. This slower rate of innovation is apparent when the data of Figs. 1 and 2 are plotted together as shown in Fig. 3. As previously indicated, the most plausible cause of the slower rate of substitution appears to be the increased investment required to effect substitution (see Reference 4).

Correlations presented in Fig. 5 based on data from Reference 5 indicate that the time between the first demonstration of a new concept in the aerospace industry and its eventual application commercially will increase in the future. Thus it may be expected that the total cycle for substitution may become longer in the future; i.e., the total time required to demonstrate a concept, apply it commercially, and then have it generally adopted by a large segment of the market appears to be increasing. This trend probably results from the general maturation process occurring in aerospace as well as increasing complexity and cost of invention and innovation.

In conclusion, it appears that the Mansfield model (Reference 1) can be applied to indicate the rate of innovation in the commercial jet aircraft engine market and should be useful in predicting market shares achieved as a function of time by a new product substitution.

References

1 Mansfield, Edwin, "Technical Change and the Rate of Imitation," *Econometrica*, Vol. 29, No. 4 (October 1961).
2 Asher, N. J., *et al.*, *Demand Analysis for Air Travel by Supersonic Transport*, Vol. 2, pp. 176–177. Institute for Defense Analysis, Arlington, Va, December 1966.
3 Conrad, G. R. and I. H. Plotkin, "Risk/Return: U.S. Industry Pattern," *Harvard Business Review* (March–April 1968).
4 Molloy, J. F., Jr., "The $12-Billion Financing Problems of U.S. Airlines," *Astronautics and Aeronautics* (October 1968), pp. 76–79.
5 Boorer, N. W., "The Future of Civil Aviation," *Futures* (March 1969), pp. 206–224.

An Innovation Index Based on Factor Analysis

A. WADE BLACKMAN, JR.
EDWARD J. SELIGMAN
GENE C. SOGLIERO

ABSTRACT

Factor analysis techniques are applied to develop an *innovation index* which indicates the relative tendency of various industrial sectors to innovate. The index is derived from various input variables which reflect the extent to which resources are allocated to achieve innovation and output variables which measure the extent to which new product and process innovation is achieved. In constructing the innovation index, each of the variables is weighted in accordance to its involvement in factor patterns which best reproduce correlations in the set of statistics. The innovation index is applied to rank-order various industrial sectors, and changes in innovation characteristics over time are indicated by index values determined for 1960, 1963, and 1970. Relationships are established between values of the innovation index and interindustry differences in the dynamics of the development of markets created through technological innovation. Application of these relationships for projecting the rate at which markets will develop for new technological innovations is discussed.

Introduction

Essential to progress in any field of activity is a standard of measure or yardstick by which to judge if conditions are improving or getting worse. In the field of economics, for example, one such measure of progress is provided by the national income and product accounts. In many areas of activity, however, such yardsticks do not exist at all or if they do exist they do not give objective measures of progress. Statistical indicators are often selected and used on the basis of their availability and/or their ability to support (or disqualify) a point-of-view. In many cases little attention is given to either the applicability of the statistical measures used or the relative weights which are placed on the various measures employed.

During the past few years, concern has increasingly been expressed over the lack of adequate yardsticks by which to assess conditions existing in the socioeconomic environment and to measure the rate at which progress is being made. As a result of this concern, NASA became the first government agency to seek to develop ways of measuring the impact of space exploration on society in general. The result of this effort was the publication by Bauer in 1966 of a volume entitled *Social Indicators* (Ref. 1) which received widespread international consideration. In 1967, the U.S. Government established the Office of Assistant Secretary of Social Indicators in the Department of Health, Education, and Welfare for the purpose of developing systems for measuring social progress. In 1969, Wilson (Ref. 2) developed an extensive system of socioeconomic indicators to measure the quality of life in the U.S. and provided rankings of individual states in a number of areas of social concern.

An area of great importance which has not been considered in past work on the

This article appeared in *Technological Forecasting and Social Change*, Vol. 4, No. 3, 1973.

development of socioeconomic indicators is the development of yardsticks for measuring the relative rates of innovation in the industrial sectors of the U.S. economy. A key requirement for continued socioeconomic progress is an economically strong industrial base. Ultimately, the maintenance of a strong economic position depends on the rate at which new innovations are developed and exploited in the market place. Although it is generally agreed that the total benefits of technological innovation exceed its total costs, it is rare that innovation occurs without exacting at least some social costs in the form of worker dislocation, creating the need for retraining, the psychological stress of change, etc. Therefore an index which would indicate periodically the innovative state of the industrial economy relative to what it has been in the past would be expected to be related to future economic and social conditions in a number of important areas.

The objective of the work described herein was to develop an index which will provide an indication of the innovation characteristics in various industrial sectors of the economy. This *index* is designed to measure *innovation* characteristics in an industrial sector on the basis of a single measure obtained by aggregating statistics related to the innovation process into a meaningful summary statistic.

Methodology

The development of a summary statistic by which to assess differences in innovation characteristics among various industrial sectors requires the following steps (see Ref. 2): (1) selection of the unit of analysis (i.e., the number of industrial sectors which are to be considered); (2) collection of a relevant set of basic statistics from which a summary statistic can be created; and (3) aggregation of the basic statistics into a meaningful summary statistic.

Selection of the industrial sectors considered in the development of the innovation indicator (Step 1) was influenced by the statistical data base selected in Step 2. The requirements established for selecting the data base included the following:

1. Statistical data were to be collected on a range of variables which could conceivably be expected to be causally related to the rate of innovation in various industrial sectors.
2. The data base was as far as possible to be internally consistent (in the definition of variables, etc.) over a wide range of industrial sectors for a given time period.
3. The data were to be consistent historically and cover a time span of at least a decade.

After surveying a number of available data sources, it was decided that the survey of research and development expenditures published annually by the Economics Department of the McGraw-Hill Company best fulfilled the above requirements and that these data (augmented as required by other data sources) would form the primary data base for the study.[1] The unit of analysis selected in Step 1 was chosen to be the industrial sectors considered in the McGraw-Hill survey.

Aggregation of the data into a summary statistic (Step 3) was accomplished through the use of common factor analysis, and detailed mathematical discussions of this technique are presented in Refs. 3–5. An approach similar to that discussed in Ref. 2 was employed to develop a single indicator from an initial set of statistics.

[1]The authors are indebted to Mr. Douglas Greenwald for supplying the McGraw-Hill data used in this study.

In factor analysis, each of n observed variables is described linearly in terms of m new uncorrelated common factors: $F_1, F_2, \ldots F_m$ and unique factors U_j, $(j = 1, 2, \ldots n)$

$$Y_1 = a_{11}F_1 + a_{12}F_2 + \cdots + a_{1m}F_m + b_1 U_1,$$
$$\vdots$$
$$Y_n = a_{n1}F_1 + a_{n2}F_2 + \cdots + a_{nm}F_m + b_n U_n,$$

where Y_j = a standardized form of a variable with known data, a_{jm} = the factor loading or weight for each factor F_m = a function of some unknown variables, U_j = a unique factor, and b_j = a unique factor weight.

The variables entering into each function, F_m, are unknown and are related in unknown (but not necessarily linear) ways. The equations relating the functions themselves are linear. Each function makes a contribution to the sum of the variances of the variables, and in general a few of the functions will account for a large amount of the total variance. The factor analysis technique provides values for the constants, a_{jm}, called loadings, which represent the extent to which each specific function is related to Y_j. Once the factor loadings or weights for each variable are determined, the initial set of statistics can be aggregated through the determination of factor scores into a single index in which each variable is weighted proportionally to its involvement in a pattern; the greater the involvement, the higher the weight. For example, an index, I_l, constructed from the first factor loadings may be expressed (see Ref. 2) as

$$I_l = \sum_{j=1}^{n} \frac{a_{jl}}{\lambda_l} Y_j,$$

where λ_l is the eigenvalue for the first factor. The computational techniques used in the determination of factor loadings and factor scores are discussed in detail in Ref. 4.

In the results presented herein, the *innovation index* was developed from the factor scores for the first factor, and in no case did the first factor account for less than 53 percent of the total variance in the data.

Results and Discussion

The overall results of this study are divided into three parts. First, the innovation characteristics of selected industrial sectors of the economy are discussed; next, the changes which have occurred in innovation characteristics from 1960 to 1970 are presented; and finally relationships between values of the Innovation Index of an industry and the dynamics of the development of markets created by technological innovation in the industry are developed.

INNOVATION CHARACTERISTICS OF SELECTED INDUSTRIAL SECTORS

The data presented in Table 1 were first used in the factor analysis model to develop values for the innovation index for various industrial sectors. The variables which were hypothesized to reflect the innovation characteristics are indicated in Table 1. The selection of these variables must be carefully considered, because the relevancy of the summary indicator is dependent upon the relevancy of the measures from which it is constructed. The selection of these measures was influenced to some degree by the data base available. Although a number of data sources were considered initially, it was difficult to obtain data which were consistent over a wide range of industrial sectors and covered a reasonable historical time span. As previously discussed, of the various sources available, the McGraw-Hill survey of business plans for research and development, which is published annually by their economics department, appeared to best meet the requirements of this study, and consequently, it was decided to base the initial values of the indicator on these data. The variables selected in Table 1 can be seen to consist of five input measures and one output measure.

TABLE 1
Data for Six-Variable Model in 1970
Source: Economics Department, McGraw-Hill

	Total R/D Expenditures, $\$ \times 10^{-6}$		Federally Financed R/D, as Percent of R/D Expenditures		R/D as Percent of Capital Spending	Estimated New Product Sales as Percent of 1974 Sales
	Estimate 1970	Planned 1974	1970	1974	1970	
Steel	140	168	0.7	0.2	7.00	8
Nonferrous metals	123	173	6.9	4.8	9.92	10
Machinery	1,716	2,211	20.0	15.0	49.45	24
Electrical machinery and communications	4,801	6,768	40.6	26.6	211.50	19
Aircraft and missiles	5,494	6,215	72.9	67.0	1,017.41	31
Autos and other transportation equipment	1,328	1,531	37.1	25.0	89.20	46
Fabricated metals and ordnance	190	240	4.5	2.8	15.32	15
Professional and scientific instruments	719	898	21.8	17.4	102.71	23
Chemicals and allied products	1,608	1,886	11.2	8.5	46.74	13
Paper and allied products	141	176	1.2	1.2	8.55	19
Rubber products	245	302	11.8	9.8	26.06	12
Stone, clay, and glass	169	191	2.0	1.0	17.07	14
Petroleum products	583	648	6.2	6.0	10.37	7
Food and kindred products	186	228	1.0	0.8	6.58	10
Textile mill products and apparel	61	78	2.0	1.6	8.93	18
Average values	1,166.93	1,447.53	15.99	12.51	108.45	17.93

Table 2 presents a correlation matrix which indicates the degree of linear relationship between the row and column variables of the matrix. If the coefficient is squared and multiplied by 100, it will express the percent variation in common for the data on the two variables. For example, the correlation coefficient of 0.82 between R/D expenditures in 1970 and R/D as a percent of capital spending means that 67 percent of the variation of the various industrial sectors of Table 1 on these two characteristics is in common. The principal diagonal of the correlation matrix in Table 2 contains communality estimates (expressed as the square of the multiple correlation coefficients) which measure the variation of a variable in common with all the others together.

TABLE 2
Correlation Matrix of Data of Table 1

Innovation Variable	1	2	3	4	5	6
1. R/D expenditures, 1970	0.99					
2. R/D expenditures planned, 1974	0.99	0.99				
3. Percent federally financed R/D, 1970	0.91	0.87	0.99			
4. Percent federally financed R/D, 1974	0.88	0.83	0.99	0.99		
5. R/D as percent of capital spending	0.82	0.75	0.88	0.94	0.98	
6. Estimated new product sales as percent of 1974 sales	0.44	0.41	0.69	0.63	0.44	0.83

The low coefficient of correlation between R/D expenditures in 1970 and new product sales as a percent of 1974 sales is noteworthy and would indicate that the creation of new products in the future is not highly related to R/D expenditures. This result would indicate that there is currently a large amount of R/D effort expended on improving existing products and/or a need exists to improve the output of R/D laboratories in terms of the number of new products created per dollar of R/D expenditure.

TABLE 3
Unrotated Factor Matrix

Variables	Factors 1	2	3	Communality
1. R/D expenditures, 1970	0.950	–0.259	–0.166	0.997
2. R/D expenditures planned, 1974	0.914	–0.300	0.270	0.998
3. Federally financed R/D, 1970	0.990	0.124	–0.003	0.995
4. Federally financed R/D, 1974	0.982	0.081	0.161	0.967
5. R/D as percent of capital spending	0.905	–0.092	0.382	0.973
6. Estimated new product sales as percent of 1974 sales	0.614	0.655	=0.155	0.830
Percent total variance	81.32	10.28	4.94	$\Sigma=96.54$
Percent common variance	84.23	10.65	5.12	
Eigenvalues	4.879	0.617	0.296	

Table 3 presents the unrotated factor loading matrix for the data in Table 1. The rotated factor loadings were essentially the same as the unrotated values. The columns define the factors and the rows refer to the variables. The loading for the row variable on the column factor is given at the intersection of the row and column. The number of columns corresponds to the number of independent uncorrelated patterns of relationships among the variables; in Table 3 there are shown three such independent patterns corresponding to three F functions in Eq. (1). The loadings indicate the degree to which the variables are involved in the factor patterns, and the square of the loading multiplied by 100 gives the percent variation that a variable has in common with an unrotated pattern. The loadings correspond to values for a_{jm} in Eq. (1). It can be seen that the first factor pattern accounts for the greatest amount of variation in the data, and the amount of variation accounted for by the other factors decreases successively. The eigenvalues shown at the bottom of the table represent the sum of the squared factor loadings in a column. The percent total variance is computed by multiplying the eigenvalue by 100 and dividing by the number of variables. It can be seen that the first factor accounts for approximately 81.3 percent of the total variation in the data and all three factors account for about 96.5 percent of the total variation. The percent common variance indicates the variation among all the variables involved in a pattern as a percent of that involved in all the three patterns and is computed by dividing the percent total variance for a particular factor by the percent total variance for all three factors.

The communality column is the sum of the squared factor loadings and indicates the proportion of a variable's total variation that is involved in the factor patterns. As discussed in Ref. 3, the communality can provide a measure of uniqueness of a variable. By subtracting from 100 the percent of variation in common with the factor patterns, a measure will be obtained which indicates the degree to which a variable is unrelated to the others; i.e., the degree to which it cannot be derived from data on the other variables. It can be seen that the communalities of all the variables shown in Table 3 are quite high, indicating that practically all of the variation in the data is explained by the three factors.

As discussed in Ref. 4, factor scores can be derived which give values for each case (i.e., each industrial sector) for the functions F of Eq. (1). Individual industrial sectors will have high or low factor scores depending on the values of the variables entering into a factor pattern. The factor scores can provide a properly weighted yardstick for evaluating the relative innovation characteristics of the various industrial sectors. Because the first factor in Table 3 accounts for 81.3% of the total variance in the data, factor scores for the first factor were used to indicate the relative innovation characteristics among the various industrial sectors.

TABLE 4
Ranking of Industrial Sectors
According to Value of Innovation Index
Determined from First Factor Scores

Industrial Sector	Innovation Index
Aircraft and missiles	3.004
Electrical machinery and communications	1.418
Autos and other transportation equipment	0.718
Machinery	0.224
Professional and scientific instruments	0.091
Chemicals and allied products	–0.057
Rubber products	–0.345
Petroleum products	–0.443
Nonferrous metals	–0.538
Fabricated metals and ordnance	–0.590
Stone, clay, and glass	–0.669
Textile mill products and apparel	–0.689
Paper and allied products	–0.694
Food and kindred products	–0.703
Steel	–0.725

TABLE 5
Data for Eight-Variable Model

	Total R/D Expenditures[a] 1970	R/D Expenditures[b] Planned 1974	Company Sponsored R/D Expenditures[a] 1970	R/D as % of Capital Spending[b] 1970	Estimated New Product Sales as % of 1974 Sales[b]	Value Added[c] 1969	No. of Mergers[c] & Acquisitions 1970	Ratio of Federal Reserve Board Index[d] 1968/1947
	$\$X10^{-6}$	$\$X10^{-6}$	$\$X10^{-6}$			$\$X10^{-6}$		
Electrical machinery and communications	4,325	6,768	2,062	211.5	19	28,275	145	3.5
Aircraft and missiles	5,173	6,215	1,107	1,017.4	31	12,829	20*	8.0
Autos and other transportation equipment	1,475	1,531	1,232	89.2	46	18,356	47*	2.9
Fabricated metals	183	240	177	15.3	15	20,841	54	3.2
Professional and scientific instruments	694	898	508	102.7	23	7,589	49	3.2
Chemicals and allied products	1,810	1,886	1,622	46.7	13	27,177	102	5.0
Paper and allied products	141*	176	140*	8.5	19	11,284	29	2.8
Rubber products	238	302	198	26.1	12	8,495	25	3.9
Stone, clay, and glass	188	191	185	17.1	14	10,049	41	2.0
Petroleum products	608	648	565	10.4	7	5,725	7	2.0
Food and kindred products	198	228	196	6.6	10	30,120	108	1.2
Textile mill products and apparel	63	78	63	8.9	18	9,672	45	1.5

* Estimated.

[a] Source: National Science Foundation.

[b] Source: Table I.

[c] Source: Statistical Abstract of the U.S. 1971.

[d] Source: Ref. 6.

Table 4 presents standardized regression estimates of the factor scores for each of the various industrial sectors. As these scores have been scaled so they have a mean of zero and about two-thirds of the values lie between +1.00 and −1.00, scores greater than +1.00 or −1.00 can be considered unusually high or low. It can be seen that the aircraft, electrical machinery and communications, and the auto and transportation equipment sectors have the highest values of the innovation index among the various industrial sectors. The paper and allied products, food, and steel sectors have the lowest values.

Although the results presented in Table 4 appeared generally plausible, it was believed that many variables important to the innovation process were not expressed in the initial formulation of the six-variable model. It was also believed that the data in Table 1 for the estimated new product sales as a percent of 1974 sales for the auto sector contained some definitional difficulties (in the sense that model changes represented new products) and tended to overrate the innovation characteristics of the auto sector. It appeared that these problems could be overcome to some extent by adding to the model additional output variables which would express other measures of innovation such as process innovation, acquisition activity, etc. The data of Table 5 were collected, and a new eight-variable model was formulated. The magnitude of the R/D expenditures by the firm was included in addition to total R/D expenditures in order to reflect the results of Leonard in Ref. 7 in which it was found that company R/D funding as a percentage of sales provided the best correlations with various measures of the rate of industrial growth. Federal R/D funding as a percentage of sales and total R/D funding as a percentage of sales did not correlate with growth measures as consistently as did company R/D funding as a percentage of sales which tends to indicate that much of past defense related R/D funding has had a relatively small effect on a number of measures of industrial growth. Value added was included as a variable to reflect the total output of a sector, and acquisition and merger activity was added as a measure of acquired technology. The Federal Reserve Index ratio was added as a measure of productivity growth. The revised, eight-factor model, contained three measures of output (viz., new product sales as a percentage of 1974 sales, value added, and the Federal Reserve Index ratio) and five measures of input.

The correlation matrix of the data of Table 5 obtained as discussed previously is shown in Table 6. It can be seen that there is a rather low correlation between total R/D spending and new product sales as a percent of 1974 sales as was obtained previously (see Table 2); however, a fairly high correlation is indicated between total R/D spending and productivity growth as measured by the Federal Reserve Index ratio. Very little

TABLE 6
Correlation Matrix of Data of Table 5

Innovation Variable	1	2	3	4	5	6	7	8
1. Total R/D expenditures, 1970	0.99							
2. R/D expenditures planned, 1974	0.98	0.99						
3. Company sponsored R/D expenditures, 1970	0.79	0.80	0.99					
4. R/D as percent of capital spending, 1970	0.83	0.76	0.37	0.99				
5. Estimated new product sales as percent	0.41	0.35	0.36	0.43	0.62			
6. Value added, 1969	0.30	0.33	0.52	−0.03	−0.01	0.92		
7. Number of mergers and acquisitions, 1970	0.29	0.38	0.57	−0.15	−0.11	0.88	0.95	
8. Ratio of federal reserve board index, 1968/1947.	0.76	0.68	0.49	0.84	0.35	0.04	−0.10	0.85

correlation appears to exist between total R/D spending and either value added or acquisition and merger activity.

TABLE 7
Unrotated Factor Matrix

Variables	Factors 1	2	3
1. Total R/D expenditures, 1970	0.702	–0.611	0.086
2. R/D expenditures planned, 1974	0.686	–0.543	0.173
3. Company sponsored R/D expenditures, 1970	0.583	–0.302	0.247
4. R/D as percent of capital spending	0.573	–0.286	–0.200
5. Estimated new product sales as percent of 1974 sales	0.309	–0.262	–0.249
6. Value added, 1969	0.295	–0.533	–0.313
7. Number of mergers and acquisitions, 1970	0.269	0.005	–0.029
8. Ratio of federal reserve board index, 1968/1947	0.536	0.006	–0.073
Percent total variance	53.81	26.10	5.82
Percent common variance	62.77	30.44	6.79
Eigenvalues	4.30	2.09	0.47

Table 7 presents the unrotated factor loading matrix for the data of Table 5. The rotated factor loadings were essentially the same as the unrotated values.

The standardized regression estimates of the eight-variable model factor scores (obtained as previously discussed) are shown in Table 8. The values shown are the scores for the first factor which accounted for 53.8% of the total variance in the data. If these results are compared with Table 4, it can be seen that slight changes in the relative rankings of the sectors have resulted; notably, the chemicals and allied product sector has moved from sixth place in Table 4 to third place in Table 8. Because the data base used in obtaining the results of Table 8 is somewhat more extensive than that used to obtain Table 4, the rankings of Table 8 are preferred.

VARIATION OF INNOVATION INDEX WITH TIME

Once values were established for the innovation index (as described in the preceding discussion), the questions arose as to how historical values for the innovation index have varied with time and if significant trends in the index for particular industrial sectors exist. In order to address these questions, it was necessary to establish a data base which was consistent over time. To achieve this consistency, it was necessary to reduce the number of variables considered from the six shown in Table 1 to the four shown in Tables 9–11 for the years 1960, 1963, and 1970, respectively, because the data base only included consistent values for these four variables for these three years. The super P factor analysis procedures discussed in Ref. 8 were then applied to the statistics for the reduced number of variables, and the first factor scores which were obtained are presented in Table 12. The data for 1970 were utilized in both the four-variable model and the six-variable model to indicate the effects of reducing the number of variables. The amount of total variance explained by the first factor was reduced from 81.3% to 72.5% as the number of variables considered was reduced from six to four. The standardized regression estimates of the factor scores for the first factor are given in Table 12 for the various industrial sectors for the years 1960, 1963, and 1970. For the year 1970, the six-variable model factor scores are compared with those of the four-variable model. Although slight differences occur in the factor scores as would be expected, both models indicate the same general order of the innovation characteristics of the various industrial sectors with only a few exceptions.

TABLE 8
Ranking of Industrial Sectors
According to Value of Innovation Index
Determined from First Factor Scores
Eight–Variable Model

Industrial Sector	Innovation Index
Aircraft and missiles	2.29
Electrical machinery and communication	1.76
Chemicals and allied products	0.60
Autos and other transportation equipment	0.29
Food and kindred products	–0.35
Professional and scientific instruments	–0.37
Fabricated metals and ordinance	–0.60
Petroleum products	–0.64
Stone, clay, and glass	–0.70
Paper and allied products	–0.75
Textile mill products and apparel	–0.75
Rubber products	–0.76

TABLE 9
Data for Four-Variable Model in 1960
Source: Economics Department, McGraw-Hill

	Total R/D Expenditures, $X10^{-6}$		1960 R/D as % of 1960 Capital Spending	Estimated New Products as % of 1964 Expected Sales
	Actual 1960	Planned 1964		
Primary metals	143.6	175.2	7.5	18
Machinery	684.8	826.9	62.0	23
Electrical machinery and communications	1,816.9	2,440.3	267.0	16
Aircraft and missiles	3,477.8	3,721.2	830.0	25[a]
Autos and other transportation equipment	849.0	849.0[a]	95.4	40
Chemicals and allied products	741.2	919.3	46.3	20
Paper and allied products	69.1	93.4	9.2	12
Rubber products	89.3	108.8	38.8	7
Stone, clay, and glass	86.1	111.2	13.9	13
Petroleum products	309.7	338.1	11.7	5
Food and kindred products	106.9	144.6	11.6	12
Textile mill products and apparel	37.0	46.6	7.0	10
Average values	700.95	818.30	116.70	16.75

[a]Estimate.

TABLE 10
Data for Four-Variable Model in 1963
Source: Economics Department, McGraw-Hill

	Total R/D Expenditures, $\$X10^{-6}$ 1963	Planned 1967	1963 R/D as % of 1963 Capital Spending	Estimated New Products as % of 1967 Sales
Primary metals	169	201	10.2	22
Machinery	1,180	1,397	95.0	20
Electrical machinery and communications	2,731	3,183	396.0	16
Aircraft and missiles	4,539	5,197	1,125.0	33[a]
Autos and other transportation equipment	918	1,029	86.5	16
Fabricated metals and ordnance	154	189	17.1	26
Chemicals and allied products	1,121	1,329	69.7	18
Paper and allied products	82	98	11.4	8
Rubber products	134	157	55.8	19
Stone, clay, and glass	98	138	16.1	14
Petroleum products	331	385	11.33	3
Food and kindred products	132	176	13.6	10
Textile mill products and apparel	239	48	6.1	15
Average Values	894.46	1,040.54	147.22	16.92

[a] 1968 value.

TABLE 11
Data for Four-Variable Model in 1970
Source: Economics Department, McGraw-Hill

	Total R/D Expenditures, $\$X10^{-6}$ Estimate 1970	Planned 1974	R/D as % of Capital Spending 1970	Estimated New Product Sales as % of 1974 Sales
Steel	140	168	7.00	8
Nonferrous metals	123	173	9.92	10
Machinery	1,716	2,211	49.45	24
Electrical machinery and communications	4,801	6,768	211.50	19
Aircraft and missiles	5,494	6,215	1,017.41	31
Autos and other transportation equipment	1,328	1,531	89.20	46
Fabricated metals and ordnance	190	240	15.32	15
Professional and scientific instruments	719	898	102.71	23
Chemicals and allied products	1,608	1,886	46.74	13
Paper and allied products	141	176	8.55	19
Rubber products	245	302	26.06	12
Stone, clay, and glass	169	191	17.07	14
Petroleum products	583	648	10.37	7
Food and kindred products	186	228	6.58	10
Textile mill products and apparel	61	78	8.93	18
Average Values	1,166.93	1,447.53	108.45	17.93

TABLE 12
Variation with Time of Values of Innovation Index for Various Industrial Sectors

	Innovation Indexes			
	Six-Variable Model		Four-Variable Model	
Industrial Sector	1970 Data of Table 1	1970 Data of Table 11	1963 Data of Table 10	1960 Data of Table 9
Aircraft and missiles	3.00	3.56	2.70	2.11
Electrical machinery and communications	1.42	2.19	1.39	0.45
Autos and other transportation equipment	0.72	0.31	0.05	0.00
Machinery	0.22	0.49	0.20	−0.19
Professional and scientific instruments	0.09	−0.20	—	—
Chemicals and allied products	−0.06	0.57	0.16	−0.16
Rubber products	−0.35	−0.51	−0.60	−0.62
Petroleum products	−0.44	−0.19	−0.41	−0.41
Nonferrous metals	−0.54	−0.61	—	—
Fabricated metals and ordnance	−0.59	−0.56	−0.59	—
Stone, clay, and glass	−0.67	−0.55	−0.63	−0.63
Textile mill products and apparel	−0.69	−0.65	−0.66	−0.66
Paper and allied products	−0.69	−0.59	−0.62	−0.64
Food and kindred products	−0.70	−0.54	−0.60	−0.62
Steel	−0.73	−0.57	—	—
Primary metals	—	—	−0.57	−0.59

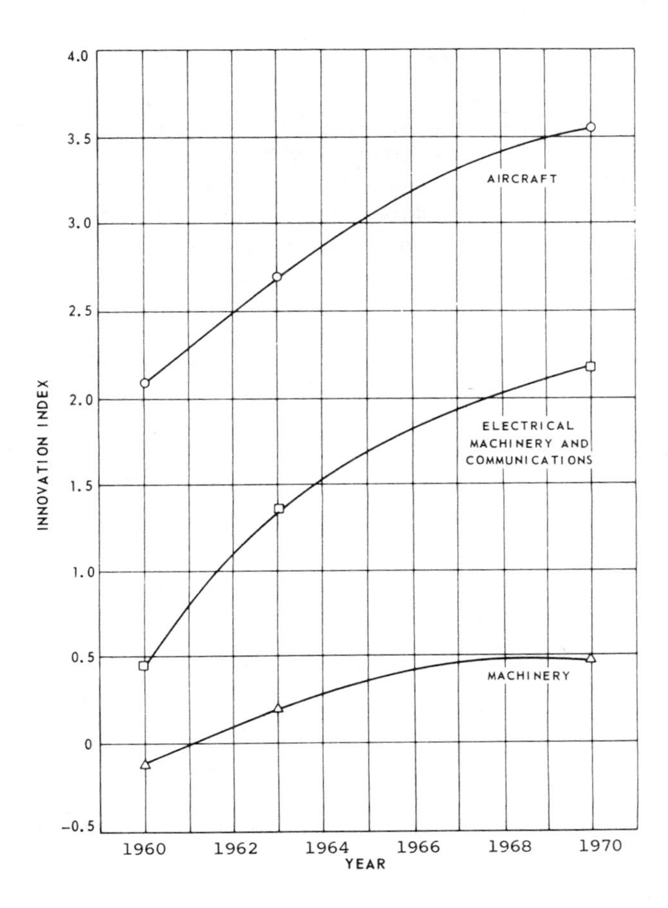

Fig. 1. Variation with time of values of the innovation index for selected industrial sectors.

Figure 1 presents a plot which shows the variation with time of the values of the innovation index for the aircraft, electrical machinery and communication, and machinery sectors. The innovation index for all three of these sectors indicates a rising trend with time. The innovation index values for the electrical machinery and communications sector exhibits the greatest percentage increase.

USE OF THE INNOVATION INDEX IN DETERMINING MARKET SUBSTITUTION RATES

A model which describes the rate at which new markets are created when new technological innovations are substituted for old products has been developed by Mansfield in Ref. 9. Mansfield obtained successful checks on the model by comparing it with historical technological substitutions which have occurred in four disparate industrial sectors; viz., railroads, coal, steel, and breweries. Subsequently, Blackman applied a modified version of the model to describe innovation rates in the commercial aircraft jet engine market (Ref. 10) and various innovations in the electrical utility and automotive sector (Ref. 11).

In the form of the model utilized in Ref. 10, the rate at which a market develops as a result of technological innovation is given by

$$\ln\left(\frac{m}{L-m}\right) = -\ln\left(\frac{L}{N_O} - l\right) + \phi\,(t - t_l),$$

where m = market share captured at time, t, by the new innovation, L = upper limit of the market share which the new innovation can capture in the long run, N_O = market share captured when t = t_l, and ϕ = constant which governs the substitution rate. The form of Eq. (3) is that of a constrained logistic equation and Fig. 2 (from Ref. 10) illustrates the typical shapes of the substitution curves. These curves were obtained by application of the substitution model to innovations in the commercial aircraft jet engine market.

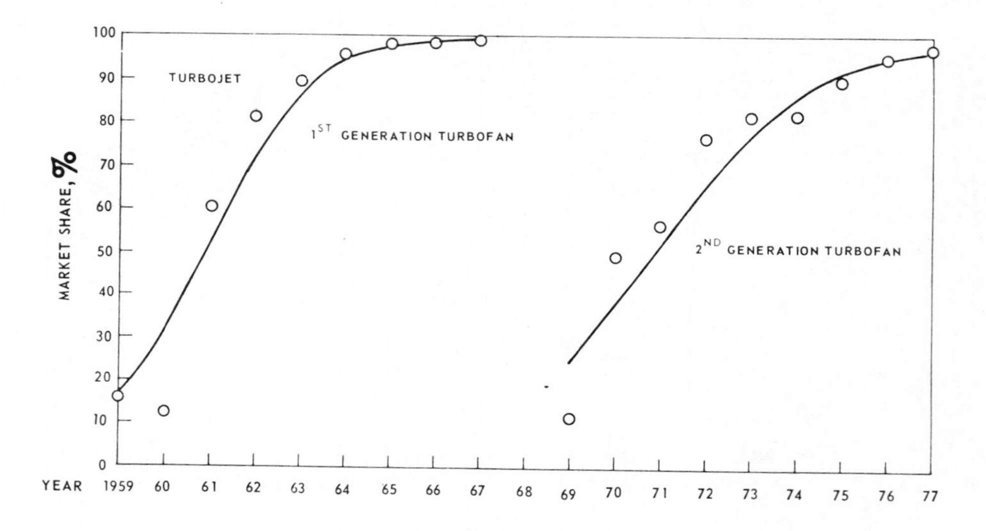

Fig. 2. Relative substitution rates of first and second generation turbofans in commercial jet engine market.

It is shown in Ref. 9, that the constant ϕ which governs the substitution rate can be expressed as

$$\phi = Z + 0.530\,\pi - 0.027S,$$

where Z = a constant representative of a given industry, π = a profitability index (see Ref. 9), and S = an investment index (see Ref. 9).

If the economic characteristics of historical technological innovations are known such that estimates can be made for π and S, it is possible to estimate the value of Z if the value of ϕ is known from historical substitution rates. Using this approach Mansfield (Ref. 9) estimated values for Z for the industrial sectors of railroads, coal, steel, and breweries and Blackman (Ref. 10) estimated a value of Z for aerospace. The magnitude of the Z values for a given industrial sector reflect the propensity of that sector to innovate; i.e., those sectors having low Z values tend to be slow in adopting technological innovations and vice-versa. It might be expected that a relationship would exist between the value of Z in an industrial sector and the value of the Innovation Index for that sector. To test this hypothesis, Z values for the steel and food and kindred products sectors were obtained from Ref. 9; Z values for aerospace were based on Ref. 10; and Z values for auto and electrical machinery sectors were computed for the technological substitutions discussed in Ref. 11. These values were then plotted versus values of the innovation index from Table 4 for the corresponding industrial sectors, and the correlation of Fig. 3 resulted. A regression analysis yielded the correlation equation shown in Fig. 3. The standard deviation of the regression coefficient was 0.0484. A multiple correlation coefficient of 0.91 was obtained which supports the hypothesis that correlation exists between the innovation index and the Z values of the substitution model.

The innovation index values from Table 8 were also found to correlate well with the Z values. The following regression equation was obtained for these data:

$$Z = 0.2221 \quad (I) - 0.3165.$$

The multiple correlation coefficient was 0.92, and the standard deviation of the regression coefficient was 0.0645.

The correlation of Fig. 3 may be used in conjunction with the substitution models described in Refs. 9 and 10 to estimate the interindustry effects on market substitution rates and to estimate the rate at which new markets can be expected to develop for new technological innovations if the economic characteristics of new technological innovation can be estimated. By identifying the industrial sector in which the new innovation will occur, a value of the innovation index can be estimated from Table 4 and/or Table 8 and a corresponding Z factor estimated from Fig. 3. The economic characteristics of the innovation can be used to estimate π and S and Eq. (4) can be solved for ϕ. By

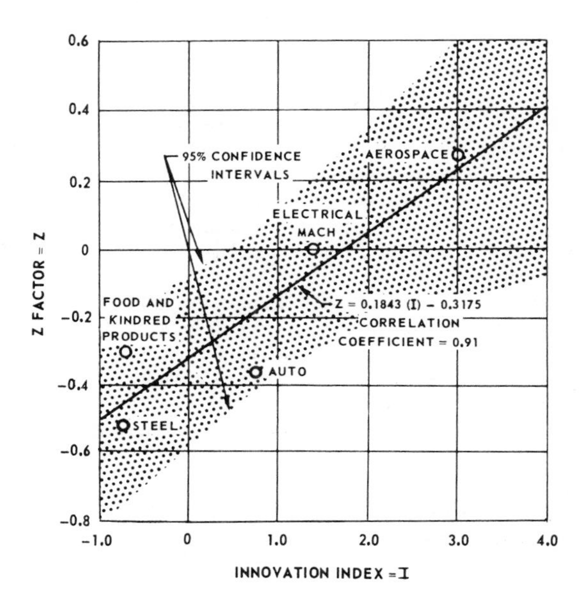

Fig. 3. Innovation index—Z factor correlation.

estimating L and N_O, Eq. (3) can then be solved for the market share of the new innovation as a function of time. Iteration procedures are sometimes necessary because the values of π and S can be a function of the technological substitution rate.

Concluding Remarks

The techniques of factor analysis provide a useful methodology for the development of an innovation index by which the relative innovation characteristics of various industrial sectors can be compared. A considerable variation appears to exist in the innovation characteristics of various industrial sectors of the economy, and to some extent, these characteristics appear to change with time.

Interindustry differences in the dynamics of the development of markets created by the substitution of new products for old products appear to correlate with the relative values of the innovation index. This correlation provides a means for projecting the rate at which market substitution will occur provided the economic characteristics of the new product innovation can be estimated.

It is believed that the results presented offer an example of how factor analysis techniques may be applied to develop an innovation index. The innovation index developed herein is somewhat limited by the lack of available data which describe the innovation process and its change over time. The development of a larger data base in the future may allow similar models to be constructed which contain a greater number of variables. Currently, the primary evidence which indicates that the innovation index developed herein does, in fact, measure interindustry differences in innovation propensity lies in the Z factor correlation. Additional checks on the model in the future would be desirable as well as the development of more extensive statistics related to the innovation process.

The application of these techniques to develop similar models which would measure innovation propensity on an international level would appear to be a useful future extension.

References

1. Raymond A. Bauer (Ed.), *Social Indicators*, M.I.T. Press, Cambridge, Mass. (1966).
2. J.O. Wilson, *Quality of Life in the United States*, Midwest Research Institute Report (1969).
3. R.J. Rummel, Understanding factor analysis, *Conflict Resolution* Vol. XI, No. 4, 444–480.
4. R.J. Rummel, *Applied Factor Analysis*, Northwestern University Press, Evanston, Ill. (1970).
5. H.J. Harman, *Modern Factor Analysis* (2nd ed.), University of Chicago Press, Chicago, Ill. (1967).
6. J.H. Halloman and A.E. Harger, America's technological dilemma, *Technology Review* **73**, 30-40 (1971).
7. W.N. Leonard, Research and development in industrial growth, *J. Pol. Econ.*, 232–256 (March/April 1971).
8. R.J. Rummel, Forecasting international relations: A proposed investigation of three-mode factor analysis, in *Approaches to Long Range Forecasting*, U.S. Government Printing Office, Washington, D.C., 94–111.
9. Edwin Mansfield, Technical change and the rate of imitation, *Econometrica* **29**, 741-765 (1961).
10. * A.W. Blackman, The rate of innovation in the commercial aircraft jet engine market, *Technological Forecasting and Social Change* **2**, 269–276 (1971).
11. A.W. Blackman, A mathematical model for trend forecasts, *Technological Forecasting and Social Change*, **3**, 441-452 (1972).

**This article also appears in Part 2 of this book.*

The Market Dynamics of Technological Substitutions

A. WADE BLACKMAN, JR.

ABSTRACT

Models are discussed which relate to the prediction of the rate of growth of markets which are created when a new product based on a new technology is substituted for an old product. A methodology is described which can be used to forecast the dynamics of the substitution process. Forecasting procedures applicable to cases where historical substitution data are available are discussed. Other procedures which can be used in the absence of historical data are also described. Specific examples of the application of both procedures are given. The methodological structure which is developed provides insights into the causes of inter- and intra-industry differences in the dynamics of market growth of new products based on new technology and can be used to establish criteria useful in evaluating and planning new technology-based products.

Introduction

The creation of new and better ways of satisfying human needs is the essence of the process of technological innovation. If broadly defined, human needs tend to be relatively invariant over time and technological innovation may be viewed as a sequential process in which new technologies better able to supply needs are substituted for old technologies in a seemingly continuous series. The changing manner in which transoceanic passenger transportation needs have been satisfied serves as an example. For many years sailing ships were the primary mode of transoceanic travel. As technological changes occurred, sailing ships were supplanted by steamships which were in turn replaced by aircraft. The process of technological innovation is of vital importance in creating an improved standard of living and a healthy economy. The motivating force behind the creation of technological innovation is the economic reward which is perceived to result from successful innovation. This reward structure induces entrepreneurs to expend a great deal of time and effort and risk large amounts of capital in producing the technological innovations which create new jobs and keep the economy healthy.

The extent of investment in technological innovation is related to the perceived rate at which a market will develop for the new technology, and the rate of market development is in turn a function of the dynamics of technological substitution. Thus the dynamics of technological substitution impacts strongly on the innovation process and thereby exerts a strong influence on the state of the economy.

Mathematical models which provide greater insight into the dynamics of technological substitution are therefore of great importance. The pioneering work in the development of models of the substitution process was done by Mansfield [1] in 1961 in which a model was developed which describes the rate at which new markets are created when

This article appeared in *Technological Forecasting and Social Change*, Vol. 6, No. 1, 1974.

new technologically improved products are substituted for old products. Mansfield obtained successful checks on the model by comparing it with historical technological substitutions which occurred in the four disparate industrial sectors of railroads, coal, steel, and breweries. Subsequently, Blackman made a revision in the model (in which the extent of substitution was defined in terms of market share captured by the new technology rather than in terms of the cumulative number of firms which employed the new technology) and applied the model to describe innovation dynamics in the commercial aircraft jet engine market [2] and innovation dynamics in the electrical utility and automotive sectors [3]. In [4], Blackman et al., developed a correlation based on factor analysis which extends the applicability of the model to other industrial sectors.

Bass [5] has advanced a new-product growth model based on the operation of innovative and imitative behavioral forces and obtained successful checks on the model in the consumer durables market sector. Nevers [6] obtained additional successful checks on the Bass model in the retail service, industrial technology, agriculture, and consumer durables sectors and conducted a detailed comparison of the Bass model with the Mansfield Model. Although it was shown that the two models can theoretically produce different results under conditions where innovative behavior and imitative behavior exert the same order of magnitude effect on sales, for all of the cases which have been examined (both in [5] and [6]) imitative effects have predominated, and the Bass model and the Mansfield model would have given essentially the same results.

In [7], Fisher and Pry describe the development of an empirical substitution model and its application to a number of various innovations. The form of the Fisher-Pry model is essentially the same as the revised form of the Mansfield model.

Both the Bass model and the Fisher-Pry model lack the ability to explain inter- and intra-industry differences in substitution rates which is an important feature of the Mansfield model and the extension of the model by Blackman et al., in [4]. Furthermore, expressing the extent of substitution in terms of market share (as in [2]) can offer some computational advantages. It is therefore believed that the Mansfield model in the revised form used by Blackman offers the most useful form of the various substitution models available.

The objective of this paper is to present a description of this model and to illustrate the use of the model through specific examples.

The discussions which follow are divided into two parts. First, a description will be given of the substitution model and, second, applications of the model will be presented. The latter discussion will describe

(1) cases in which the substitution model was used in conjunction with historical substitution data to forecast future substitution rates; and
(2) cases which illustrate the use of the model for making projections of substitution rates in the absence of an adequate historical data base.

Description of the Substitution Model

A detailed development of the substitution model is presented in [2]. It is shown that the rate at which a market developed as a result of technological innovation is given by

$$\ln\left[\frac{m}{L-m}\right] = -\ln\left(\frac{L}{N_0} - 1\right) + \phi\,(t - t_1), \tag{1}$$

where m = market share captured at time, t, by the new innovation

L = upper limit of the market share which the new innovation can capture in the long run

N_0 = market share captured when $t = t_1$

ϕ = constant which governs the substitution rate.

The form of eq. (1) is that of a constrained logistic equation and a linear plot of m vs. time exhibits a typical S-shaped substitution curve.

SUBSTITUTION METHODOLOGY BASED ON HISTORICAL DATA

The general form of Eq. (1) may be expressed as

$$\ln \left[\frac{m}{L-m}\right] = C_1 + C_2\,(t - t_1), \tag{2}$$

where C_1 and C_2 are constants and t_1 can be taken to represent the year in which the new technology product first captures a portion of the market.

If an estimate is made of L and historical data are available which give market shares captured by the new-technology product as a function of time, values for the constants C_1 and C_2 in Eq. (2) may be estimated through the use of regression analysis, or alternatively, the constants may be estimated graphically. If Eq. (2) is plotted on semi-logarithmic paper, a straight line will result, and the constants may be estimated by determining the slope of the line and the zero intercept.

Thus if an adequate historical data base exists, the model can be used to forecast the rate at which substitution will occur in the future.

It is not necessary that the data base be extensive as discussed in [7] and [8]. When a new-technology product is first introduced, it is less well-developed than the older product it is replacing, and the new product is likely to undergo improvements and cost reductions as it gains market share. If the new product is economically viable after it has gained a small market share, it is likely to become more competitive as time progresses, and once a substitution has begun, it is highly probable that it will eventually take over the available market. It is demonstrated in [8] that predictions of substitution may be made when a new technology product has substituted for as little as two percent of the total market.

The value estimated for L affects only the longer-range part of the forecast, which occurs as substitution nears completion. If uncertainty exists in the estimate of L, high and low values which represent the range of uncertainty may be estimated and the effects of uncertainty indicated in the forecast.

SUBSTITUTION METHODOLOGY IN THE ABSENCE OF HISTORICAL DATA

In [1], Mansfield examined historical substitutions which have occurred in four disparate industrial sectors; viz, railroads, coal, steel, and breweries. Using regression analysis, an empirical expression was developed for the constant ϕ (see Eq. (1)), which controls the substitution rate. Very high correlation coefficients were obtained. The form of this expression is

$$\phi = Z + 0.530\,\pi - 0.027\,S, \tag{3}$$

where Z = a constant representative of a given industry

π = a profitability index (see [1])

S = an investment index (see [1])

If the economic characteristics of historical technological innovations are known such that estimates can be made for π and S, it is possible to estimate the value of Z if the value of ϕ is known from historical substitution rates. Using this approach, Mansfield [1] estimated values for Z for the industrial sectors of railroads, coal, steel, and breweries; and Blackman [2] estimated a value of Z for aerospace. The magnitude of the Z values for a given industrial sector reflect the propensity of that sector to innovate; i.e., those sectors having low Z values tend to be slow in adopting technological innovations and vice-versa.

In [4], the techniques of factor analysis were applied to develop an Innovation Index which measures the relative propensity towards innovation of various industrial sectors of the U.S. economy. The index is derived from various input variables which reflect the extent to which resources are allocated to achieve innovation in selected industrial sectors and output variables which measure the extent to which new product and process innovation is achieved.

It was hypothesized in [4], that a relationship might exist between the value of Z (see Eq. (3)) for an industrial sector and the value of the Innovation Index for that sector. To test this hypothesis, Z values for the steel and food and kindred products sectors were obtained from [1]; Z values for aerospace were based on [2]; and Z values for auto and electrical machinery sectors were computed for the technological substitutions discussed in [3]. Regression analysis was used to obtain a relationship between these values and values of the Innovation Index, I, for the corresponding industrial sectors. The following regression equation was obtained:

$$Z = 0.2221\,(I) - 0.3165. \tag{4}$$

The multiple correlation coefficient was 0.92, and the standard deviation of the regression coefficient was 0.0645.

The correlation of Eq. (4) may be used in conjunction with Eq. (3) to estimate the inter-industry effects on market substitution rates and to estimate the rate at which new markets can be expected to develop for new technological innovations if the economic characteristics of the innovations can be estimated. By identifying the industrial sector in which the new innovation will occur, a value of the Innovation Index can be obtained from Table 1 or computed using the procedures given in [4], and a corresponding Z factor can be estimated from Eq. (4). The economic characteristics of the innovation can be used to estimate π and S, and Eq. (3) can be solved for ϕ. By estimating L and N_0, Eq. (1) can then be solved for the market share of the new innovation as a function of time.

It should be noted that intra-industry differences in substitution rates are a function of the profitability and investment characteristics of a new-technology product vis-a-vis an old product, because the value of Z is fixed for a selected industrial sector. Inter-industry differences, however, are a function of profitability and investment characteristics as well as differences in the propensity towards innovation of the various industrial sectors.

TABLE 1

Ranking of Industrial Sectors According to Value of Innovation Index Determined from First-Factor Scores of the 8-Variable Model of [4]

Industrial Sector	Innovation Index
Aircraft and missiles	2.29
Electrical machinery and communication	1.76
Chemicals and allied products	0.60
Autos and other transportation equipment	0.29
Food and kindred products	–0.35
Professional and scientific instruments	–0.37
Fabricated metals and ordinance	–0.60
Petroleum products	–0.64
Stone, clay, and glass	–0.70
Paper and allied products	–0.75
Textile mill products and apparel	–0.75
Rubber products	–0.76

Application of the Substitution Model

In the discussions which follow, applications will first be described in which the substitution model was utilized with an existing data base to forecast the rate of technological substitution in future time periods. Next, cases will be described to illustrate the use of the model in the absence of an adequate historical data base.

MARKET SUBSTITUTIONS BASED ON HISTORICAL DATA

In the following sections, specific cases drawn from the marine, automotive, electrical appliance, and aircraft market sectors will be discussed in which the general form of the growth model, expressed by Eq. (2), was curve-fitted to a historical data base and used where appropriate to forecast future substitution rates.

Marine Market Examples[1]

Four examples of historical technological substitutions which have occurred in the marine market sector will be presented chronologically in the following sections.

Substitution of steam for sail propulsion. Although a number of small steam vessels had operated successfully since Fulton operated the North River Steamboat in 1807, it was not until 1820 that steam propulsion had reached the point where it could be considered suitable for ocean transportation, and on that date scheduled Trans-Atlantic service began. From the very start, steamships were a success. Westbound trips across the Atlantic by steam averaged 17 days, eastbound trips averaged 15.4 days. By contrast, sailing ships averaged 34 days westbound and 22.1 days eastbound. Steamships also offered the advantage of reliable scheduling because of their dependability.

The results of a regression analysis using data from [9] on the growth of steamships since 1820 are shown in Figs. 1 and 2. A correlation coefficient of 0.97 was obtained, and the substitution model agreed reasonably well with historical data for this innovation. As

[1] The author is indebted to Mr. C. E. Jones, Jr., formerly of United Aircraft Research Laboratories, for the data used in these examples.

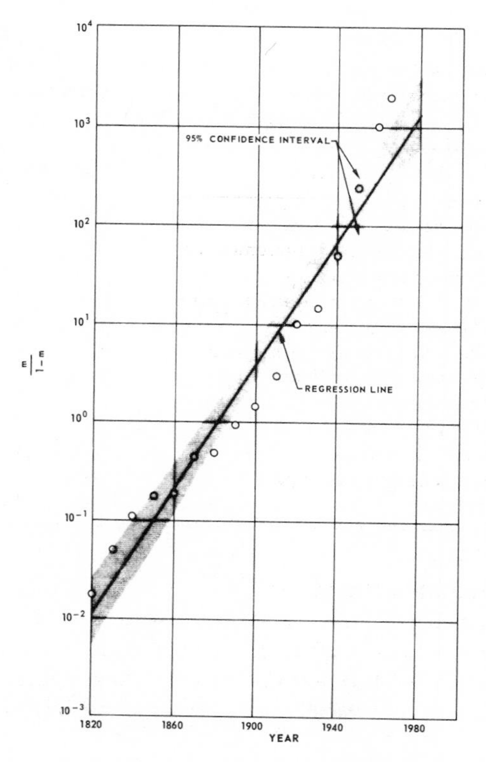

FIG. 1. Substitution of steam for sail in the U.S. commercial marine market. m = Market share of steam owered ships.

shown in Fig. 2, the ϕ-parameter, as determined by the regression analysis, was 0.0727, and the standard deviation of the regression coefficient (indicated in parenthesis under the regression equation) was 0.0046.

Substitution of oil for coal. Coal was the primary fuel used on Atlantic crossings until

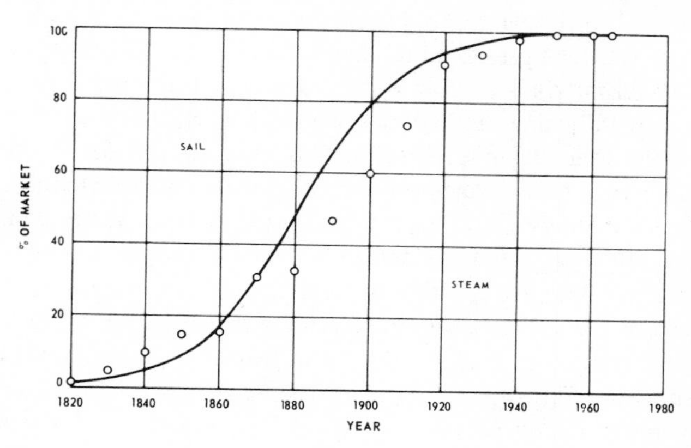

FIG. 2. Dynamics of the displacement of sail by steam propulsion in the U.S. commercial marine market. —— Model predictions; ○ historical data; $\ln\left[\frac{m}{1-m}\right] = \underset{(0.0046)}{-4.4167 + 0.0727\,t'}$; **$m$ = Market share of steam powered ships; t' = Year−1820; correlation coefficient = 0.97.**

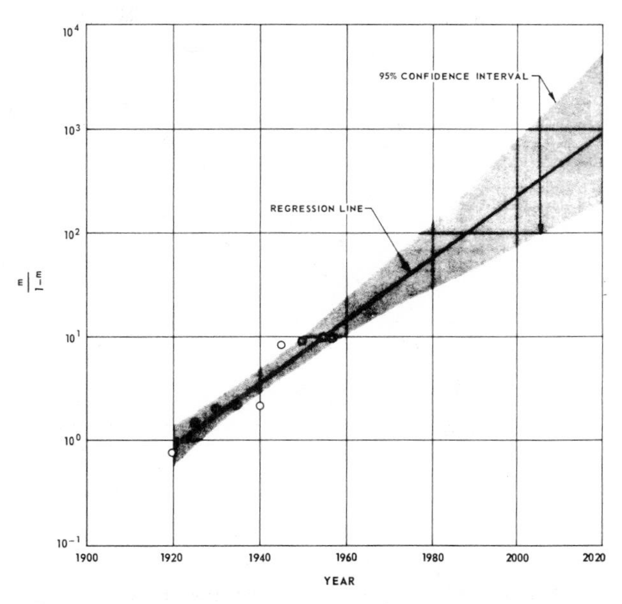

FIG. 3. Substitution of oil for coal in the U.S. commercial marine market. m = Market share of oil fueled ships.

1902. At this time, the reduced labor costs, more efficient fuel storage, and increased plant efficiency associated with burning fuel oil began to be appreciated, and as a result oil was substituted for coal as an energy source for generating steam. By World War II virtually all vessels engaged in the Atlantic trade used oil.

Based on historical data given in [9], a regression analysis for this innovation (Figs. 3 and 4) produced a ϕ-parameter of 0.0699. Again, very good agreement between the substitution model predictions and historical data was obtained as indicated by the correlation coefficient of 0.95.

Substitution of Fiberglass for Wood in Small Craft Structures. Fiberglass was introduced in the marine industry in 1948, since that time it has been rapidly accepted

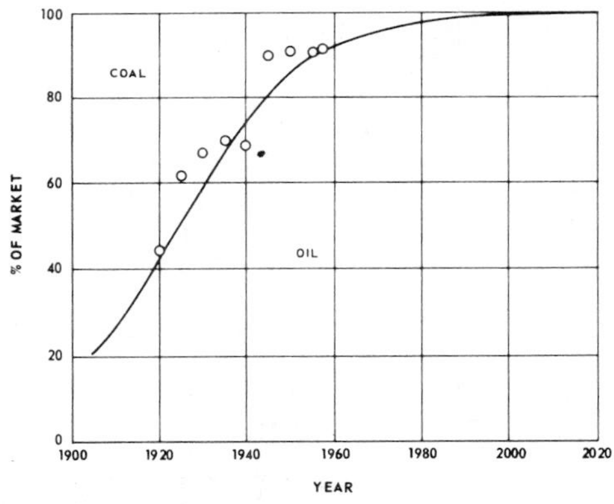

FIG. 4. Dynamics of the displacement of coal fuel by oil in the U.S. commercial marine market. ——— Model predictions; ○ historical data; $\ln\left[\frac{m}{1-m}\right] = -1.3560 + \underset{(0.0087)}{0.0699}\, t'$; **$m$ = Market share of oil fueled ships; t' = Year–1902; correlation coefficient - 0.95.**

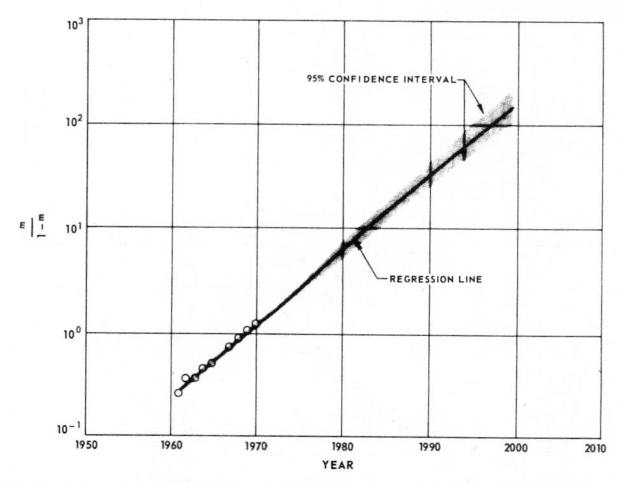

FIG. 5. Substitution of fiberglass for wood in the U.S. recreational marine market. m **= Market share of fiberglass.**

everywhere because fiberglass hulls can be less expensive than equivalent wooden hulls, if produced in large volumes. Fiberglass hulls have the additional advantages of being noncorrosive, leak and rot resistant, and easy to maintain. Fiberglass is generally preferred for the larger hulls because it is cheaper than aluminum; however, in "mini-boats," such as rowboats, prams, and dinghies, aluminum is more economical.

A regression analysis, using data from [10] and [11] starting with 1948, the first production year, showed excellent correlations, as is indicated in Figs. 5 and 6. The ϕ-parameter derived from this analysis was 0.1635.

Substitution of Inboard/Outdrive (I/O) Engines for Inboard Engines. Strictly speaking, the inboard boat will never be fully replaced by the inboard-outdrive boat; the inboard engine merely will be relegated to larger and more expensive craft. On the average, the ratio of current sales of I/O boats to inboard boats is about four-to-one. The rapid growth of I/O boats is attributed primarily to the development of interest in

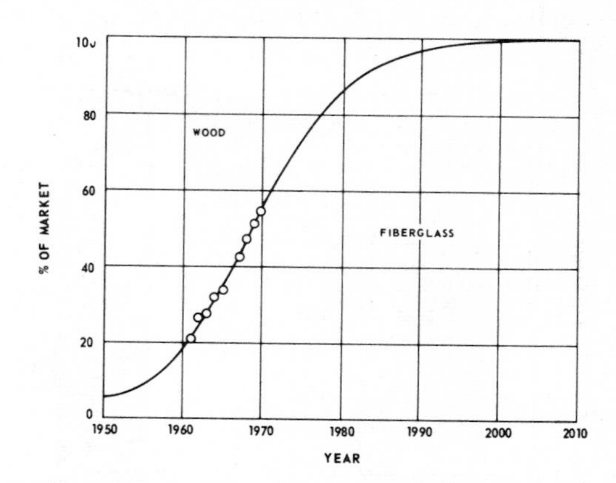

FIG. 6. Dynamics of the displacement of wooden hulls by fiberglass hulls in the U.S. recreational marine market. ——— **Model predictions; ○ historical data;** $\ln\left[\frac{m}{1-m}\right] = -3.3911 + \underset{(0.0049)}{0.1635}\, t'$; m = **Market share of fiberglass; t' = Year−1948; correlation coefficient = 0.99.**

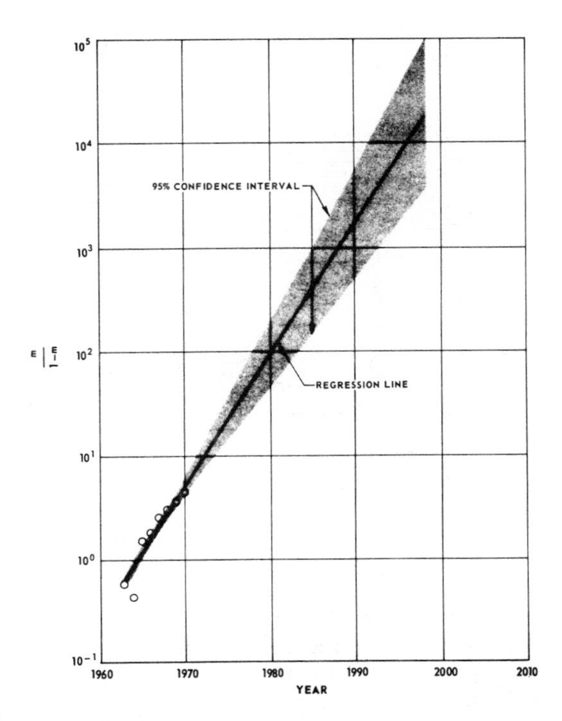

FIG. 7. Substitution of inboard/outdrive engines for inboard engines in the U.S. recreational marine market. m **= Market share of inboard/outdrive engine.**

waterskiing, which requires a considerable amount of power, particularly if two or more people are towed. The I/O boats have captured a large share of this market, even though I/O units are much more expensive than outboards, because people have found them to be a convenient way of getting higher horsepower while circumventing the complexities of the inboard boat. Compared with an inboard boat, an I/O boat does not require separate rudders and shafting; these are part of the outdrive unit. The I/O boat is easier to beach because the outdrive unit can be tilted like the outboard, and it is convenient to transport by trailer. The I/O boat claims higher propulsive efficiencies than the inboard (as much as 15% in some cases), because the thrust line of its propeller is not inclined at an angle relative to the keel, as is the usual case with the inboard.

Figures 7 and 8 present the results of a regression analysis of this substitution using historical data from [12]. It can be seen that correlation is excellent, and the ϕ-parameter is 0.2904.

Automotive Market Examples[2]

The results of curve-fitting the growth model (as described previously) to historical data (from [13] and [14]) for three technological innovations in the U.S. automotive industry are shown in Figs. 9 through 11, respectively, for the introduction of power brakes, power steering, and air conditioning. The vertical axis values in each plot represent the annual percentages of autos which included each of the indicated customer-

[2] The author is indebted to Mr. W. R. Davison of United Aircraft Research Laboratories for the data used in these examples.

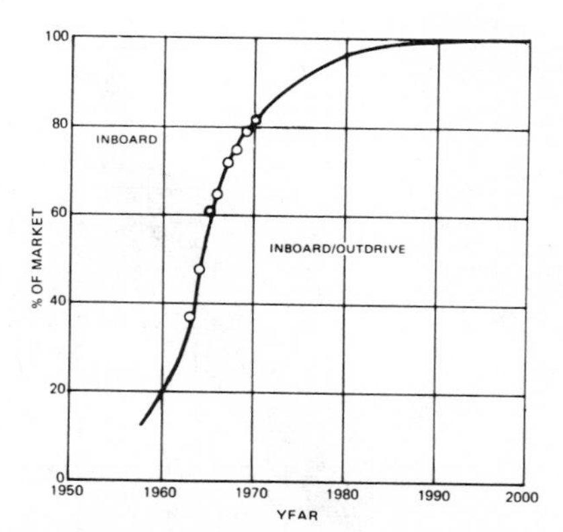

FIG. 8. Dynamics of the displacement of inboard engines by inboard/outdrive engines in the U.S. recreational marine market. ——— **Model predictions; ○ historical data;** $\ln\left[\frac{m}{1-m}\right] = -1.8562 + \underset{(0.0223)}{0.2904}\, t'$; m = **Market share of inboard/outdrive engines; t' = Year−1958; correlation coefficient = 0.98.**

option accessories. It can be seen that the statistical correlation coefficients obtained in each case are quite high, which indicates good agreement between the historical data and the form of the model equation.

The ϕ values, which control the market dynamics for the three innovations, range from a low of approximately 0.1 for power brakes (Fig. 9) to a high of approximately 0.3 for automotive air conditioning (Fig. 11). The ϕ value of approximately 0.25 for power steering (Fig. 10) is intermediate between these two.

The market dynamics of the above three innovations would be expected to be controlled basically by consumer acceptance. It is interesting to compare a situation in which the market dynamics are not controlled by consumer acceptance but are controlled by government regulation. A case in point was the introduction of automobile seat belts which was forced by government regulation in the early and mid-1960's. The market dynamics for this case are shown in Fig. 12 (based on data from [15]). It can be seen that

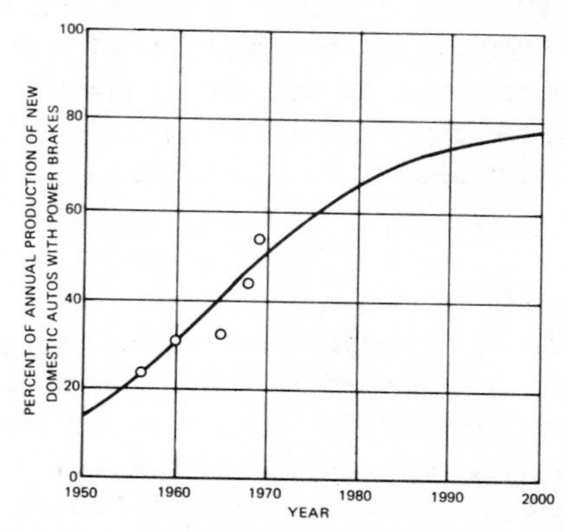

FIG. 9. Dynamics of the acceptance of power brakes on new U.S. autos. $\ln\left[\frac{m}{0.8-m}\right] = -1.504 + \underset{(0.0287)}{0.1018}\, t'$; m = **Market share;** t' = **Year−1950; correlation coefficient = 0.898;** ——— **Model predictions; ○ historical data.**

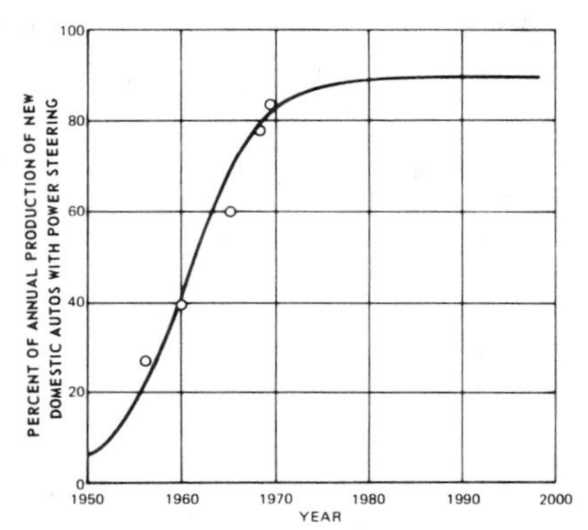

FIG. 10. Dynamics of the acceptance of power steering on new U.S. autos. $\ln\left[\frac{m}{0.9-m}\right] = -2.639 + \underset{(0.04421)}{0.2544}\, t'$; m = **Market share;** t' = **Year−1950; correlation coefficient = 0.961;** ——— **Model predictions;** ○ **historical data.**

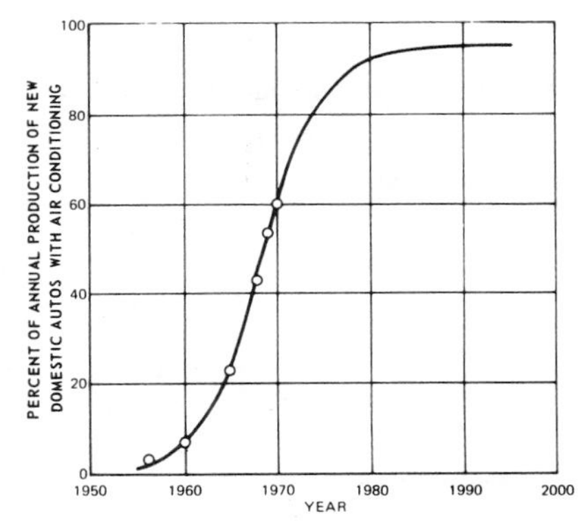

FIG. 11. Dynamics of the acceptance of air conditioning on new U.S. autos. $\ln\left[\frac{m}{0.95-m}\right] = -4.078 + \underset{(0.0159)}{0.3031}\, t'$; m = **Market share;** t' = **Year−1955; correlation coefficient = 0.995;** ——— **Model predictions;** ○ **historical data.**

the rate of introduction is much more rapid than in the other three cases, as is indicated by a ϕ value of approximately 2.3.

Electrical Appliance Market Examples

Figures 13 through 23 present curve-fits for the growth model and historical market data (obtained from [16]) which pertain to electrical appliances. The form of the growth model used in these cases is slightly different from the form used previously. Instead of expressing the extent of market build-up in terms of market shares based on annual sales, it is expressed in terms of the number of homes using the appliance divided by the total number of electrically wired homes, because this was the format used in the data base. In all of the cases shown, the correlation coefficients tended to be fairly high, even though the number of data points used in the correlations are not as numerous as might be desired.

The ϕ values range from a low of approximately 0.05 for electric water heaters (Fig. 14) to a high of approximately 0.44 for color TV (Fig. 23).

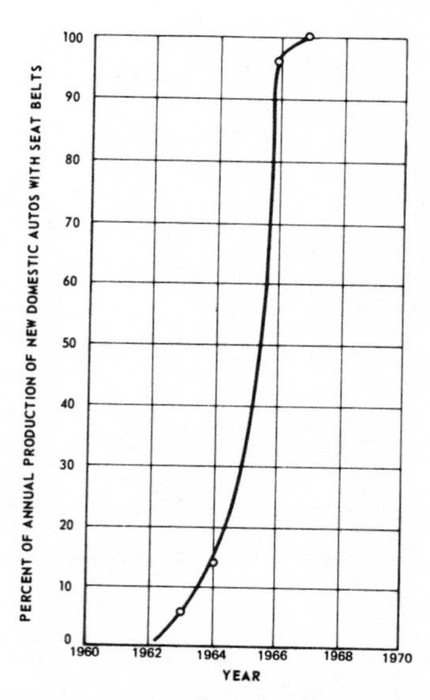

FIG. 12. Dynamics of the acceptance of seat belts in U.S. automobiles. ———— **Model predictions; ○ historical data;** $\ln\left[\frac{m}{1-m}\right] = \underset{(0.240)}{-5.738 + 2.312\, t'}$; m = **Market share;** t' = **Year−1961; correlation coefficient = 0.989.**

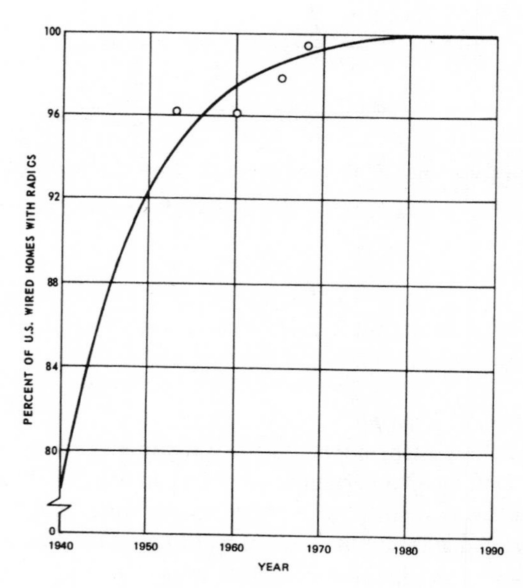

FIG. 13. Dynamics of the acceptance of radios in U.S. ———— **Model predictions; ○ data;** $\ln\left[\frac{P}{1-P}\right] = \underset{(0.06207)}{-232.91 + 0.12072\, t}$; $P = \frac{\textbf{number of homes in U.S. using appliance}}{\textbf{number of electrically wired houses in U.S.}}$; t = **Year; correlation coefficient = 0.81.**

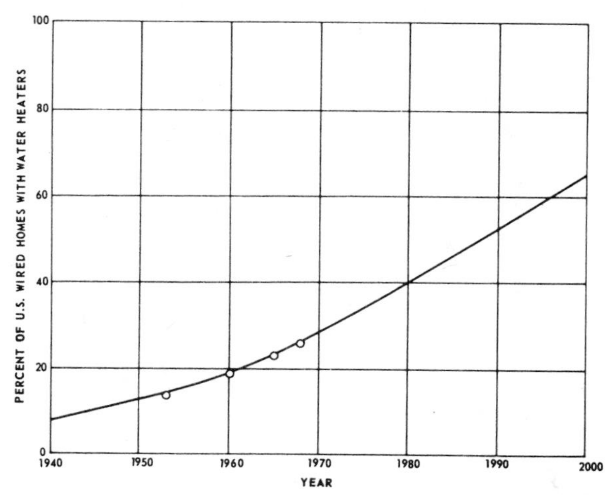

FIG. 14. Dynamics of the acceptance of electric water heaters in U.S. ——— Model predictions; ○ data; $\ln\left[\frac{P}{1-P}\right] = -105.24 + \underset{(0.00075)}{0.05295}\, t$; $P = \frac{\text{number of homes in U.S. using appliance}}{\text{number of electrically wired homes in U.S.}}$; t = **Year; correlation coefficient = 0.99.**

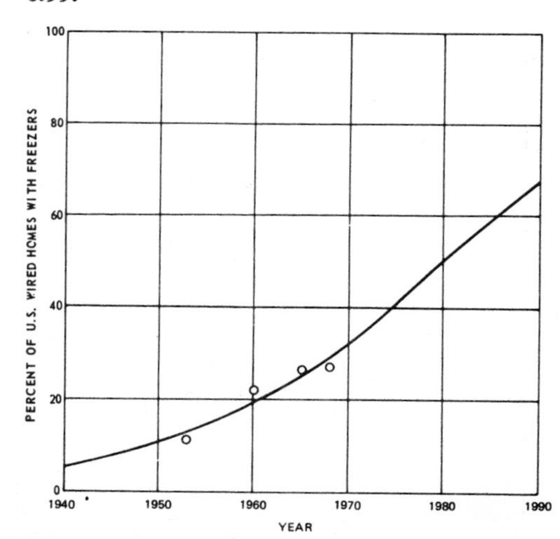

FIG. 15. Dynamics of the acceptance of freezers in U.S. ——— Model predictions; ○ data; $\ln\left[\frac{P}{1-P}\right] = -142.79 + \underset{(0.01538)}{0.07212}\, t$; $P = \frac{\text{number of homes in U.S. using appliance}}{\text{number of electrically wired homes in U.S.}}$; t = **Year; Correlation coefficient = 0.96.**

Commercial Aircraft Engine Market Examples

To check the applicability of the model expressed by Eq. (2) to the commercial aircraft engine market, the rates of displacement of the commercial turbojet by the first generation turbofan and the rate of displacement of the first generation turbofan by the second generation turbofan were investigated. As described in [2], data for the free-world market shares of turbojets and first generation turbofans were used in a regression analysis and the results are presented in Fig. 24, which show that excellent agreement between the model and the data was obtained, as indicated by a multiple correlation coefficient of 0.97.

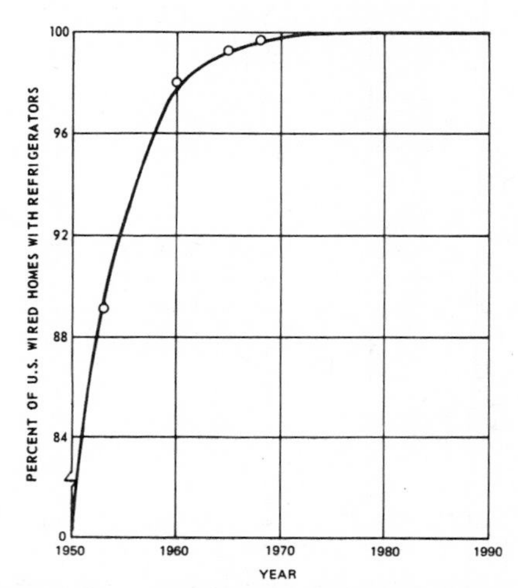

FIG. 16. Dynamics of the acceptance of electric refrigerators in U.S. ——— Model predictions; ○ data; $\ln\left[\frac{P}{1-P}\right] = -471.68 + \underset{(0.00722)}{0.24261}\, t$; $P = \frac{\text{number of homes in U.S. using appliance}}{\text{number of electrically wired homes in U.S.}}$; t=Year; **correlation coefficient = 0.99.**

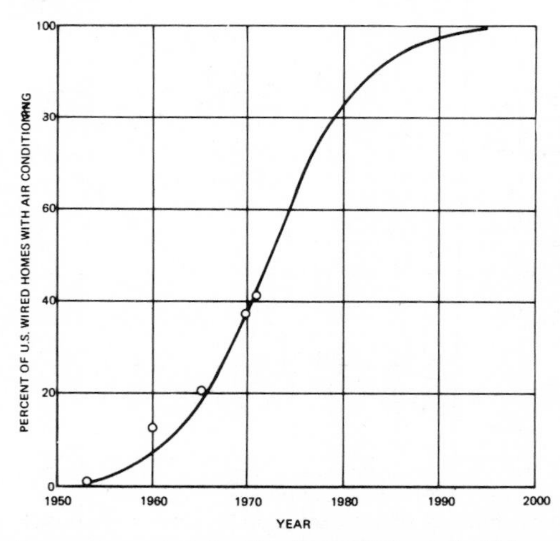

FIG. 17. Dynamics of the acceptance of electric air conditioners in U.S. ——— Model predictions; ○ data; $\ln\left[\frac{P}{1-P}\right] = -409.055 + \underset{(0.0283)}{0.2074}\, t$; $P = \frac{\text{number of homes in U.S. using appliance}}{\text{number of electrically wired homes in U.S.}}$; t = Year; **correlation coefficient = 0.97.**

The model was next used to investigate the displacement dynamics of the first generation turbofan by the second generation turbofan. Sales forecast data were utilized in a regression analysis as before to examine the extent to which eq. (2) correlated the data. The results are shown in Fig. 25. Again, good agreement was obtained between the model and the data. A multiple correlation coefficient of 0.96 resulted.

Figure 3 of [2] presents the two sets of data on the same plot. It can be seen that the rate at which the second generation turbofan is being substituted for the first generation turbofan is occurring at a slower rate than the rate at which the turbojet was displaced by

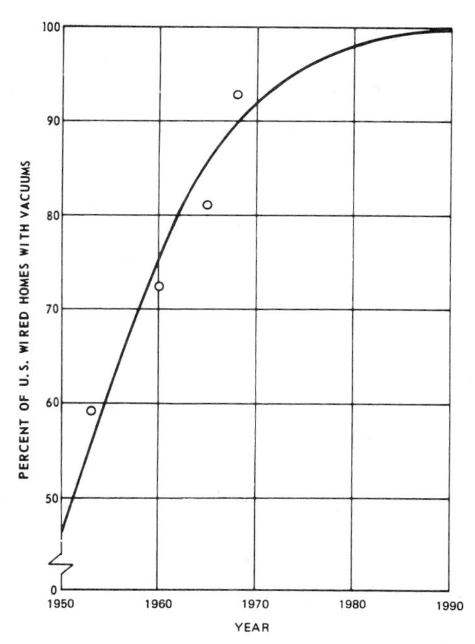

FIG. 18. Dynamics of the acceptance of vacuums in U.S. ——— Model predictions; ○ data; $\ln\left[\frac{P}{1-P}\right] = -259.33 + \underset{(0.03467)}{0.13290}\,t$; $P = \frac{\text{number of homes in U.S. using appliance}}{\text{number of electrically wired homes in U.S.}}$; t **= Year; correlation coefficient = 0.94.**

the first generation turbofan. As discussed in [2], these differences are believed to result because the investment levels associated with aircraft using the second generation turbofan are greater than investments associated with aircraft using the first generation turbofan.

Market Substitutions in the Absence of Historical Data

Many situations of practical interest occur in which projections are required of the rate at which a new product concept, having superior technological and economic characteristics, can be expected to substitute for an existing product in a specified market sector if the new product concept is introduced to the market at some specified future time period.

The methodology which applies to this situation has been previously discussed. In an effort to indicate the applicability of this methodology, historical substitutions were examined retrospectively. Two of the previously discussed marine market substitutions (which occurred at different time periods and which involved substitutions having different economic characteristics) were used as a basis for checking the methodology. To perform the check, the ϕ value, which was obtained from historical data for the substitution of fiberglass for wooden hulls (Figs. 5 and 6), was used with Eq. (3) to determine a value of the industry constant Z from computed values of π and S. This value of Z was then used in Eq. (3) with values of π and S computed for the substitution of inboard-outdrive engines for inboard engines (Figs. 7 and 8) to compute a value of ϕ for this substitution. The computed value was then compared with the value obtained from the historical substitution data in order to check the methodology. The two constants, π and S, of Eq. (3) had to be determined by an assessment of the costs and benefits associated with each of the two substitutions.

The profitability index, π, can be expressed as

$$\pi = \frac{ROR}{ROR_{req}},$$

where

ROR = rate-of-return associated with the innovation

ROR_{req} = minimum rate-of-return required for investments

If π is greater than 1.0, the product substitution is considered profitable and if π is less than 1.0, it is unprofitable. The following expression may be used to relate the present value of the net benefits received from a product to its purchase price.

$$Y_o = \sum_{k=1}^{j} \frac{B_{o,k}(1-T)}{(1+ROR_{req})^k} - \sum_{k=1}^{j} \frac{(E_{o,k}+D_{o,k})(1-T)}{(1+ROR_{req})^k} + \sum_{k=1}^{j} \frac{D_{o,k}}{(1+ROR_{req})^k}, \quad (5)$$

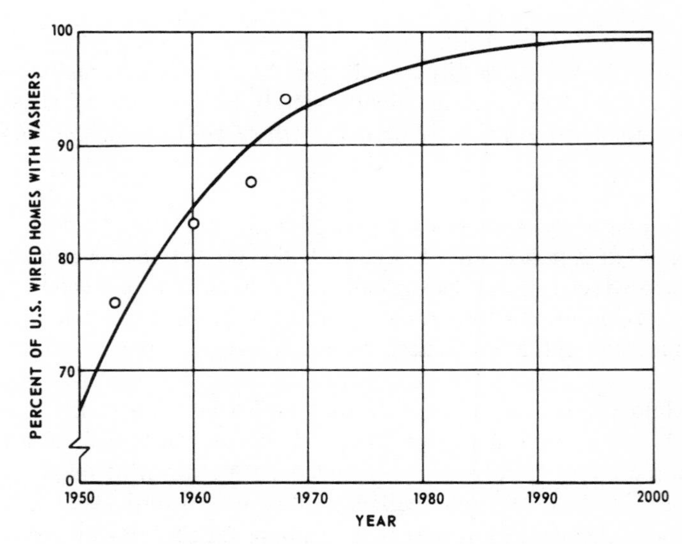

FIG. 19. Dynamics of the acceptance of washers in U.S. ——— Model predictions; ○ data; $\ln\left[\frac{P}{1-P}\right] = -189.38 + \underset{(0.02955)}{0.09750}\, t$; $P = \frac{\text{number of homes in U.S. using appliance}}{\text{number of electrically wired homes in U.S.}}$; t = Year; correlation coefficient = 0.92.

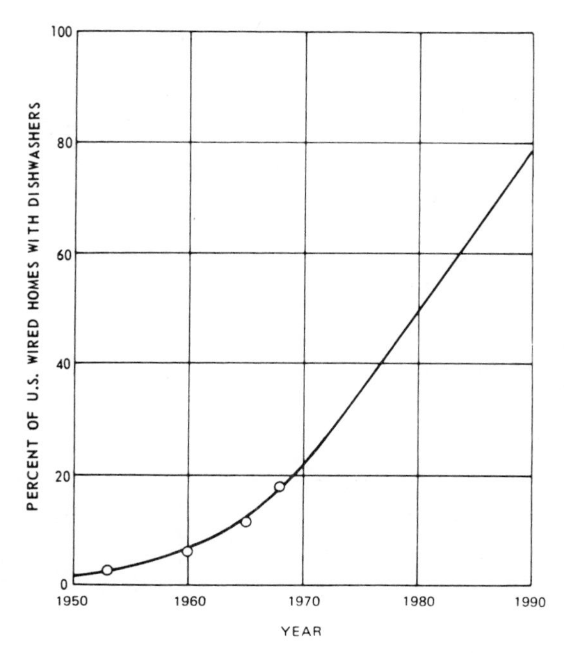

FIG. 20. Dynamics of the acceptance of dishwashers in U.S. ——— Model predictions; ○ data; $\ln\left[\frac{P}{1-P}\right] = -257.13 + \underset{(0.00782)}{0.12985}\,t$; $P = \frac{\text{number of homes in U.S. using appliance}}{\text{number of electrically wired homes in U.S.}}$; **$t$ = Year; correlation coefficient = 0.99.**

where

Y_o = initial purchase price of the old product
B_o = annual benefit to the user of the old product
T = tax rate
E_o = annual operating expense using the old product
D_o = depreciation expense of the old product
j = life of the product, years.

Based on a known initial purchase price, life cycle expenses, and tax rate, eq. (5) can be used to determine the annual benefits received from the old product (B_o). Assuming that the annual benefit of the new product must equal the annual benefit of the old product and that the initial cost (Y_N), annual operating expenses (E_N), and depreciation (D_N) of the new product are known (or can be estimated), the ROR of the new product can be determined from

$$Y_N = \sum_{k=1}^{j} \frac{B_{o,k}(1-T)}{(1+ROR)^k} - \sum_{k=1}^{j} \frac{(E_{N,k}+D_{N,k})(1-T)}{(1+ROR)^k} + \sum_{k=1}^{j} \frac{D_{N,k}}{(1+ROR)^k} \qquad (6)$$

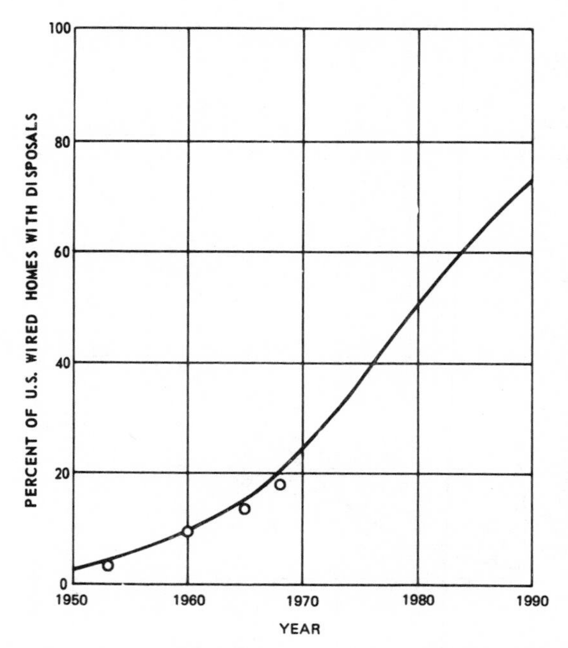

FIG. 21. Dynamics of the acceptance of disposers in U.S. ——— Model predictions; ○ data; $\ln\left[\frac{P}{1-P}\right] = -241.59 + \underset{(0.01320)}{0.12202}\, t$; $P = \frac{\text{number of homes in U.S. using appliance}}{\text{number of electrically wired homes in U.S.}}$; **$t$ = Year; correlation coefficient = 0.99.**

Equations (5) and (6) were formulated on the basis of tax schedules applicable to business firms for which depreciation and expenses are tax deductible. For non-business situations, depreciation expenses should not be included and an appropriate non-business tax rate should be used.

The investment index, S, can be determined by estimating the average initial investment in the new product and then estimating the total assets of an "average" firm introducing the new product.

$$S = \frac{\text{initial investment} \times 100}{\text{total assets of an average firm}}$$

In developing the industry constant, Z, from the value of ϕ, determined from the historical data for the substitution of fiberglass for wooden hulls, it was assumed that a representative purchase situation would involve a choice of buying either a 24-ft wooden boat priced at \$7,470 or a fiberglass boat priced at \$8,710 (based on pricing data from [17]). It was assumed that each boat had a 15-year life and zero salvage value. The purchaser's ROR_{req} was assumed to be 8%, based on the cost of borrowing money. It was assumed that the boats would be used for recreational purposes only and that no business income would be derived from their use. The operating, maintenance, and tax expenses for the fiberglass boat were estimated to be \$394 annually, and the corresponding

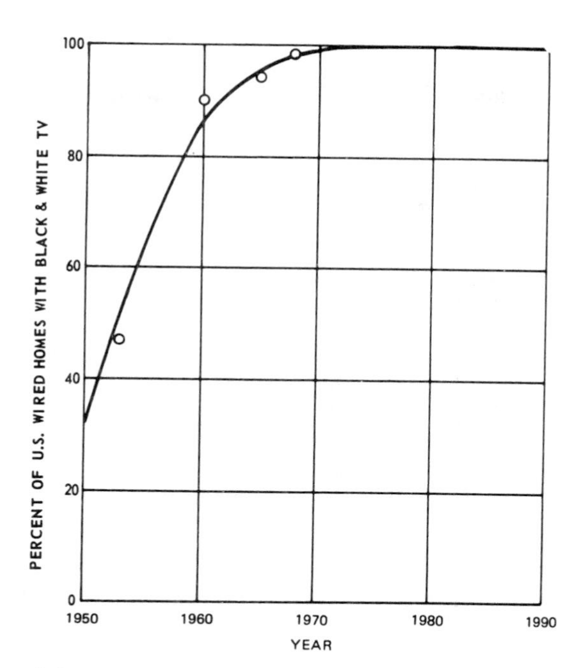

FIG. 22. Dynamics of the acceptance of black and white T.V. in U.S. ——— Model predictions; ○ data; $\ln\left[\frac{P}{1-P}\right] = -502.20 + \underset{(0.03260)}{0.25715}\, t$; $P = \frac{\text{number of homes in U.S. using appliance}}{\text{number of electrically wired homes in U.S.}}$; t = **Year; correlation coefficient = 0.98.**

expenses for the wooden boat were estimated to be \$670 annually. The effect of the purchase on personal income taxes was neglected. Equations (5) and (6) were solved using the above assumptions, and a π value of 1.25 was determined for the substitution. The value of S was based on the difference in the purchase prices of the two boats divided by the total assets of a representative purchaser which were assumed to be three times his annual salary of \$15,000. For these assumptions, S was determined to be 2.75.

By substituting the ϕ-value based on historical data of 0.1635, the π-value of 1.25, and the S-value of 2.75 into Eq. (3), Z was calculated to be -0.424.

A ϕ-value was then computed for the substitution of the inboard-outdrive (I/O) engine based on the above value of Z. Using Eqs. (5) and (6), as previously described, a value of π was computed to be 1.35. For these calculations, it was assumed that a prospective buyer would have the choice of either buying a 24-ft I/O boat for \$7,470 or a similar inboard boat for \$8,800. It was assumed that each boat had a 15-year life and zero salvage value. The purchaser's ROR_{req} was assumed to be 8%, and it was assumed that the boats would be used for recreational purposes only and no business income would be derived from their use. For each boat, the operating, maintenance, and tax expenses were assumed to be \$670 annually. The effects of the purchase on personal income taxes were neglected.

As no positive incremental investment would be required to purchase the I/O boat when compared with the inboard boat (because its purchase price was less than the inboard boat), the value of S was assumed to be zero.

By substituting the value of Z of -0.424, the value of π of 1.35, and the value of S of

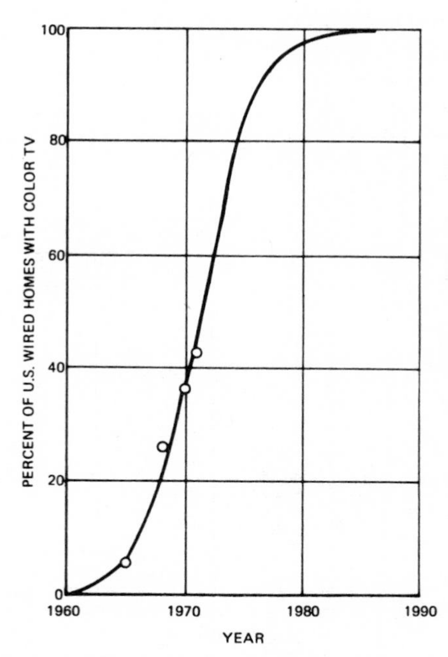

FIG. 23. Dynamics of the acceptance of color T.V. in U.S. ——— Model predictions; ○ data; $\ln\left[\frac{P}{1-P}\right] = -870.64 + \underset{(0.0732)}{0.4417}\, t$; $P = \frac{\text{number of homes in U.S. using appliance}}{\text{number of electrically wired homes in U.S.}}$; **$t$ = Year; correlation coefficient = 0.97.**

zero into Eq. (3), a value of ϕ of 0.292 was calculated. This value compared favorably with the value of ϕ of 0.290 which was determined from the historical substitution data and serves to check the calculation procedures.

This close agreement is, however, not believed to be indicative of the uncertainty level which can be expected in the calculation of substitution dynamics in the absence of a historical data base. In general, it is believed that uncertainty levels of the order of 10–20 percent should be more representative, although it is difficult to arrive at good estimates, because checks on the procedures have been made mainly through the use of hindsight. In cases where the procedures have been applied to obtain projections of substitution dynamics for a new product prior to its introduction to the market and these projections have been compared with independently determined projections based on expert opinion, agreement has been very good, although more data are needed before a good assessment can be made of the accuracy of prediction.

The calculation procedures should, therefore, be used with some caution when historical substitution data are not available. Nevertheless, it is often necessary to make estimates of the substitution dynamics of new technology-based products before they have been introduced to the market, and the methodology described is believed to be the best analytical technique available for making such estimates. This methodology is perhaps best suited to the formulation of inputs to venture analysis procedures (such as those described in [18]) in which quantitative evaluations are made of the effects of input uncertainty on the determination of the worth of the venture under consideration.

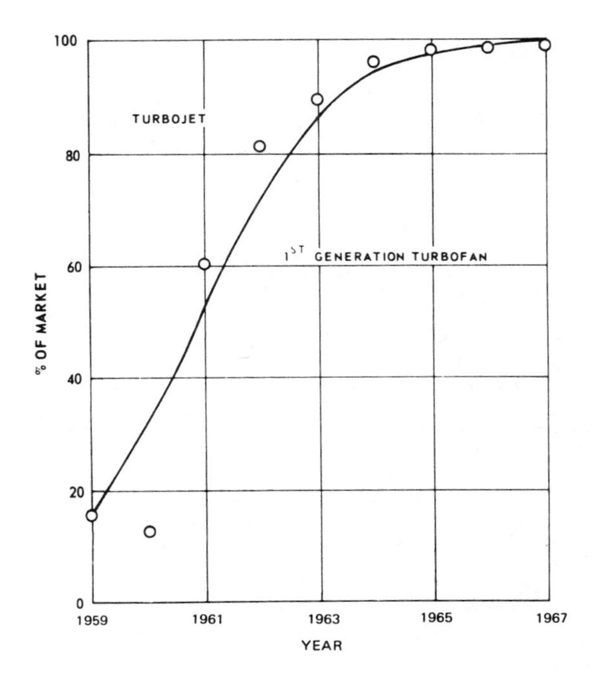

FIG. 24. Displacement dynamics of turbojets by first generation turbofans in commercial jet engine market. ——— Model predictions; ○ historical data; $\ln\left[\frac{m}{1-m}\right] = -1.6506 + \underset{(0.0810)}{0.8671}\, t'$; m = **Market share of 1st generation turbofan;** t' **= Year−1959; correlation coefficient = 0.97.**

Concluding Remarks

The dynamics of the process by which a new technology is substituted for an old technology is of importance in a number of planning-related activities which include new venture analysis, the definition of new markets, resource allocation for research and development, technology forecasting and assessment, etc.

The methodologies described herein provide a structure for predicting the dynamics of technological substitutions. If a technological substitution has been initiated, forecasts based on historical data may be formulated to indicate the rate at which the substitution will proceed in the future.

If a technological substitution has not been initiated, projections may also be made of the rate of substitution which can be expected if a new product, based on the new technology, is introduced at a specified time in the future and the new product has specified economic characteristics. Although the uncertainty associated with projections of this type can be expected to be greater than the uncertainty associated with forecasts based on a historical data base, they are often of great importance in the evaluation of new product ventures, and the procedures described herein are believed to be the best available analytical techniques for making such estimates. Projections made without a historical data base are perhaps most useful as inputs to those venture analysis procedures in which quantitative evaluations are made of the effects of input uncertainty on the determination of the worth of new product ventures.

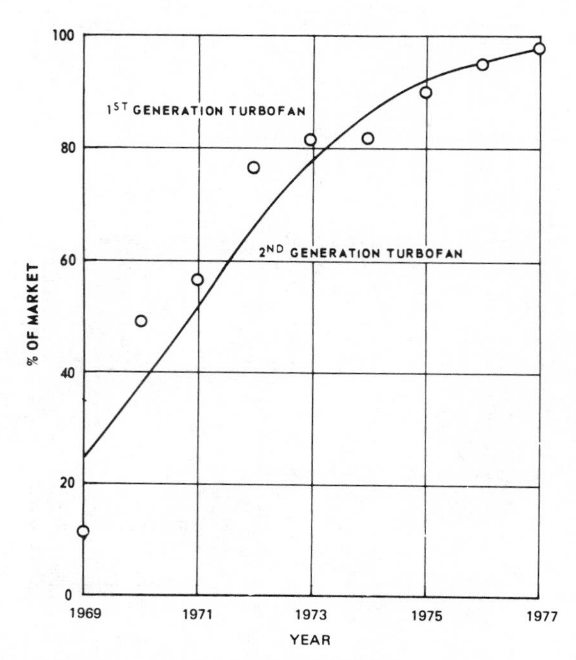

FIG. 25. Displacement dynamics of first generation turbofan by second generation turbofan in commercial jet engine market. ——— Model predictions; ○ historical and forecast data; $\ln\left[\frac{m}{1-m}\right] = -1.0971 + \underset{(0.0644)}{0.5823}\, t'$; m = **Market share of 2nd generation turbofan;** t' = **Year−1969; correlation coefficient = 0.96.**

The substitution forecasting methodologies discussed herein provide insights into the causes of inter- and intra-industry differences in the dynamics of market growth of new products based on new technology and can be helpful in establishing criteria for selecting new products which can be expected to achieve high rates of market growth in the future.

References

1. Edwin Mansfield, Technical change and the rate of imitation, *Econometrica,* **29,** No. 4, 741–765 (October 1961).
* 2. A. W. Blackman, The rate of innovation in the commercial aircraft jet engine market, *Technol. Forecasting Social Change,* **2,** 269–276 (1971).
3. A. W. Blackman, A mathematical model for trend forecasts, *Technol. Forecasting Social Change,* **3,** 441–452 (1972).
* 4. A. W. Blackman, E. J. Seligman, and G. C. Sogliero, An innovation index based on factor analysis, *Technol. Forecasting and Social Change,* **4,** 301–316 (1973).
5. F. M. Bass, A new product growth model for consumer durables, *Management Science,* **15,** No. 5 (January 1969).
6. J. V. Nevers, Extensions of a new product growth model, *Sloan Management Review,* **13,** No. 2, 77–91 (Winter 1972).
7. J. C. Fisher and R. H. Pry, A simple substitution model of technological change, *Technol. Forecasting Social Change,* **3,** No. 1, 75–88 (1971).
8. R. C. Lenz, Jr. and H. W. Lanford, The substitution phenomenon, *Business Horizons,* **15,** No. 1, 63–68 (February 1972).

**This article also appears in Part 2 of this book.*

9. *Historical Statistics of the United States–Colonial Times to 1957,* A Statistical Abstract Supplement Prepared by the U.S. Department of Commerce, Bureau of the Census.
10. *Annual Market Research Notebook–The Marine Market 1970,* Boating Industry Association, Chicago, Ill.
11. *Boating Statistics, CG-257,* Annual Report Prepared by U.S. Coast Guard.
12. *The Boating Business, 1970,* Prepared by Boating Industry Magazine, Chicago, Ill.
13. Anon: *Automotive Industries,* Detroit, Michigan (March 15, 1969).
14. Anon: *1971 Automobile Facts and Figures,* Automobile Manufacturer's Association, Detroit, Michigan (1971).
15. Anon: *Annual Issues of Ward's Automotive Yearbook,* Ward's Communications, Inc., Detroit, Michigan (1964 through 1967).
16. *Statistical Abstract of the U.S.,* U.S. Department of Commerce, 92nd edition, (1971), and earlier editions.
17. *Motor Boating and Sailing,* Hearst Corporation (December 1971).
18. A. W. Blackman, New venture planning: The role of technological forecasting, *Technol. Forecasting and Social Change,* **5,** No. 1, 25–49 (1973).

The International Diffusion of New Technology

M. BUNDGAARD-NIELSEN

ABSTRACT

An analysis of the factors influencing the international diffusion of new technology is presented. As a result of the analysis, a model of the international diffusion process is proposed and tested statistically using diffusion data for two major innovations in the Steel Industry.

Introduction

The importance of international transfer or diffusion of new technology is widely recognized in the literature of technological forecasting. As an example one might cite Kosobud [3], who has shown that the spectacular economic growth of post-war Japan—to a very large extent—was due to the active management of the transfer of new technology from other countries. The purpose of this paper is to present a quantitative analysis of the parameters that influence the diffusion process, and to discuss the results of a case study of the diffusion of two major innovations into the European Steel Industry.

The Decision to Adopt New Technology

Observations of the number of adoptions of a particular new technology versus time display the characteristics of an S-shaped curve. Thus, the diffusion process may be described by a simple mathematical model—the well-known Pearl Reed curve, i.e.,

$$A(t) = A(\infty)/(1 + \exp[-C(t-t_0)]) \tag{1}$$

where $A(t)$ is the number of adoptions of the new technology that have occurred in year t after the introduction of the new technology. $A(\infty)$ is the final number of adoptions—a figure that frequently has to be estimated or extrapolated, since empirical data may not be available. C is the rate constant of the diffusion process and t_0 is the time when half of the final number of adoptions has occurred—the half-time.

Although the outcome of the diffusion process, i.e., the number of actual adoptions, is represented by a simple mathematical model, such as Eq. (1), a more thorough understanding of the decision-making involved rests upon a proper knowledge of the factors influencing the parameters of the simple model, i.e., the rate constant C as well as the time t_0. Mansfield [4] showed that the rate constant C was positively correlated with the profitability of the new technology and negatively influenced by the relative capital investment needed to introduce the new technology. In addition, Mansfield, who was

---

This article appeared in *Technological Forecasting and Social Change*, Vol., 9, No. 1, 1976.

studying diffusion processes in the US-industry, tested the possibility that the rate constant should increase with the growth rate of the industry into which diffusion took place and with the year of introduction. The idea behind these two hypotheses being that a rapidly expanding industry would adopt new technology more readily, and thus one should expect a higher rate constant, and similarly, the development of better communication channels as time passes should increase the speed of adoption. However, neither one of the hypotheses was found to be statistically significant. Later on, Bundgaard-Nielsen [2] showed that late adopters in the petroleum industry tend to adopt new technology faster than earlier adopters, and that the rate constant and the half time were dependent upon the technological background of the adopting firm as well as upon the strategic pressure upon the firm.

Studies of international diffusion of new technology are comparatively rare. One recent exception is a study by Swan [6] on the diffusion of a consumer product–synthetic rubber. Swan tested the hypothesis that the rate constant in the simple mathematical model, Eq. (1), was correlated with the time of adoption in a given country, the growth of the rubber industry in the country and the size of the rubber import per capita to the country. Swan found all three factors to be statistically significant. As observed quite correctly by Swan, the diffusion of synthetic rubber internationally is characterized by not requiring investments in new production equipment nor does it require that new working methods are adopted. Thus, the resistance towards diffusion is virtually exclusively on the consumer side, and very little resistance is to be found towards the diffusion in the rubber industry itself.

The purpose of this paper is to report on the results of a study of the international diffusion of new technology into the steel industry, where no barriers towards the diffusion are to be found on the consumer side since the product, steel, is unchanged. However, considerable resistance towards the new technology is to be expected in the industry itself since the adoptions studied require extensive investments in addition to the introduction of new technological concepts.

As shown by Bundgaard-Nielsen [2] the rate constant C is positively influenced by the technological background of the adopter, i.e., an advanced technological background facilitates the adoption of a new technology. This in turn implies that late adopters will adopt faster than earlier ones provided information concerning the new technology is available to all potential adopters–a provision which implies that late adopters are in a better position to assess the new technology than earlier ones–and consequently, the rate constant C should increase with t_0–the latter being a measure of the point in time when the average firm in a given country adopts the new technology. As a new technology develops, it becomes more flexible and more compatible, and since the decision to adopt a new technology depends upon how compatible it appears to be with the currently applied technology, one should also for this reason expect the rate constant to increase with time of adoption–here measured by t_0.

The growth rate, r, of the industry into which diffusion occurs may as pointed out by Mansfield [4] influence the rate constant quite simply because rapidly expanding firms may be more willing to experiment with new production methods etc. However, the opposite may also be the case, in particular, it is conceivable that if the growth rate is brought about as a result of a recently completed investment plan increasing the production capacity, then it is rather unlikely that the industry is going to scrap new and very durable equipment simply because a new–and in the long run–more profitable production process has just been introduced.

Concluding, one may then state that the rate constant C may be positively as well as negatively correlated with the growth rate of the industry—the sign depending among other factors upon the most recent capital investments undertaken by the industry. In mathematical form we obtain:

$$C_i = C(t_{0i}, r_i) \tag{2}$$

or assuming a linear functional relationship:

$$C_i = at_{0i} + br_i + c \tag{3}$$

where a, b and c are constants, while C_i, t_{0i} and r_i are the rate constant, the half-time and the industrial growth rate for the industry into which the new technology diffuses. The index i refers to country i.

The European Steel Industry

The simple model given by Eq. (1) was used to calculate rate constants and half-times for the diffusion of two major innovations into the steel industry of six European countries. The final number of adoptions, $A(\infty)$, was taken to be the number that minimized the residual sum of squares in estimating the parameters of Eq. (1). The actual data used are shown in the Appendix together with the $A(\infty)$-values, while the estimated rate constants and half-times are shown in Table 1.

It has been shown by Martino [5] that estimates of rate constants and half-times are affected systematically by errors in upper limits. Consequently, the procedure used to obtain the $A(\infty)$'s in this investigation was combined with a sensitivity analysis of the effect of errors in the $A(\infty)$'s. The analysis showed that a 5% error in $A(\infty)$ would lead to errors in C_i and t_{0i} of roughly the same relative size with one exception where the 5%

TABLE 1

Diffusion of New Technology into the European Steel Industry (1952–1968)

Technology	Country i	Rate constant C_i	Half-time t_{0i}	Level of significance
Continuous casting of steel	Germany	0.27	13.8	1%
	Italy	0.40	15.5	2%
	U.K.	0.21	12.7	1%
	France	0.18	4.5	1%
	Austria	0.21	9.2	1%
The use of oxygen converters	Germany	0.53	11.6	1%
	Italy	0.56	14.3	10%
	Sweden	0.41	10.1	1%
	U.K.	0.79	10.2	1%
	France	0.50	10.9	1%
	Austria	0.17	4.0	1%

relative error in $A(\infty)$ produced a 25% relative error in the half-time—an exception which, however, would have had no effect upon the trend and significance of the subsequent conclusions of this study. Next the rate constants for the two technologies were correlated with the half-times and growth rates providing estimates for the constants a, b and c in Eq. (3). The results are shown in Table 2, where the figures in the parentheses indicate standard errors. The growth rates for the national steel productions, r_i, were taken as averages from 1952–1968.

Figure 1 shows a scatter-diagram and indicates as does Table 2 a satisfactory correlation, i.e., one may conclude that Eq. (3) is a useful model of the relation between the rate constant and its influencing factors—half-time and growth rate for the steel industry in country i. The results in Table 2 show that the rate constant is positively influenced by half-time—as one would expect. As for the impact of growth rate upon the rate constant, the results indicate a positive influence for continuous casting and a negative influence for oxygen converters. A possible explanation is that the high growth rate in some of the national steel industries in Europe after the war was obtained due to the installment of new and durable production equipment, in particular, open hearth and electric furnaces for steel-making. These processes represent large investments, and clearly no firm wants to replace newly installed capital-intensive production processes with a new design, such as oxygen converters immediately after a heavy capital outlay, even if the latter promises a better economy in the long run. Thus, the high growth rate had in effect a retarding influence upon the diffusion of oxygen converters. This was not the case for the continuous casting process, simply, because the latter did not replace durable and expensive equipment, it merely extended the productive capacity of existing equipment. Thus, this latter result confirms the commonly held opinion, that a fast growing industry is more likely to invest in new processes than a stagnant industry—a rule, which as this study has indicated, has some important exceptions.

Concluding Remarks

There are two major conclusions to be drawn from this study:

(1) It is possible to describe the rate constant of diffusion as a function of some measure of the time of adoption in a given country and the relevant industrial growth rate in the country.

(2) While half-time seems to influence the rate constant of diffusion positively, the

TABLE 2

The Rate Constant in The Simple Diffusion Model

Technology	Eq. 3, The rate constant
Continuous casting of steel	$C_i = 0.012\, t_{0i} + 0.025\, r_i - 0.014$ (0.004) (0.008)
The use of oxygen converters	$C_i = 0.05\, t_{0i} - 0.06\, r_i + 0.34$ (0.01) (0.02)

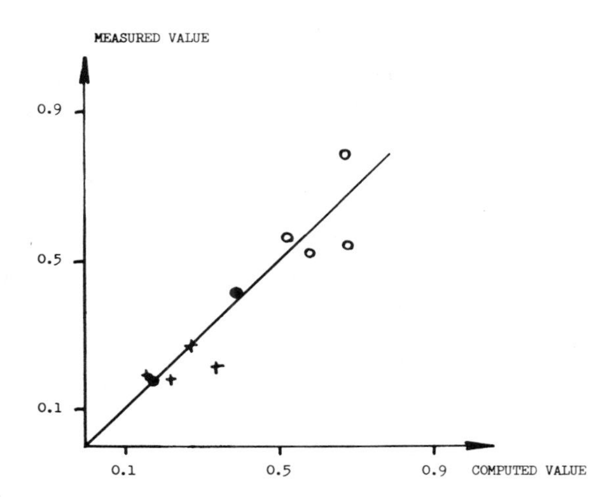

Fig. 1. Scatterdiagram. 1, ○, the use of oxygen converters; +, the continuous casting of steel.

impact of growth rate is more complex and may lead to an increased as well as decreased rate of diffusion, depending upon the particular situation prevailing in the industry.

Clearly, our knowlege concerning the factors influencing the international diffusion of new technology is far from complete. However, empirical studies as the one reported in this paper may help to increase our understanding of this subject, which will certainly gain in importance in the years to come. In particular, one might point out that the results presented in this paper indicate that major industrial countries, such as Italy and Germany are late but fast adopters, while more specialized industrial countries, such as Austria and Sweden are early adopters, but with a lower rate constant than the former countries. This observation combined with the demonstrated industrial growth rate dependence might serve as a first step towards the development of an international innovation index, whereby the innovation propensity of different countries could be characterized in a manner similar to the factor based innovation index developed by Blackman et al. [1].

References

* 1. A. W. Blackman, E. J. Seligman, and G. C. Sogliero, *Technol. Forecast. Soc. Change* **4**, 301–316 (1973).

† 2. M. Bundgaard-Nielsen and P. Fiehn, *Technol. Forecast. Soc. Change* **6**, 33–39, (1974).

3. R. Kosobud, *Technol. Forecast. Soc. Change* **5**, 395–406 (1973).
4. E. Mansfield, *Econometrica* **29**, 741–766 (1961).
5. J. Martino, *Technol. Forecast. Soc. Change* **4**, 77–84 (1972).
6. P. L. Swan, *J. Ind. Econ.* **22**, 61–69 (1973).

**This article also appears in Part 2 of this book.*

†This article also appears in Part 4 of this book.

Appendix

The Adoption of Continuous Casting of Steel

Years since introduction (1952)	Number of adoptions				
	Germany	Italy	U.K.	France	Austria
0	0	0	2	0	1
2	1	0	2	2	2
4	1	0	2	3	2
6	2	1	2	3	3
8	2	1	2	4	3
10	5	1	5	5	5
12	7	2	9	5	5
14	8	5	11	5	5
16	13	12	11	5	7
18		17			
$A(\infty)$	18	20	14	6	8

The Adoption of Oxygen Converters

Years since introduction (1952)	Number of adoptions					
	Germany	Italy	Sweden	U.K.	France	Austria
0	0	0	0	0	0	2
2	0	0	0	0	0	4
4	0	0	1	0	0	5
6	1	0	1	1	0	5
8	3	0	2	5	2	7
10	9	0	5	14	5	7
12	13	2	6	27	10	7
14	18	5	8	29	10	7
16		7	8		12	8
$A(\infty)$	24	10	9	31	13	9

Data: The Diffusion of New Technology, a study of ten processes in nine industries, Industriens Utredningsinstitut Stockholm, 1969.

Part 3.
Economic Analyses

Part 3. Economic Analyses

Introduction

The following three articles echo the theme of the preceding chapter that technological substitution has a strong economic basis. However, there are important differences between the approach taken by the authors of these articles and those of Mansfield, Blackman, and others described previously.

In the first paper, Stern, Ayres, and Shapanka take into account a general fact applicable to virtually all substitutions: alternate technologies do not compete merely on the basis of price, but rather on the basis of the "utility" they provide. Unfortunately, while it is comparatively easy to obtain data on the various performance characteristics representing the qualitative aspects of the technologies involved, there are difficulties of a fundamental nature in determining a "unit of utility".[1] A second problem inherent in "real world" substitution forecasting relates to the definition of an application or the identification of the relevant "market". For one thing, when a new product appears, the market ultimately available to it is not necessarily the market currently held by its nearest competitors.

In an attempt to come to grips with these problems, Stern et al. present a simulation model that allows some of the factors otherwise ignored in technology substitution to be evaluated and incorporated explicitly and quantitatively. The model is based on utility analysis, wherein for each of the competing technologies the relevant attributes and performance characteristics are first identified and, subsequently, the indicators that reflect the relative importance of each identified attribute to the principal users of the product. By combining the physical ratings of the properties it then becomes possible to obtain an overall utility factor for each product. The model itself is composed of numerous variables in addition to the utility adjusted price, e.g., obsolescence constant, the annual rate of growth of the *combined* market representing the demand for both the new and the old product, the learning time (i.e., the number of years to perfect the new product), etc. An application of the model is illustrated in the case of the substitution of glass by polyethylene for household and industrial chemicals. Clearly, the model is best suited to cases where competing technologies are rather precisely defined, where a good deal of current technical and economic data are available, and where an in-depth analysis is desired.

In the second article, Ayres and Shapanka present a new methodology for the integration of information on technological substitution into the well-known model of input-output analysis and illustrative applications of their approach to examples of substitution drawn from the aluminum and the iron and steel industries. As a forecasting

[1] On some of the difficulties involved, see *Price Indexes and Quality Change* (Zvi Griliches, ed.), Harvard University Press, Cambridge, 1971; H. Houthakker, Compensated changes in quantities and qualities consumed, *Review of Economic Studies* **19** (3), 155–64 (1952); and Sherwin Rosen, Hedonic prices and implicit markets: Product differentiation in pure competition, *J. Politic. Econ.* **82** (1), 34–55 (1974).

model, the input-output table has substantial advantages over other forecasting approaches. The coefficients and the variables of an input-output model represent real, measurable physical or monetary quantities, i.e., commodities, energies, services, labor, and capital. When any of these quantities or the relationships between them are likely to be altered by new developments, the effect of these changes can be specifically traced in the input-output table with greater clarity than in any other forecasting model. However, a serious problem in the use of the input-output technique is that it often aggregates into a single industry sector a number of producers whose outputs are not mutually substitutable for one another. At the same time, it ignores, in some cases, certain substitution possibilities between industries. As a consequence of aggregation, errors are likely to be introduced in predicting the impact of changes in the final demand or requirements for many physical materials. Obviously, it would be of considerable advantage to combine the methodology for forecasting technological substitutions and that of input-output analysis.

The approach presented by Ayres and Shapanka achieves this objective by means of the following four steps. First, the more important technological trends that can be expressed as substitutions are identified. Next, the information is translated into a form such that it can be used in the input-output model, i.e., sector inputs and outputs and ratios, and the historical rates of change of the variables are estimated. In the third step, the future time-path of the substitution process is projected by means of the simple trend extrapolation model based on the assumption that the substitution exhibits an S-shaped curve. The fourth and final step is to modify the existing input-output coefficients in conformity with the projections. In this manner, it becomes possible to obtain a considerably more reliable forecast based on input-output analysis.

The third article by Sahal takes as its point of departure both the Mansfield-Blackman type of models of technological substitution as well as those based on trend extrapolation. But in this work the assumption that the path of substitution takes the form of an S-shaped curve is regarded as "extraneous" and, at best, unnecessary. Technological substitution is considered as an economic phenomenon, subject to both profit as well as technical considerations. The role of the technological background is stressed in the decision to adopt an innovation–specifically, the scale of production of the firms under study. The combined importance of these considerations is borne out in the empirical analysis of the substitution of mechanical cornpickers by field-shelling technology. It is found that we can dispense with the assumption that the diffusion of an innovation follows an S-shaped curve, without any loss in the explanatory power of the models. The production scale of the firms, i.e., farm size, is suggested to be the most important determinant of technological substitution.

A Model for Forecasting the Substitution of One Technology for Another

M. O. STERN, R. U. AYRES and A. SHAPANKA

ABSTRACT

Most technological changes can be described as a substitution of one material, process or product for another. Each such substitution, if successful, normally tends to follow an S-shaped (or "logistic") curve: that is, it starts slowly as initial problems and resistances have to be overcome; then it proceeds more rapidly as the competition between the new and the old technology grows keener and the new technology gains an advantage; and finally, as the market for the new technology approaches saturation, the pace of substitution slows down. Sometimes, when the process is completed, the old technology continues to retain some specialized portion of the total market (i.e., a sub-market) for which it is particularly well adapted. In forecasting the course and speed of the substitution process, especially when it has already begun and partially taken place, the simplest approach is to project a function having the appropriate S-shaped curve, using historical data to determine the free parameters of the function. While useful, especially where data are not available for a more sophisticated study, the simple curve-fitting techniques fail to take into account several important factors that affect economic and management decisions on the part of producers and intermediate users (as well as "final" consumers) and thereby influence the course which the substitution process is likely to take. To overcome this limitation, a simulation model has been developed at IR&T which allows some of these factors to be evaluated and incorporated explicitly and quantitatively. The model is described and its application is illustrated in the case of the substitution of plastic for glass in bottles. It is most applicable where the competing technologies are rather precisely defined, where a good deal of current technical and economic data are available, and where an in-depth analysis is desired. Because this particular forecast was made before the sudden precipitous increase in petroleum prices, which upsets the price relationships assumed in the forecast, there is discussion of the vulnerability of forecasts to political and other contra-economic developments.

Market Substitution Analysis

This paper presents a method of forecasting the course of market penetration where a new product or technology competes with and gradually supplants an older one. We call

---

This article appeared in *Technological Forecasting and Social Change*, Vol. 7, No. 1, 1975.

The research reported here was supported, in part, by the National Science Foundation, under Contract No. NSF-C-652. A full report of this research is available through NSF(RANN) or directly from IR&T under the title "Materials-Process-Product Model, A Feasibility Demonstration Based on the Bottle Manufacturing Industry," IRT-305-FR, revised July 1974.

this process *substitution* because the new product or technology over a period of time replaces or, in a sense, "substitutes for" the old one in at least some portion of the market. Innumerable historical examples come to mind, such as the substitution of synthetic fibers for cotton, wool, etc., in the textile and clothing industries; of plywood for board lumber in a number of building applications; of coal for wood, and later of oil for coal, as fuel; and so on. In most cases, the new product or technology does not completely replace the old in all applications. Usually, the old product continues to retain a hold on some portion of the market or on a sub-market, as, for example, in the case of synthetic fibers which are blended with natural fibers to produce materials which have advantages over either type of fiber alone.

Long-term economic planning demands that we obtain some idea, early in the game, of how a technological change is likely to proceed. The more complex and interdependent our economy becomes, the more essential it is that changes be, to some extent, foreseen so that their impact on different industries can be evaluated before a crisis arises. But, at the same time, this growing complexity increases the difficulty and uncertainty of the forecasting process. The need is for forecasting techniques which will take into account what is known and project, on the basis of this knowledge, a "most likely" course of events in the future.

This is a task which individuals directly involved in an industry are not always able to do effectively; they tend to be preoccupied by the problems of daily management, production, short-term market planning, etc., and thus may be more aware of perturbations caused by these problems than they are of the ultimate potential of a new product or process. Businessmen and engineers–especially in senior executive positions–often take a negative attitude toward the commercial possibilities of a new technology because the costs and risks of development and implementation are more apparent and immediate than the long run benefits. Indeed, the difficulties may be borne, to some extent, personally, while the rewards, necessarily deferred, tend to accrue to successors in their jobs.

At the same time, however, what a businessman demands of a technological forecast in terms of specificity and accuracy tends to be quite rigorous. An academic researcher doing a 25-year forecast can (and must) tolerate errors in timing of plus or minus 5 years and a considerable degree of vagueness with regard to details. A businessman who is trying to plan production and marketing strategy for the next five years obviously cannot. Therefore, while the long-range forecast can proceed on the basis of relatively speculative information on a new technological possibility, a short-range forecast should probably not be attempted in the absence of information from a pilot plant or a test marketing program.

The techniques of technological forecasting span a range from the so-called Delphi method, which employs the structured opinions of a group of experts, to semi-empirical mathematical models which attempt to quantify the variables that determine the course of a changing technology. We are concerned in this paper largely with the latter approach; for we believe that presently existing mathematical models are sufficiently developed to permit the testing of different assumptions regarding the rate of change of price, technical performance, technological diffusion, market growth, etc.

History of Work on Substitution Analysis

What we now call "substitution analysis" is a recent outgrowth of input-output analysis pioneered more than twenty years ago by Wassily Leontief [1]. Examination of the complex ways in which the outputs of one industry become the inputs of other industries is essential to an understanding of how the economy works; and this, in turn, is fundamental groundwork for technological forecasting. Work on input-output analysis has been considerably refined in the last two decades. At the present time, active work is under way at the Bureau of Labor Statistics, U.S. Department of Labor, the Office of Business Economics, U.S. Department of Commerce, the Economic Commission for Europe (ECE), and various other organizations (including IR&T).

One of the early problems in these studies was that of how to balance an input-output table as the various elements change over time. An input-output table theoretically must balance: that is, the total of sales must continue to equal the total of purchases, not only in the initial year but in later years also. In practice, when data are collected and assigned to the cells of an input-output table, it is never possible to balance the table precisely, and an "unallocated" row and column are introduced to handle the discrepancies. When an input-output table is updated, and more recent production levels are used, then the changes in interindustry coefficients result in an increase in the size of the discrepancy. R. A. Stone [2] proposed a heroic assumption, namely that the coefficients in each row and each column be adjusted upward in a fixed proportion to eliminate the unallocated transactions. This method of balancing tables is called the RAS method. Stone further proposed that forecasts of changes in coefficients be based upon the same procedure: the rate of increase of the discrepancy for each row and each column should be estimated, and assumed to continue constantly into the future. When used as a device for extrapolation of interindustry coefficients, the RAS assumption has two parts: first, it assumes that the trends in coefficients occur at a constant rate over time; second, it assumes that every coefficient in a row and in a column is changing at the same rate.

This is a statistician's method; it is quick, simple and convenient. But later empirical tests show it to be substantially in error when technologies grow at every uneven rates (as they generally do). It has been found necessary to examine particular technologies and to take into account their individual characteristics and probable rates of growth or decline. This leads to the concept of market substitution analysis.

A model describing the process of market substitution analysis was presented by Edwin Mansfield [3] of the University of Pennsylvania in 1961. Other substitution studies have been carried out by Chow (1967) [4], Blackman (1971) [5], Fisher and Pry (1971) [6], and by Ayres, Noble and Overly at IR&T (1970) [7]. A more elaborate market substitution model has recently been developed by M. O. Stern of IR&T. Rather than attempting to review the advantages and disadvantages of all of these approaches, we will limit the historical discussion to a review of two of the simpler methods, namely those of Fisher-Pry and Ayres-Noble-Overly (IR&T). Then we will go on to a discussion of the more sophisticated Stern model which is the main subject of this paper.

Projecting the S-Shaped Curve

The two methods we will now review are curve-fitting techniques based on empirical observations of historical data on substitutions. The general picture shown by these data is that the "path" of a substitution (i.e. the percentage of the market acquired by the new product, as a function of time) tends to take the form of an elongated S-shaped curve. The slope of the curve before the first inflection point is rather flat, reflecting the fact that during the first few years a new product, process or service must overcome the ignorance or resistance of consumers, performance "bugs" in early models, contract arrangements based on the old technology, production diseconomies due to small scale, problems of financing, developing, installing and learning to use new processing equipment and methods, etc.

During later phases of the substitution, the rate of increase of market penetration increases sharply as the product becomes accepted, as production processes are improved, as economies of scale are achieved, as new contracts are made, and as the "learning period" for the industry comes to an end. And, finally, after the second inflection point the curve flattens again: the new product is "mature," that is to say, it is no longer changing rapidly and has already exploited most of its scale economies; also, it is now challenging the "defender" mainly in those markets or applications for which the latter is best suited.

An obvious forecasting approach, then, when a substitution process is already under way, is to anticipate the usual S-shaped curve and try to project its probable shape and time-span on the basis of available data, either for the substitution in question or for substitutions judged to be similar which have taken place in the past.

The basic forecasting equation of the *Fisher-Pry model* [6] is derived from the hypothesis that the fractional share f of the market held by a "challenger" (A) increases at a rate proportional to the fraction not yet captured from a defender (B). This fraction is $(1-f)$. The relation postulated above can be expressed as a differential equation

$$\frac{1}{f}\frac{df}{dt} = k\,(1-f) \tag{1}$$

where the constant k is defined as the annual growth rate of the fractional market share of the new product during the first few years of substitution (obtained from historical data). Let $t = u$ define the time at which $f = 0.5$, i.e. the substitution is halfway completed and the challenger and defender have equal market shares.

The differential equation above can be integrated to yield the expression

$$f/(1-f) = \exp k\,(t-u), \tag{2}$$

whence

$$f = \frac{1}{1 + \exp k(u-t)} \tag{3}$$

This is one version of the so-called "logistic" curve. Note that f has two parameters, k and u. However, for purposes of fitting to empirical data the actual markets Y_A and Y_B must be identified. By assumption the total market is the sum of the sales of A and B, viz.,

$$Y_A + Y_B,$$

where A is the challenger, B the defender. Hence

$$f = Y_A/(Y_A + Y_B).$$

On reflection, it can be seen that the market as a whole is, itself, changing in time and that observed historical changes in Y_A simultaneously reflect both the substitution of A for B and the changing sum of A and B together. In effect, there are four parameters rather than two, to select. This obviously increases the complexity of the statistical fitting problem in practice, but does not change its nature.

The Fisher-Pry approach bypasses such questions as comparative prices or comparative utilities (usefulness for specific purposes) of the competing materials or products and assumes that a substitution process—once it has been initiated—will continue to completion, following a predetermined path. This kind of forecast is probably most useful in dealing with substitutions between "categories" of products which comprise a broad range of applications. In this type of situation it is difficult to determine realistic comparative prices or comparative utilities (discussed below) or any of the other parameters required by the more sophisticated models. The absence of complicating detail does not exempt the Fisher-Pry model from problems, however, since it is often difficult to determine exactly what the "total" joint market really is and therefore what fraction of it should be assigned to each of the competitors at any given time. Thus the Fisher-Pry model is inherently more adaptable to retrospective analysis than to *ex ante* projections.

The *Ayres-Noble-Overly (IR&T) model* is superficially similar to the Fisher-Pry model described above, but with the addition of a price mechanism which incorporates some of the dynamics of the substitution. Before we take this up, however, let us look at a class of problems that is inherent in all substitution analyses, but that has been largely ignored in previous work—namely, the relative usefulness or adaptability of two competing products or technologies for particular applications, and the extent to which new applications evolve, in turn, as a consequence of improved performance.

Utility Analysis

When one material or technology competes with and gradually replaces another, the new material or technology is never an exact equivalent of the old; hence, the unit prices of the two are not directly comparable. For example, in the case of the substitution of plastic for glass in bottles, which we will consider in more detail later, it is clearly not valid to compare the average price of, let's say, a milk bottle made of glass with one of the same size made of plastic. In the first place, the lighter weight of polyethylene (PE) means lower shipping costs and easier handling which affect the price of milk to the consumer. But, even more important, the lightness, unbreakability and moldability of PE bottles permits customers to carry and manipulate a gallon container of milk—with a molded handle—about as easily as they could formerly carry and manipulate a single quart container made of glass. What is actually happening in the market place shows that the competition is *not* between one-quart glass bottles and one-quart PE bottles for milk, since in reality one-quart glass bottles have largely been replaced by two-quart waxed paper containers and gallon-sized PE bottles.

The above example illustrates a general fact applicable to virtually all substitutions: alternate technologies do not compete on the basis of price alone, but rather on the basis of maximum "utility" provided per dollar of cost. Or, looked at the other way around, cost-effectiveness implies minimum cost per unit of "utility." But what is "utility"?

Unfortunately, while it is comparatively easy to obtain data on prices per unit of weight, per unit of contained volume, or per unit of some other explicit performance measure, it is not so easy to determine a "unit of utility." Still, if a forecast is to bear much relation to reality, an attempt has to be made.

The second critical problem inherent in "real-world" substitution forecasting relates to the definition of an application or "market." The difficulty can be illustrated, again, by observing that only a few decades ago the class of "household containers" consisted essentially of wooden crates, cardboard boxes, paper bags, tin-plated steel cans and glass bottles. The advent of new and hybrid materials has vastly increased this range. Restricting consideration to the smaller category of airtight dimensionally stable containers, one still has to add to the list: aluminum cans, tin-plated steel cans with aluminum tops, polymer-coated steel cans, PVC or PE bottles, and (soon) laminated plastic-glass bottles. Each of these tends to be preferred for certain applications—and as the number of distinguishable alternative types of containers continues to grow, so does the number of distinguishably different *markets.*

When a new product appears, the market ultimately available to it is not necessarily the market currently held by its nearest competitors. When detergents began competing with soap they quickly replaced soap for most uses in washing machines—but not for personal washing. The two markets are now distinguishably different.

Clearly a general procedure for forecasting substitutions must also attempt, somehow, to come to grips with this problem. The differentiation of markets is closely related to the question of assessing utility-in-use. The procedure currently used by IR&T does not attempt to forecast the proliferation of new markets, because we share the general inability to adequately determine utility values of hypothetical products or processes for which there is, as yet, no body of actual experience. We feel more comfortable in dealing with situations where the set of distinguishable applications has already been established by a market process of sorts. Our procedure is as follows.

Step 1—Identify the relevant attributes and performance characteristics of each of the competing products or technologies. This means first compiling a list of these attributes, which can be done by a team of investigators who are familiar with the properties of the products or processes. Once this list is complete, each product or technology is "rated" by the team to determine the extent to which it possesses each attribute. The following rating scale was used in the following analysis:

Rating Scale for Attributes

3—the product possesses this attribute to a high degree
2—the product possesses this attribute to a considerable degree
1—the product possesses this attribute to a limited degree
0—the product does not possess this attribute.

Step 2—Identify indicators which reflect the relative importance of each identified attribute to the principal users of the product. Again, a quantitative rating scale is normally used to translate verbal evaluations (e.g., obtained by a survey of consumers, manufacturers and related literature) into numerical values. The rating scale is adjusted to take into account evidence that indicates that human beings tend to perceive order-of-magnitude (i.e., geometric) differences on an arithmetic (or logarithmic) scale.[1] In other words:

Rating Scale for the Importance of Properties

16—critical
8—very important
4—somewhat important
2—of little importance
0—of no importance

[1] This is well known to be the case for perceptions of sound or light intensity, for instance.

TABLE 1

Attribute/Importance Ratings

	Physical Attributes	Physical Rating of Materials: PE	PVC	GL.	Perceived Importance of Attributes to Various Markets (Average): Liquor	Wine	Soft Drinks	Food	Dairy	Med. & Health	Cos-metics	Hshld. Chems.	Beer
Of concern to consumer bottler	Unbreakability	3	3	1	5.3	5.3	5.3	5.3	5.3	4.0	6.4	8.0	5.3
	Lightweight	3	3	0	8.0	8.0	8.0	8.0	8.0	8.0	8.0	8.0	8.0
	Transparency	1	3	3	4.0	6.0	2.0	4.0	2.0	4.0	6.0	3	4.0
Of concern to consumers	Tradition	1	1	3	2.3	4.0	2.3	2.3	2.3	2.3	2.0	2.0	2.3
	Reusable in home	2.5	2.5	2	0	0	0	3.0	2.5	2.0	0	2.0	0
	Convenient dispensing	3	3	2	0	0	0	2.0	0	2.0	2.0	4.0	0
	Chemical inertness	2	2	3	10	8.0	8.0	16.0	8.0	16.0	16.0	4.0	8.0
	Heat resistance	1	.5	3	0	0	0	16.0	0	0	0	0	16.0
Of concern to bottlers	Flexibility of color and shape	3	3	2	6.0	0	3	3	2.0	3	6.0	3.0	2.0
	Non-permeability	1	1	3	8.0	8.0	8.0	16.0	8.0	8.0	8.0	8.0	16.0
	Adaptable to bottling eqpt.	2	2	2	3.0	3.0	8.0	8.0	8.0	8.0	4.0	8.0	4.0
	Pressure resist-ance (rigidity)	1	1	3	1.0	1.0	16.0	14.0	1.0	1.0	1.0	1.0	16.0
Concerning social policy	Disposability	2	2	2	12.0	12.0	12.0	12.0	12	12	12.0	12.0	12.0
	Recycleability	1	1	3	1.0	1.0	2.0	2.0	1.0	1.0	1.0	1.0	2.0

TABLE 2

Aggregate Utility Scores

	Beer	Liquor	Wine	Soft Drinks	Food	Dairy	Med. & Health	Cos-metics	Hshld. Chems.
PE	118.2	94.2	75.9	95.2	148.7	82.4	104.3	117.2	97.0
PVC	118.2	102.2	87.9	99.2	148.7	86.4	112.3	129.2	103.0
Glass	202.2	96.2	89.3	126.2	232.2	81.2	110.9	124.4	83.0

Step 3—After completion of Steps 1 and 2, *the final step is to multiply the physical ratings of the products by the importance ratings of the properties in order to obtain an overall utility factor for each product.* Then the utility factors are normalized so that the utility factor of the old product equals one. By applying these factors to the unit price, it is possible to obtain what we call the "utility-adjusted price" of the new product in relation to its predecessor or competitor at a given point in time. As an example, see Table 1, which illustrates the method as applied to the substitution of plastic for glass in bottles.

Tables 1–3 illustrate the application of these three steps to the utility analysis of plastics versus glass for bottles in various markets (applications). The scores can be compared down, among the three materials for each market, rather than across markets. (Comparing across markets would reflect the rigidity of requirements for each market rather than the suitability of the materials.) Table 2 shows the utility scores after normalization, taking glass to have a "standard" utility of 1. These values will be used in the implementation of the Stern model, discussed later.

The Ayres-Noble-Overly Model

The Ayres-Noble-Overly model [7] was the first of which we are aware that explicitly introduced a utility-adjustment for prices. The approach is simple and can be illustrated by the substitution of aluminum for copper in electrical applications. Suppose product A (in this case, aluminum) is challenging and gradually supplanting product B (copper) in a specified market (electrical equipment). Let Y_A represent the (dollar) output of the industrial sector which produces product A (e.g., the aluminum industry). Let P_A represent the utility-adjusted[2] unit price of the material produced (aluminum). Y_B and

TABLE 3

Normalized Utility Scores

	Beer	Liquor	Wine	Soft Drinks	Food	Dairy	Med. & Health	Cos-metics	Hshld. Chems.
PE	0.58	0.98	0.85	0.75	0.64	1.01	0.94	0.94	1.17
PVC	0.58	1.06	0.98	0.79	0.64	1.06	1.01	1.04	1.24
Glass	1.0	1.0	1.0	1.0	1.0	1.0	1.0	1.0	1.0

[2] The utility analysis in the case of aluminum versus copper has to take into account the following factors: that aluminum is lighter than copper; that its electrical conductivity is less; that, because of its greater electrical resistance, a greater proportion of the electrical energy is lost by heat; that, because of the latter, more heat-resistant types of insulation have to be used; and that the ductility and certain other physical characteristics of the two metals affecting workability are different.

P_B, then, would signify, respectively, the (dollar) output of the copper industry and the unit price of the copper used in electrical manufactures.

Assuming that all the relevant measures exist, we can define the ratios

$$Y = \frac{Y_A}{Y_B} \quad P = \frac{P_A}{P_B}$$

in such a way that Y varies from zero to infinity over a period of time as the substitution process goes to completion. It is convenient to introduce a function, f, the fractional penetration, as before:

$$f = \left(\frac{Y_A}{Y_A + Y_B}\right) = \frac{Y}{1 + Y}. \tag{4}$$

The actual replacement, of course, takes place gradually over a period of time, due to a number of "inertial" factors in the system to which we have already referred.

We postulate the following simple relationship between output ratio and price ratio:

$$Y = Y_0 P^{-m}. \tag{5}$$

We also assume that the price ratios tend to change rather smoothly in time, following an exponential decay, viz.

$$P = P_0 \exp\left(-\frac{k}{m} t\right), \tag{6}$$

where t represents elapsed time in years starting from an arbitrarily selected and initial year (e.g., 1900).

It follows, then, that

$$Y = Y_0 P_0^{-m} \exp(kt). \tag{7}$$

Thus, substituting (7) into (4):

$$\begin{aligned} f &= \frac{1}{1 + Y_0^{-1} P_0^{m} \exp(-kt)} \\ &= \frac{1}{1 + \exp[k(u-t)]}, \end{aligned} \tag{8}$$

where $\exp(ku) = (P_0^m / Y_0)$ or,

$$u = [\ln (P_0^m / Y_0)] / k \tag{9}$$

by definition. Eq. (8) is the same logistic growth function appearing in the Fisher-Pry model. The parameters have the same interpretation. Actual fitted values of k and u for some exemplary cases, based on IR&T studies, are listed in Table 4.

As with the Fisher-Pry model, this is essentially a simple curve-fitting methodology. Clearly, the basic assumption regarding the relation between prices and market shares is unrealistic. It is possible–indeed, commonplace–for a new product to begin penetrating a market even *before* its price is truly competitive. Moreover, the substitution may proceed to completion or to an equilibrium point even though the two utility-adjusted prices are

TABLE 4

Fitted Parameters for Substitution Function (Listed in Increasing Order)

Substitution	k	u
Electricity for Labor	0.021	1971
Telephone for Mail	0.031	1916
Knits for Woven Goods	0.115	1992
Vinyl for Hardwood Floor	0.150	1966
Aluminum for Copper Wire	0.153	1986
Air for Rail Travel	0.164	1957
PVC for Cast Iron Pipe	0.183	1966
Nylon for Wool	0.188	1963
Fiberglass for Steel in Autos	0.219	1988
Vinyl for Leather Goods	0.227	1975
Computers for Labor	0.548	1971–1975

Source: IR&T.

still approximately equal. In order to take into account the underlying dynamics of such situations, a more sophisticated model has been developed at IR&T by M. O. Stern [8].

The Stern model, because it takes into account more factors in the dynamics of technological change than do the simpler models, requires more specific current data on the technologies involved. The competing products or technologies need to be quite specifically defined. It is necessary to know something of the age, condition and rate of obsolescence of the capital equipment used in the old technology, since, obviously, the new one can take over more rapidly if the old capital equipment will soon have to be replaced. It is necessary, too, to have some idea of how long it will take to perfect the new technology and train its users in the industry affected; for, if the new process is highly complex and its application requires a radical change in the users' way of doing things, obviously the substitution will proceed more slowly than if the changeover is comparatively simple. Another element to be taken into account is the price elasticity of demand, which determines the extent to which price competition is a critical factor in the market for the product.

The Stern model, like those previously discussed, exhibits a general "S" shape; but, unlike the two simpler models we described, its inflection point is not necessarily at the point $f = 0.5$. In addition, although continuous, $f(t)$ is not differentiable at one point—the discontinuity in the first derivative is generally too slight to be observable in practice.

We have attempted to structure the model in such a way that most of the input it requires is obtainable from observations that are either readily available or derivable, with one exception: like all substitution models (whether explicitly or implicitly), this one requires us to adjust unit prices of the competing products or technologies for differences in their "utility" in the manner referred to earlier. Some of these differences are qualitative, hard to measure physically and harder to translate into dollars. This requirement also forces us to subdivide markets in which the substitution is played out into submarkets for which the ratio of utilities of the competing products is roughly constant. This leads to data problems, because it may require a degree of disaggregation not readily at hand. As already mentioned, however, the assumption is basic to the substitution process, and it is better to face it explicitly than to sweep it under the rug. From now on, when we speak of equivalent units, we mean that we are comparing units adjusted for equal utility.

Qualitative Description of the Stern Model

We consider a "market" Y (units/year) for materials, processes or products (called "products" for generality). Two products, i, the defender, and j, the challenger, are battling for domination of the market $Y = Y_i + Y_j$ (conceptually, the model can readily be extended to three or more products). For flexibility, we assume a time origin $t = 0$, which can be in the past, present or future. The challenger makes his first appearance at $t = t_1 \geqslant 0$. Before this time, all we need to know is $Y = Y_i(t)$, and $P = P_i(t)$, the market price of a unit of Y_i.

At $t = t_1$, product j makes its initial entry; it does so on the basis that its ultimate cost of production per equivalent unit is estimated to be substantially lower than that of product i, so that it can ultimately undersell i and drive it out of the market (based on prices adjusted for relative utility values). It usually takes a substantial "learning period" before this ultimate cost is approached. During this learning period, several things may take place:

(1) Processes for producing product j are shaken out and certain obstacles to low-cost production are removed by research and development.

(2) The customer is introduced to j and undergoes an educational process or habit-changing process regarding j's usefulness relative to i; i.e., he learns to perceive or accept that, at an equivalent utility-adjusted price, both are, on the average, equally useful.[3]

(3) An initial market Y_j for j is created that justifies its production at a level where scale economies permit its cost to become competitive.

(4) Competitors become aware of product j and increasingly consider replacing obsolescent capital for producing i by new capital for producing j.

We therefore postulate that between time t_1 when product j first appears, and a later time t_2 when its cost of production becomes competitive, j is placed on the market below cost,[4] in order to create an initial entry and test possible market acceptance, before a major investment is made in production scale-up. Various pricing policies might be pursued for j at this time. For example, it could be priced at P_i, physical unit for unit, which might permit it to have a price edge from the utility point of view. Analytically, it is also permissible for t_1 to equal t_2; in that case, j is introduced at an initial price equal to or lower than that of its competitor i. Its further price descent then is determined by the learning dynamics.

In any case, the penetration in the period $t_1 \leqslant t \leqslant t_2$ is not modeled specifically. Instead, we assume that the market price of a unit as well as the total market Y remain on their trend prior to t_1; marketing penetration of j, abetted by suitable but unspecified pricing policies, is governed by learning dynamics, and takes advantage of growth in Y, and obsolescence of production in Y_i. The latter helps Y_j to the extent that "word gets around" about a serious challenge to i, discouraging investment in new production facilities for i. In certain cases (e.g., durable goods), obsolescence of the i inventory may determine the replacement dynamics more than obsolescence of production capital.

---

[3] In practice, we must be content with such a definition; the case that all customers will be indifferent with respect to i and j is too much to hope for, even with careful utility adjustments. At best one can hope for equal demand for both.

[4] Or at least below its shadow sales price, i.e., one permitting a reasonable venture profit at which Y_j would be sold between $t_1 < t < t_2$ if it dominated the market.

At time t_2, a certain degree of market penetration Y_j, determined by the dynamics of learning, market growth, and obsolescence in the production or inventory of i has taken place. At this point, we postulate that the shadow price for j (adjusted for relative utilities of i and j) becomes competitive, i.e., $P_j(t_2) \leqslant P_i(t_2)$,[5] and thereafter it decreases further, tending toward its ultimate low value. We postulate that the real market price P is a weighted average of what, following t_2, will be the separate shadow prices for equivalent units of i (whose price is assumed to follow its prior trend) and j. As long as $Y_j << Y_i$, $P(t)$ will "ride the coattails of $P_i(t)$," allowing Y_j to cash in on the extra profit. As Y_j becomes $\cong Y_i$, competition in the production of j will force the market price P down toward P_j, and drive i out of the market except for possible specialized applications.

If the product represented by market Y exhibits marked price elasticity of demand, the decrease in P contributes to growth of the market Y, and accelerates the penetration of Y_j. If the obsolescence in Y_i inheres in its production capital (rather than its stock), we assume that this obsolescence is also accelerated by the price elasticity, in that the strategy for marketing j will lean toward optimizing present worth of future profits by forcing the price down toward P_j, rather than riding the coattails of P_i. In that case, shutdown of older, marginal production capital for i will be further hastened.

Mathematical Description of the Stern Model

Capitalized Latin alphabet letters were, for the most part, already introduced in the previous section and represent dimensioned quantities. Lower-case Latin alphabet letters are dimensionless numbers or representations (ratios) of the capitalized ones. Greek letters represent inverse time-dimensioned dynamic constants.

The crucial parameters to be estimated are as follows:

- t_1 The year in which the new product is introduced.
- t_2 The year in which the utility-adjusted price of the new product becomes equal to that of the old product.
- λ Obsolescence constant; the average lifetime (in years) of capital equipment in the old segment of the industry.
- κ The annual rate of growth of the *combined* market (as distinguished from the parameter k defined for the Fisher-Pry and Ayres-Noble-Overly models described earlier).
- σ Learning time; the number of years required for the new product and its manufacturing processes to be largely perfected, and for this know-how to be disseminated throughout the industry.
- m The price elasticity of demand.
- $\overline{P}_j(\infty)$ The utility-adjusted price of the new product at some future time when its utility has reached a maximum and its monetary cost has reached a minimum.

Other quantities used in the mathematical presentation (including those referred to in the previous section) are as follows:

$Y_i(t)$ Yearly market for defender i (units per year).
$Y_j(t)$ Yearly market for challenger j (units per year).
$Y(t) \doteqdot Y_i(t) + Y_j(t)$ Total market for products i and j.

[5] The $<$ sign applies generally where the challenging product is less costly from the outset, in which case $t_1 = t_2$ and $Y_j(t_2) = 0$.

$P_i(t)$ Market price per equivalent unit (i.e., utility-adjusted) of product i.
$P_j(t)$ Shadow market price per equivalent unit of product j.
$P(t)$ Actual market price per equivalent unit of either i or j.
$f(t) = Y_j(t)/Y(t)$ Market share of j.
$r_i(t) \equiv P_i(t)/P_i(t_2)$ Price of i at time t relative to that at t_2.
$r_j(t) \doteq P_j(t)/P_i(t_2)$ Shadow price of j relative to that of i at t_2.
$r(\infty) \doteq P_j(\infty)/P_i(t_2)$ Price of j in the distant future relative to that of i at t_2.
$r(t) \doteq P(t)/P(t_2)$ Average market price relative to that at $t = t_2$.

We now write down and discuss the relationships postulated, one by one, and turn to their solution, $Y_j(t)$ or $f(t)$.

$r_i(t)$ must be given or assumed.

We postulate that $r_i(t)$ is at least piecewise continuous and differentiable. If defender i has dominated for a period of years without technological change, $r_i(t)$ is likely to be stable up to $t = t_1$. For $t \geqslant t_1$ one may want to introduce a price behavior based on a defensive strategy against product j.

$$Y(t) = \frac{Y(0)\,(P(0))^m}{(P(t_2))^m}\,(r(t))^{-m} e^{\kappa t}, \text{ for all } t. \tag{2}$$

This relationship expresses the assumption that the market is changing at the rate κ/year, and that the price elasticity of demand is a constant m. It is so normalized that at some chosen time origin $t = 0$, usually taken in the past, the market is $Y(0)$. If the market has been changing at a variable rate, a more complicated function can be used, or κ can be chosen to represent the growth rate (at constant price) during the period of substitution.

$$r(t) = r_i(t), t \leqslant t_2. \tag{3}$$

$$\begin{aligned} r(t) &= (Y(t))^{-1}\,(Y_i(t)r_i(t) + Y_j(t)r_j(t)) \\ &= r_i(t)\,(1 - (Y_j(t)/Y(t))\,(1 - r_j(t)/r_i(t))) \end{aligned} \tag{4}$$

for $t > t_2$; we require $r_j(t_2) \leqslant r_i(t_2)$.

We postulate that before $t = t_2$, while j may make a market penetration, it will not be based on any "orderly" strategy that can be modeled, and that the market price $r(t)$ will remain $r_i(t)$, although this function may exhibit reactions to the challenger's strategy. For $t > t_2$, we assume that *the market price for equivalent utility-adjusted units of i and j is the same,* and is given as a market-share-weighted average of the two shadow prices r_i and r_j. While such weighting appears plausible, it cannot be derived from economic theory; it could be checked empirically, but this is a difficult task since shadow prices are not directly observable, but can at best be imputed from extrapolation or venture analysis. In practice, the market price for equivalent units of i and j may differ; in that case, $r(t)$ is taken as the *average* market price.

$$r_j(t) = r(\infty) + (1 - r(\infty))\,[\exp \sigma'(t_2 - t)]\,, \text{ for all } t;$$

where σ' = elasticity-modified learning time constant; $\sigma' \doteq \sigma(2 - r(\infty))^{ml}$.

This relation causes the equivalent unit shadow price for j to decrease from a value greater than unity at $t < t_2$, through 1 at t_2, to $r(\infty)$ at $t \to \infty$, decaying with a characteristic modified learning period $\sigma' = \sigma(2 - r(\infty))^{ml}$, where the modifying factor $(2 - r(\infty))^{ml}$ takes into account possible pricing strategies for the challenger j that yield greater present venture worth by lowering price and increasing market penetration and volume at the expense of profit margin per unit sold. If some production costs for j are available as a function of time, these can be used to estimate values for σ and t_2.

Finally, we postulate, for all $t > t_1$,

$$\frac{dY_j}{dt} = (1 - Be^{-\sigma'' t})(\frac{dY}{dt} + \lambda' Y_i) \doteq S(t)(\frac{dY}{dt} + \lambda' Y_i) \tag{6}$$

where $B \doteq \exp(\sigma'' t_1)$; σ'' represents market penetration time $[\sigma'' \doteq \sigma(1 - r(\infty))]$; and λ' represents elasticity-modified obsolescence time constant $[\lambda' \doteq \lambda(2 - r(\infty))^{mn}]$.

This equation says that the rate of increase with time of the market for j is proportional to the rate of increase of the total market Y, and to the rate of obsolescence $\lambda' Y_i$ of the capacity for production of i. The degree (or proportionality factor $\leqslant 1$) to which dY_j/dt can take advantage of market growth dY/dt and obsolescence $\lambda' Y_i$ is given by the market penetration factor $S \equiv 1 - Be^{-\sigma'' t}$, which increases from zero at $t = t_1$ to 1 as $t \to \infty$.

There are several assumptions inherent in Eq. (6):

(1) It is not obvious that the learning constant σ used in Eq. (6) should be the same as that used in Eq. (5), although to avoid introducing additional difficult-to-calibrate parameters we have assumed this to be the case. In practice, we often distinguish two learning time constants; one (σ') is associated with production shakeout, scale economy and pricing strategy; the other (σ'') with market penetration.

(2) The rate of obsolescence $\lambda' Y_i$ for stock of (or manufacturing capital for) defender i presupposes either an exponential decay (which is never strictly the case, though mathematically convenient) or a situation close to static equilibrium for the value Y_i, which holds only if $\kappa << 1$.

(3) The use of $\lambda' = \lambda(2 - r(\infty))^{mn}$ rather than λ is based on the notion that a high price elasticity, coupled with diffusing knowledge regarding the low value of $r(\infty)$, causes some of the older plants for manufacturing product j to be shut down prematurely before the end of their normal life. The parameter n is a (usually positive) number, not necessarily equal to l in Eq. (5).

In theory, appropriate values for l and n might be estimated from a rather complex venture analysis in which both challenger and defender use appropriate strategies to maximize their present worth. In practice, such an analysis goes beyond our present scope and, while promising to afford additional insight into the substitution dynamics, would have limited practical applicability. Thus l and n will be considered available for improving fit in those cases where data are available.

Solution of the Equations

The solution can best be divided into two regimes. The first regime obtains for $t_1 \leqslant t \leqslant t_2$, so that $r(t) \doteq r_i(t)$. The second regime prevails for $t \geqslant t_2$.

In the first regime, only Eq. (1), (2), (3) and (6) come into play. Substituting from (2) for $\dot{Y} = dY/dt$ and Y in (6) yields a differential equation in Y_j, where $r_i(t)$, hence $Y(t)$, is assumed given (we define $\dot{r}_i = dr_i(t)/dt$:

$$\dot{Y}_j + S(t)\lambda' Y_j = S(t)Y(t)(\kappa + \lambda' - m\dot{r}_i(t)/r_i(t)), \tag{7}$$

which, subject to the boundary condition $Y_j(t_1) = 0$, is solved by

$$Y_j(t) = e^{-\lambda'\int S dt} \int_{t_1}^{t} e^{\lambda'\int S dt} SY(\kappa + \lambda' - m\dot{r}_i/r_i)dt. \tag{8}$$

This equation can readily be solved (at least numerically) if $r_i(t)$ is given.

In the second regime, Eqs. (4) and (5) must be introduced as well. We shall first deal with the case that $t_1 < t_2$. In that case, some market penetration has taken place by the time t_2, i.e., $Y_j(t_2) > 0$; we then postulate that $P_j(t_2) = P_i(t_2)$, so that $r_j(t_2) = r_i(t_2) = r(t_2) \equiv 1$. It can easily be shown that dY/dt becomes discontinuous at t_2, because $Y_j(t_2)$ is finite, and introduces an abrupt (albeit usually small) change in the slope of $r(t)$.

The discontinuity is derived by requiring Y_j to be continuous at $t = t_2$. From Eq. (4), we can solve for Y_j in terms of $r_i(t)$, $r_j(t)$, $r(t)$ and $Y(t)$:

$$Y_j = \frac{(r_i - r)\, Y}{r_i - r_j}. \tag{9}$$

But from (2), $r(t)$ can also be expressed in terms of Y, and r_i and r_j are assumed given (Eqs. (1) and (5)). Thus we can find dY/dt for $t = t_2 - \epsilon$, (from Eq. (2)) and for $t + \epsilon$ (from Eq. (9)) and let $\epsilon \to 0$ in order to obtain the discontinuity in dY/dt at t_2 in terms of Y_j. From the direction $t_2 + \epsilon$,

$$(\dot{Y})_{t_{2+}} = Y_j(m\dot{r}_i + m\sigma'(1 - r(\infty))) + Y(\kappa - m\dot{r}_i), \tag{10}$$

with all variables evaluated at $t = t_2$. It is easy to check that $(\dot{Y})_{t_{2-}} = \dot{Y}(\kappa - m\dot{r}_i)$, so that, if $\dot{r}_i$ is continuous at t_2, the discontinuity in $\dot{Y}$ at t_2 is given by the coefficient of Y_j in (10).

We next replace $Y_j(t)$ and $\dot{Y}_j(t)$, obtained from (9), in (6) in order to derive an ordinary nonlinear differential equation of first order for Y:

$$\dot{Y} = \frac{Y}{r_i - r_j} \; \frac{\sigma'(r_i - r)(r_j - r(\infty)) + [\frac{\kappa}{m} r + S\lambda'(r - r_j)](r_i - r_j) - \dot{r}_i(r - r_j)}{r_i - \frac{m-1}{m} r - S(r_i - r_j)}, \quad t \geqslant t_2. \tag{11}$$

As expected because of the singularity, this equation becomes indeterminate at $t = t_2$, so Eq. (10) is used for $\dot{Y}(t_2)$, allowing the integration to proceed from there, with the help of Eqs. (1), (2) and (5).

In case $t_1 = t_2$, $Y_j(t_2) = 0$, and $\dot{Y}_j(t_2) = 0$ also (see Eq. (6)). Thus, $\dot{Y}$ is continuous and must be put equal to $Y(\kappa - m\dot{r}_i)$ at $t = t_2$, while we put $r_i(t_2) = r(t_2) = 1$; $r_j(t_2)$ can have any value $\leqslant 1$; to make it smaller, we merely replace $C = \exp(\sigma' t_2)$ in Eq. (5) by a suitable $C' < C$.

To sum up, we need the following inputs to run the model:

(1) Historical data, to the extent available, on $Y(t)$, $Y_i(t)$, $Y_j(t)$, $P_i(t)$, $P_j(t)$, $P(t)$;
(2) Values or estimates for t_1, t_2, $r(\infty)$;
(3) Estimates for parameters m, l, n, σ, κ and λ (in the case of l, n only if $\neq 1$).

Application of the Stern Model to Forecasting the Substitution of Plastic for Glass in Bottles

BACKGROUND

This forecast is presented as an illustration of how the foregoing theory can be applied in a specific case. The analysis was made before the political situation in the Near East precipitated an extreme and unprecedented jump in the price of petroleum; hence, the price relationships between plastic, which is a petrochemical product, and glass are likely now to be disrupted. The forecast is nevertheless presented in its original form for two reasons:

(1) It is a valid illustration of the application of a useful forecasting technique; and the fact that the forecast results now have to be reexamined because of the sudden impact of external factors does not negate the usefulness of the method itself.

(2) The study offers a valuable object lesson in the vulnerability of any technological forecast to political and other contra-economic factors. This vulnerability can best be minimized by explicitly considering several alternative contingencies. That is to say, each technology should be carefully evaluated to identify its foreseeable points of vulnerability (e.g., raw material supplies, product safety legislation, etc.). Several forecasts should be performed incorporating various assumptions regarding these different possibilities. By this method the forecaster is able to both determine the sensitivity of the substitution to certain exogenous factors, and to "bracket" the probable future path of the substitution.

The background information underlying the forecast is as follows:

In 1970 the United States consumed 43,780 million bottles, of which 5,421 million were plastic. Most of these plastic bottles were made of high-density polyethylene, a much smaller number of low-density polyethylene, and a small but growing number of polyvinyl chloride. Of the 551 million pounds of resin used, 75% was polyethylene, 17% was polyvinyl chloride, and 8% was polystyrene and polypropylene [9].

High-density polyethylene (HDPE) is the dominant resin used for bottles today. HDPE reached significant volume when it was introduced for household bleaches and liquid detergents in 1958. Since then, this material has also been used to package various drugs, cosmetics, and toiletries, and has gained a small foothold in milk and foods packaging as well.

High-density polyethylene has very good resistance to impact, chemicals, alcohols, and water vapor. It is limited in some applications because it is permeable to oxygen and oils and is not available in crystal clear form. Although the price of PE bottles has increased in recent years, HDPE remains the lowest cost resin for blow-molded bottles. The long-term historical trend to lower resin prices for bottle grade HDPE has been from 44¢ per pound in 1958 to 18¢ per pound in 1970. (Increased petroleum prices may alter this trend.)

PVC is potentially the most direct competitor for glass. It can be made in rigid, impact-resistant, crystal-clear form. It has good chemical resistance to alcohols and oils, low permeability to water vapor and gases, and is opaque to ultraviolet light. The basic resin is available in large quantity because it is widely used for a variety of non-packaging products, ranging from floor coverings to garden hoses and raincoats. The cost of unmodified PVC has been about the same as that of bottle grade HDPE (20¢ per pound in 1970). However, the modifiers required for PVC brought the cost at that time to about 30 to 33¢ per pound after compounding.

The most likely candidates for PVC packaging in food are vegetable oils, vinegar, wine, salad dressings, and seasoning products. Nonfood products include toiletries and cosmetics (hair lotions, mouthwashes, shampoos, etc.) and chemicals such as household cleaners.

Machinery technology has kept pace with resin technology, so that today many proprietary PE formulations can be blow-molded on equipment which is readily available. However, a good deal of equipment is custom designed for specific applications; and in fact, the development of packaging "systems" has been an important factor which has enabled resin producers and packaging companies to successfully enter new markets. Conversely, the need to develop new in-plant filling equipment has slowed the growth of PE bottles.

Since polyethylene has led the way in molded plastic bottles, the existing blow-molding technology has aided other resins used for bottles, for instance, polystyrene. Some resins, however, cannot be formed on PE equipment; one of these is PVC. Most of the PVC blow-molding capacity is in the form of proprietary in-plant machinery owned by major resin suppliers. Assuming that PVC is accepted in food packaging applications on a limited scale, exploitation of the potential will have to await availability of machinery to turn out the billions of units which might be demanded. Thus, in the short run, lagging forming technology may delay PVC acceptance on a wide scale; but in the long run PVC should establish itself as second only to polyethylene in the bottle field.

The substitution of plastic for glass bottles is primarily a result of customer preference and the economics of transportation rather than the relative cost of the bottles themselves. The ready acceptance by consumers of plastic bottles in the household chemicals market has convinced bottlers that consumers prefer plastic bottles because of their lightness and unbreakability. Bottlers themselves prefer plastic bottles because they can save on shipping costs–plastic bottles being 8 to 20 times lighter than glass bottles of equal size. Because the strength-to-weight ratio is much higher for plastic than for glass, bottlers can also economize by using larger capacity bottles.

It was noted earlier that one of the key problems in the application of substitution analysis is that of market differentiation. Thus, the market for transparent containers (i.e., "bottles") was essentially homogeneous, although clearly differentiated from the markets for metal cans, paper containers, etc. The advent of plastic as a competing material has accelerated the process of differentiation *within* the bottle market. There are now nine recognized sub-markets for bottles in which plastics do or could compete with glass: beer, wine, liquor, soft drinks, food, dairy, medicinal and health, toiletry and cosmetic, and household chemicals. The consumption of containers in each market in 1970 is shown in Table 5.

Up to the present time, there are four markets into which plastics have not penetrated at all: beer, wine, liquor, and soft drinks. For wine and liquor this is due primarily to tradition, since there appear to be plastics available which can meet the physical requirements. In beer and soft drinks, however, plastics have not yet been able to overcome problems of heat tolerance (beer is pasteurized in the bottle), pressure resistance, and permeability.

Plastics presently have captured nearly 6% of the market for food bottles. They are apparently prevented from capturing a much larger segment of the market at this time by

TABLE 5

Number of Bottles Produced for Major Markets [8]

Market sector	Total glass bottles 1970 ($\times 10^6$)	Total plastic bottles 1970 ($\times 10^6$)
Beer	6,760	–
Liquor and Wine	2,730	neg.
Soft Drinks	7,600	–
Food	11,375	700
Dairy	125	1,400
Medicinal and Health	3,560	540
Toiletry and Cosmetic	2,200	1,480
Household Chemicals	630	2,360

their inability to withstand the temperatures used in retorting and the pressures involved in vacuum packing. Permeability is also a problem with some foods. For dairy products, glass bottles have all but disappeared. The active competition is now between plastic and paper.

Plastics have 15% of the market for bottles for medicinals and health care products, but are used mainly for tablets, capsules and other dry substances. Permeability becomes a problem with liquid products; many drugs lose potency rapidly in plastic bottles.

In toiletries and cosmetics, plastics reached 40% of the market by 1970. Their share came primarily from hair care products; however, there seems no physical reason why plastics cannot be used for other cosmetics as well. An exception to this is perfume, where permeability and chemical inertness are crucial.

For household chemicals, plastics are now the dominant packaging material, with 79% of the market by 1970. There are no apparent physical limits to the application of plastics here.

In addition to the sectors discussed above, which purchase bottles, the glass, plastic fabrication, and resin production sectors will also be affected by the substitution. The likely substitution path (as it appeared at the time of the forecast) is shown in Fig. 1.

Shipments of glass bottles to three markets have ceased to grow and are beginning to drop off. These are: household chemicals, medicinal and health products, and toiletries and cosmetics. In one market, dairy, glass has almost disappeared. The glass industry is shutting down individual bottle lines and may soon have to shut down some of its furnaces if it cannot expand other markets for glass products. The most probable action of the defender (glass) is that it will soon attempt to compete with plastic by introducing an improved glass bottle. Such a bottle, lightweight (20% lighter than normal) with a thin transparent plastic coating, is now being marketed. The reduced weight reduces shipping costs, and the plastic coating renders it shatterproof but not unbreakable. It is too soon to tell whether this innovation can reverse the trend toward plastic.

UTILITY ANALYSIS

The general problems of utility analysis were discussed earlier in this paper, and these problems were illustrated by examples drawn from this particular application. The specifics of this analysis as developed in this study are as follows:

A list of the relevant desirable attributes of bottles was compiled from conversations with bottlers, bottle manufacturers and consumers, from marketing research studies and

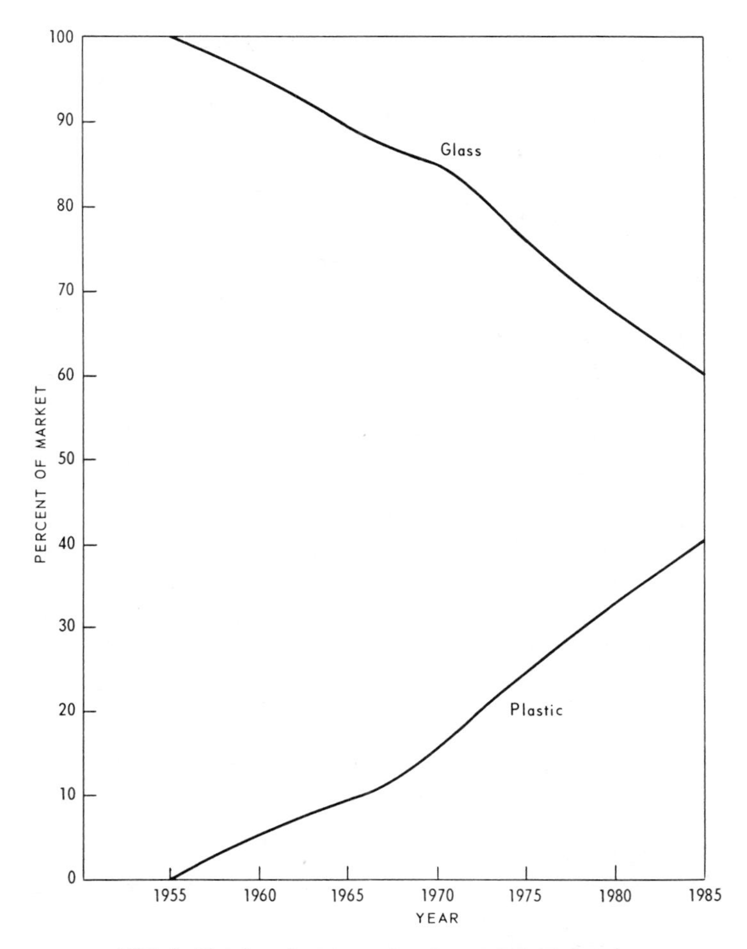

FIG. 1. Total market shares for glass and plastic bottles.

from product literature. The attributes were grouped into four categories for analytical purposes.

1. Attributes of Concern Primarily to Bottlers

The following properties are those which affect the mechanics of filling and storage of products.

Chemical inertness—Is the container non-reactive with the product? Will it corrode, deteriorate, or affect the taste, color or aroma of the product? Also subsumed under this heading are the container's barrier properties with respect to ultraviolet light; most products require some degree of protection.

Heat resistance—Does the container maintain its shape and strength at temperatures above room temperature? Is it resistant to thermal shock?

Non-permeability—Is the container permeable to moisture, gases or odors? Will it protect the product from oxidation and from loss of moisture, aroma, or potency?

Pressure resistance (rigidity)—Does the container maintain its shape under atmosphere pressure or under stress?

Flexibility of color and shape—This is the only property in this group which does not relate to physical performance. It is a merchandising concern which some bottlers mention as being of some importance.

Adaptability to existing bottling equipment—Does the nature of the bottling material require modifications to filling lines, etc.?

2. Attributes of Concern Primarily to Consumers

The following properties are those which primarily affect consumers' buying in a conscious explicit manner.

Transparency—Can the product be seen clearly through the container?

Tradition—Is the container the traditional one for the product?

Reusability—Is the container reusable in the home for other purposes after the product has been consumed?

Convenient dispensing—Does the container incorporate a convenient mode for dispensing the product?

3. Attributes Which Concern Both the Bottler and the Consumer

The following properties directly affect the utility of the container both to the bottler and the consumer.

Unbreakability—Can the container withstand impact?

Lightweight—Is the container light enough to be shipped economically and handled easily?

4. Attributes Which Are of Social Concern

The following properties are those which affect society at large rather than individual bottlers or consumers, i.e., they are externalities.

Disposability—Is the packaging material easily handled by at least one of the conventional solid waste disposal methods?

Recycleability—Is the packaging material amenable to being recycled back into the resource flow?

The next phase of the analysis was to determine the relative importance to each group of bottlers of the attributes described in the first section. This was done primarily through telephone interviewing of bottlers. The rating scales described in the general section on utility analysis earlier in this paper were used for converting the verbal assessments of the container materials into quantitative ratings.

The rating procedure was carried out by a group of persons having diverse background. After general discussion of the interviews and the list of properties, the ratings were decided upon by consensus. (More elaborate techniques could obviously be used at this point.) "Adaptability to bottling equipment" and "disposability" drop out of the analysis, in effect, because the three materials are rated approximately equally with respect to these attributes.

PROJECTIONS

Figure 1 shows a projection to 1985 of the market shares of plastic and glass for the total bottle market. This projection was obtained by simple trend extrapolation, since it covers too many diverse applications to be projected with the model.

Figure 2 shows the model projection for the household chemicals market. Figure 3 shows a similar model projection for the liquor bottle market [8].

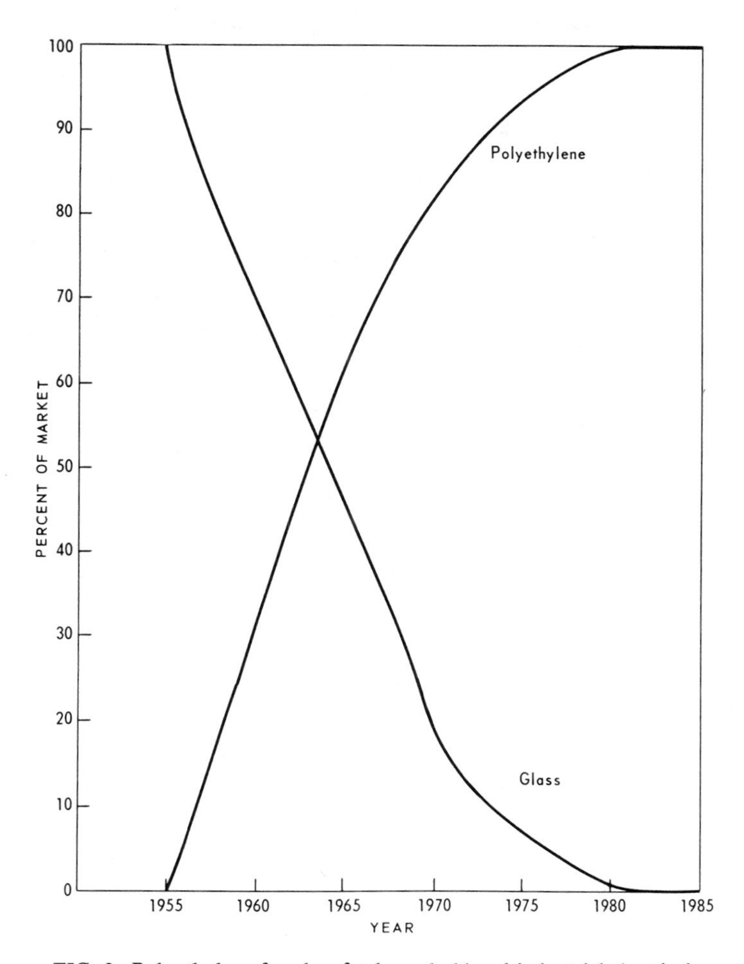

FIG. 2. Polyethylene for glass for household and industrial chemicals.

Comparison of Figs. 1, 2 and 3 leads to the observation that plastic bottles move rather slowly from one market to another, so that growth of the overall market share is slow. However, once plastic bottles enter a particular market the substitution proceeds quite rapidly.

MODEL PARAMETERS

The model parameters for the PVC for glass in liquor bottles substitution were:

t_1 = 1970 λ = 10 years

t_2 = 1970 plus six months m = 0.2

σ = 3 years $P(\infty)$ = 0.6

κ = 0.027 (2.7%)

The model parameters for the polyethylene for glass for household chemicals substitution were:

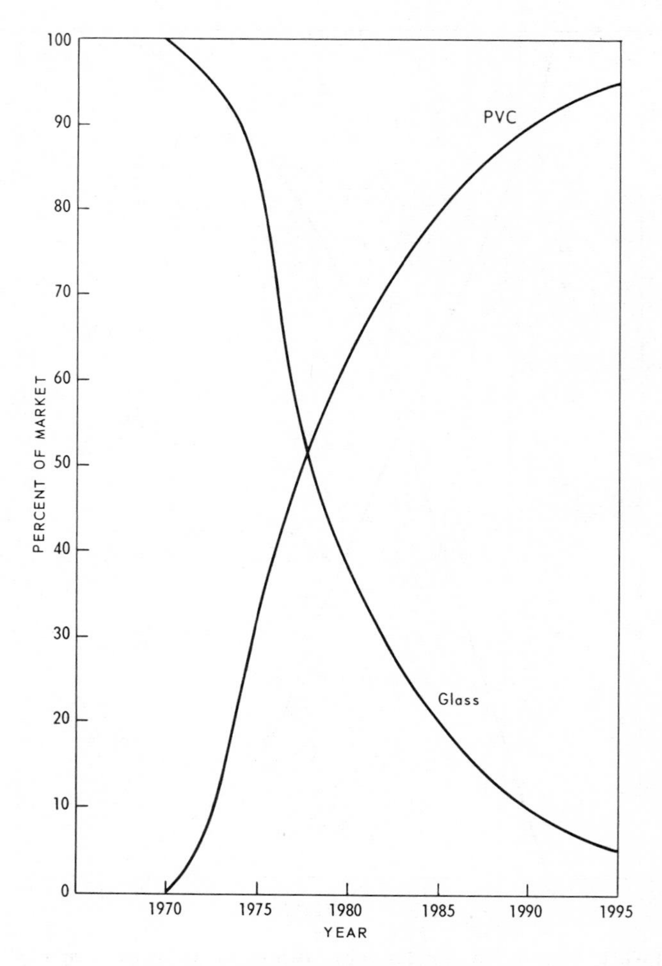

FIG. 3. PVC for glass in liquor bottles.

$t_1 = 1955$
$t_2 = 1955$
$\sigma = 6$ years
$\kappa = 0.024$ (2.4%)

$\lambda = 10$ years
$m = 0.2$
$P(\infty) = 0.4$

The authors gratefully acknowledge the assistance of Jim Saxton with information on technology and L. W. Ayres with the computations.

References

1. Wassily Leontief, *Structure of the American Economy, 1919–1939,* 2nd ed., Oxford Univ. Press, New York, 1951.
2. University of Cambridge, *A Programme for Growth: Input-Ouput Relationships 1954–1966*, Sec. 3, Chapman and Hall, London, 1963.
3. E. Mansfield, Technical Change and the Rate of Imitation, *Econometrica* **29** (Oct. 1961).

4. G. Chow, Technological Progress and the Demand for Computers, *Amer. Economic Rev.* **57**, 1117 (Dec. 1967).
* 5. A. W. Blackman, The Rate of Innovation in the Commercial Jet Engine Market, *Technol. Forecasting Soc. Change* **2**, 268–276 (1971).
6. J. C. Fisher and R. H. Pry, A Simple Substitution Model of Technological Change, *Technol. Forecasting Soc. Change* **3**, 75–88 (1971).
7. R. U. Ayres, S. Noble, and D. Overly, Technological Change as an Explicit Factor of Economic Growth, in *Effects of Technological Change on and Environmental Implications of an Input-Output Analysis of the United States, 1967–2020,* Part I. For Resources For the Future, Inc., and the Commission on Population Growth and the American Future.
8. R. U. Ayres, J. Saxton, and M. Stern, *Materials-Process-Product Model for the Bottle Manufacturing Industry,* prepared for the National Science Foundation; International Research and Technology Corporation, IRT-305-EP., December 1972.
9. Arsen Darnay and William E. Franklin, The Role of Packaging in Solid Waste Management 1966 to 1976. Prepared for U.S. Department of Health, Education and Welfare, 1961; Midwest Research Institute.

**This article also appears in Part 2 of this book.*

Explicit Technological Substitution Forecasts in Long-Range Input-Output Models

ROBERT U. AYRES and ADELE SHAPANKA

ABSTRACT

This paper presents a new methodology for making long-range forecasts of inter-industry patterns of trade that take into account explicit technological developments. Patterns of trade are statistically quantified by means of input-output tables that show the flow of goods and services between industrial sectors and between industry and "final demand" (i.e., individual consumers, government, and capital investment). This paper describes methods of building an input-output model for forecasting purposes and shows how the effects of technological change can be built into the model.

1. Introduction

The businessperson or bureaucrat trying to make long-range plans to meet anticipated needs ten or twenty years hence requires a more precise set of tools than a projection of some broad economic index such as Gross National Product. He needs specific information about the outlook for particular industries, and, of course, for the industries and consumers that buy from him. It is of little help to him to learn that GNP may double in the next fifteen years if, in fact, his share of that total is likely to drop to a small fraction of what it is now, or, conversely, to rise substantially.

In the past, such estimates have been made largely on a "hunch" basis. The classic entrepreneur says to himself, "I have faith in the future of the widget industry. Think of all the people out there who need widgets". And he decides to float a large stock issue to double his plant capacity. If he guesses right he may become wealthy; or his market may vanish for a variety of reasons, one of which might be that another product, the gidget, has supplanted his widget.

"Seat-of-the pants" forecasts, such as these are becoming less and less practicable. Resource scarcity, among other things, can drive up the prices of the materials of which the widget is made and, perhaps, make some of them ultimately difficult to obtain. The competing product, the gidget, might be made of materials in more abundant supply; technological changes, which now appear with greater and greater rapidity, may make the widget obsolete; or the main customers for the widget may be in an industry about to decline.

Not only businesses but governments and other institutions need information on what is likely to happen to inter-industry patterns of trade over various future periods. The State of Michigan, for example, as well as the City of Detroit, might well be concerned with the probably condition of the automobile industry a decade or so from now. Will the demand be primarily for small cars or large ones? Will the automobile power plant be different from that of today? Will the trend be away from luxury cars toward simpler, more functional ones? Or the reverse? Answers to these questions affect employment, industrial concentration, energy requirements, city planning, transportation, and many other matters.

Long-range forecasts that do not take into account changing conditions–and, especially, technological changes–can go absurdly awry. Take, for example, the forecasts of public transportation needs in the Chicago area that were made in 1916, 1927, 1930 and 1937 [1]. All these forecasts essentially projected future transit usage on the basis of historical rates of growth. As Fig. 1 illustrates, all were greatly in error. They failed to take into account the growing use of the private automobile, although the automobile had already made significant inroads into transit usage at the time these estimates (especially the later ones) were made. The projections were also upset by the 1929–1939 depression and the effects of World War II.

To avoid such expensive errors as those in the Chicago transit case, businesses and governments need three kinds of information:

1. Knowledge of the intricate patterns of inter-industry trade that now prevail: of who is buying from whom and who is selling to whom. This knowledge is quantified in what are known as "input-output" tables.
2. Knowledge of what factors are likely to alter these patterns of trade in the future. This means knowledge of new technological developments in the laboratory and in industrial plants, and careful appraisals of present resources of minerals, fuels, water, etc.
3. Long-range forecasts of what the patterns of trade between industries are likely to be at various future dates. For this purpose, an input-output model is constructed which takes into account the technological knowledge referred to in point 2.

In this paper, we discuss, first, the input-output tables that provide the basic data from which the projections start. Then we describe the manner in which information on technological changes is integrated into the model citing examples drawn from the aluminum and the iron and steel industries to illustrate the approach.

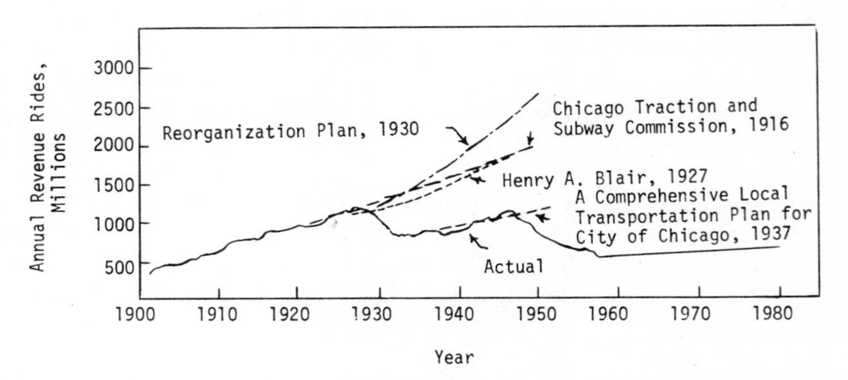

Fig. 1. Transit Use Projections–Chicago. Actual use of transit facilities from 1901 to 1960 and estimated to 1980 compared with various projections of estimated use (Chicago).

What will probably seem most novel about this presentation to many readers will be the treatment of technological changes. There has been a tendency for the economic fraternity and the scientific and engineering fraternity to avoid peering into each other's domains; each has preferred to look upon the other as having its own mystique and its own initiates. But this separation is a handicap to both. Economists need to find ways of incorporating technological innovation into their theoretical structure, for production and trade are vitally influenced by technology–have been in the past and will certainly be in the future. Conversely, scientific and engineering achievements are dependent upon economic considerations for their full development and application.

So awesome have been the social and economic transformations brought about by technology in the past that there has been a tendency on the part of those outside the scientific and engineering communities to treat these developments almost as if they were unpredictable "acts of God". But they are not unpredictable at all. Quite the contrary. Technological changes that have the potential of upsetting prevailing patterns of trade between industries cast their shadows years in advance. It takes a long time–some years, usually–for a new process or product to be worked through the research and development phases to the point where plants are ready to be built to introduce it into the economy. Once the new process or products enters the market and begins actively to compete with established ones, more years are likely to pass before it has a very substantial effect on established patterns of trade. These events simply do not, and cannot, take place overnight. There is ample time to analyze their potential, and the problem of the forecaster is not so much how to spot the developments likely to have significant economic effects, but rather how to apply this knowledge specifically to forecasts of trade patterns. This paper will describe a methodology that we have developed for the purpose of facilitating this step.

2. Input-Output Tables

The input-output table has become, by now, a fairly familiar tool of economic analysis. Essentially, it divides the economy into sectors corresponding to industry groupings, and then shows the flow of products and services between each individual sector and each other sector. The basic classification used in determining these sectors is some kind of aggregation of the industrial categories defined in the Standard Industrial Classification (SIC) Code [or, in countries other than the United States, the International Standard Industrial Classification Code (ISIC)]. For our forecasts, we used a set of 185 industry categories developed by Clopper Almon, Jr. for the Interindustry Forecasting Model of the University of Maryland (INFORUM) [2]. These are listed in Appendix I.

Input-output tables are in matrix form, i.e., each column and each row represents an industrial sector. Table 1 shows how such a table is structured, using a few industry sectors chosen at random.

The sectors designated by the rows are, for convenience, referred to as the i sectors and those designated by the columns as j sectors. The sectors in the rows and columns are identical, but the i and j designation enables us to keep track of the direction in which goods and services are flowing; namely, from i to j.

An *input* or A matrix, shows for each of the j sectors what fraction of its total inputs is derived from each i sector. These fractions, designated by the symbol A_{ij}, are known as "input coefficients". If we include the "value added" segment, all the input coefficients in each j column add up to 1.

An *output,* or B matrix, shows for each of the i sectors what fraction of its total

TABLE 1

Example of Input-Output Matrix

j sectors– / i sectors	Indust. Chemicals	Plastics Resins	Rubber Prod.	Alumi-num	Farm Mach'y	. . .	Final Demand	Total
Indust. Chemicals								1.0
Plastics & Resins								1.0
Rubber Products								1.0
Aluminum								1.0
Farm Mach'y								1.0
. . . .								
Value Added by Sector								
Total	1.0	1.0	1.0	1.0	1.0			

outputs is sold to each j sector and to "final demand". ("Finald demand" is defined as sales to individuals for private consumption, sales to government, and capital investment in new plant and equipment.) These fractions, designated by the symbol B_{ij}, are known as "output coefficients". If we include the final demand segment, all the output coefficients in each row of the table add across to 1.

These coefficients represent established patterns of trade between the sectors. Underlying their calculation is the assumption that these patterns are not random but are a result of the unique production technologies in each sector as well as established consumption habits. While the coefficients are comparatively stable over short periods of time, new technological, political, or social developments can bring about substantial changes in the coefficients over longer periods—say, a decade—a fact which the forecaster needs to take into account. It is worth emphasizing, incidentally, that output coefficients (but not input coefficients) are, in principle, *price independent.* That is to say, if the basic assumption of homogeneity within the sector were rigorously valid, the allocation of outputs from a given sector to others involved presumably remain unchanged regardless of the price of the product. Thus output coefficients have a significant advantage in that they are "inflation-proof".

An input-output table is originally constructed from data on the physical flow of materials, energy, and services, derived (in the U.S.) from the quinquennial Census of Manufactures and its annual updates, plus a variety of trade and other sources. The work is difficult because of gaps that have to be filled by estimates, and duplications that have to be eliminated. Since the physical units of different kinds of items are incommensurable, all items are finally priced out and converted to dollar values (or to values in the appropriate unit of currency for the country involved). The labor involved in preparing an input-output table by present methods is so great that the table is generally five or more years out of date before it is completed–a serious disadvantage in an era of rapid change. Prof. Clopper Almon, Jr. of the University of Maryland has done valuable work in updating these tables [2], using a variety of current trade and production information, and "balancing" the tables by forcing row and column sums to be consistent.[1]

Input-output tables expressed in dollar terms and including the "value added" and "final demand" terms are designed to be in balance: that is, for each industry sector, the total of the inputs plus value added (column totals) is equal to the total of the outputs including final demand (row totals). The total of final demand for all industries combined is equal to the total of value added for all industries combined, and the total is the Gross National Product (GNP).

The structure of the input (A) matrix is expressed by the following equations [3]

$$\begin{aligned} (1-A_{11})x_1 - A_{12}x_2 - \cdots - A_{1n}x_n &= y_1, \\ -A_{21}x_1 + (1-A_{22})x_2 - \cdots - A_{2n}x_n &= y_2, \\ -A_{n1}x_1 - A_{n2}x_2 - \cdots + (1-A_{nn})x_n &= y_n, \end{aligned} \qquad (1a)$$

where A_{ij} = input coefficient, representing the fraction of all inputs to the jth sector derived from the ith sector. (Thus, A_{12} = fraction of all inputs to sector 2 derived from sector 1.) $x_1, x_2, \ldots, x_n$ = total output of sector 1, sector 2, etc. $y_1, y_2, \ldots, y_n$ = final demand (as previously defined) for the product of sector 1, sector 2, etc. n = number of sectors.

Input coefficients are calculated from dollar values rather than physical quantities because the physical quantities of different input products are incommensurable and because "value added" (wage and capital costs), which is included in the denominator of the fraction, is necessarily a dollar quantity. Output coefficients, on the other hand, can be directly calculated from either dollar or physical quantities and are identical for both. This fact is important in input-output analysis.

In matrix notation, the equations stated above can be expressed in the form

$$(I-A)x = y, \qquad (1b)$$

$$(I-A)X = Y, \qquad (1c)$$

where I is the identity matrix and A is the matrix of input coefficients (either in physical or dollar units); x is a vector of physical outputs; X is a set of dollar outputs; y is a vector of final demands for physical quantities, while Y is a set of final demands in dollar terms. Evidently, $X_i = p_i x_i$ and $Y_i = p_i y_i$, where p is the vector of unit prices. The sum Y_j for all sectors is defined as the GNP.

The inverse of the dimensionless matrix, $(I-A)^{-1}$, satisfies the following relationship:

$$(I/A)^{-1}\ (I-A) = 1,$$

whence

$$X = (I-A)^{-1}\, Y. \qquad (2)$$

That is to say, the inverse can be multiplied by final demand to give the output levels of each industry in dollar terms. A term of the inverse, such as "coal to steel" represents the dollar output of coal that is required for one dollar of final demand for steel. When an

[1] This is the so-called RAS method.

input-output matrix is given, even with several hundred sectors, the inverse matrix can be readily calculated with the aid of an electronic computer. Then, if we multiply the steel column of the inverse matrix by the final demand for steel, we obtain the output level of each industry that is attributable to purchases by the steel sector.

The input-output matrix can be "bordered" by other vectors of coefficients to yield useful results. For instance, it may be convenient to introduce vectors of labor coefficients, $k_1 \ldots k_n$, and wage coefficients, $w_1 \ldots w_n$, where k_j represents the number of man-hours per unit of physical output and w_j represents the wages per man-hour of labor. The total of all wages and salaries paid in the economy is given by the sum of the products

$$W = \sum_{j}^{n} w_j k_j x_j. \tag{3}$$

The value-added per (physical) unit of output by an industry, plus an additional term to reflect the annualized cost of capital, can also be defined as the unit value (i.e., the price) of its output minus the costs (i.e., prices) of all purchased inputs. For the jth industry, this can be expressed:

$$v_j = p_j - \sum_{i=1}^{n} A_{ij} p_i, \tag{4a}$$

or in vector notation

$$V = p[I-A]. \tag{4b}$$

This equation can be inverted (multiplying both sides by the inverse matrix) to yield

$$P = v[I-A]^{-1}. \tag{5}$$

These relationships can be used to determine the price consequences of changes in the cost structure of a given industry due, for instance, to environmental regulation.

3. Input-Output Forecasting Model

As a forecasting model, the input-output table has substantial advantages, as we will show. But, in its present state of development, it is far from a perfect instrument. In addition to the labor of preparing it and the fact that it is out of date before it becomes available, there are also a number of difficulties in its use:

1. It often aggregates into a single industry sector a number of producers whose outputs are not mutually substitutable for one another. At the same time, it ignores, in some cases, certain substitution possibilities *between* industries.
2. As a result of aggregation, spurious inter-industry relationships are sometimes introduced (e.g., "cycles" where industries appear to be mutually dependent in a closed loop).
3. Again, due to aggregation, errors may be introduced in predicting the impact of changes in final demand or requirements for many physical materials, when these are a small fraction of the inputs to an industry.

On the other hand, the input-output model has a great advantage over other forecasting approaches in that the coefficients and independent variables represent real, measurable physical and/or monetary quantities, i.e., commodities, energy, services, labor, and capital. When any of these quantities, or the relationships between them, are likely to be altered by new developments, the effect of these changes can be specifically traced in the model—an advantage possessed by no other forecasting device of which we have knowledge.

The input-output table—the starting point from which the forecast is made—represents a static condition, i.e., it portrays the patterns of trade prevailing at a particular moment of time. The problem of forecasting is to project the various elements of the table into the future, taking into account the probability that the input and output coefficients will change. Economists, as a rule, regard final demand (Σy_2) as one of the more predictable factors, and the usual procedure, therefore, is to begin with a forecast of final demand made from an econometric projection that takes into account changes in such factors as population, average productivity, probable trend of personal income or GNP, etc. If the coefficients A_{ij} and B_{ij} remain fixed, assuming no significant changes in technology, the inputs and outputs for each industry can be computed from the projected final demand, using Eq. (1a). But, of course, the coefficients do not remain fixed. They, too, have to be projected, using a variety of procedures that we will discuss shortly.

A given production technology at a particular time is associated with a given set of matrix elements. As technology changes, so must the matrix elements. If alternative production technologies are available and the switch from one to another can occur as a matter of management choice, then there should be one such matrix for each possibility; the mathematics obviously become more complex and a linear programming approach would normally be used to select among the various possibilities.

Several kinds of external constraints may also be imposed on the input-output quantities to facilitate certain types of analysis. For instance, the capacity of each industry may be assumed to be fixed at a given point in time. A projected "final demand" vector—along with other assumed quantities—might conceivably imply a set of intermediate outputs exceeding the capacity of one or more industries. This is obviously an impossible situation. By imposing capacity constraints, a set of possible trade patterns—as well as impossible ones—can be determined. In the same way, some final demand vectors may imply labor requirements greater than the available labor force. Other constraints, e.g., resource availability or pollution output, can also be imposed. Again, the range of possible and impossible outcomes can be explored with the model.

It can be seen that, at the present stage of development, the input-output forecasting model hangs on an econometric projection—namely, of "final demand" or total GNP—and amounts to an analysis of the inter-industry trade patterns that would prevail under the assumed conditions of the projection.

If it were practical to do so, we might define enough time-dependent relationships to permit the model itself to "generate" all future states of the economy. Starting from a fully specified starting point (say, 1963), the I–O model(s) might predict outputs and prices for a succession of years, while a long-range econometric model might determine final demand (consumption), capital spending, imports, wage rates, and so on from assumed growth in the labor force, government expenditure, and exports. For instance, expenditures for capital goods could be tied to the relationship between capacity, labor supply and output. Existing industrial capacity could be assumed to depreciate at some predetermined rate, based on economic and engineering considerations. Spending for new plant and equipment also tends to depend on the nonutilized fraction of existing capacity, and on the availability of labor. As either of these drops, capital spending will increase, and vice versa. The quantitative relationships could be determined by means of econometric analysis of historical data. A model prepared for the Environmental Protection Agency, based on the Interindustry Forecasting Model of the University of Maryland (INFORUM), has "internalized" the capital sectors in practice, treating them essentially as intermediate rather than as final demand [4].

In principle, the linkage between econometric and input-output models can also be extended with respect to income and final demand, since personal income is the sum total of all wages (plus transfer payments, interest, and dividends), and final demand can be closely correlated with income. Up to the present, however, this has not yet been done satisfactorily, although several input-output modelers are working along these lines.

Wherever it is not feasible to let a model package "generate" its own future environment, of course, a set of exogenous projections must be used. This is particularly necessary with regard to technological changes, as we will see shortly.

4. Treatment of Technological Changes

Technological changes have an economic impact that can emerge in an input-output forecast in either of two ways: (1) by creating new goods and services causing changes in final demand, or (2) by creating new materials or new methods of processing or fabrication, causing shifts in inter-industry coefficients.

Technological change, insofar as it affects final demand, involves an exercise in social as well as technological forecasting, and will not be addressed by this paper. We will be concerned only with technological changes that will alter inter-industry trade patterns, such as, for example, the substitution of aluminum for copper in electrical applications or the substitution of plastic for metal in a number of industries. On these kinds of changes, the literature is more extensive [5–16].

The latter approach, called "*substitution analysis*", assumes that a technological change can be defined as the substitution of a new material, production process or product for an old one.

A particular substitution may occur much more rapidly in some industries than in others. For example, computers were fully applied to petroleum refining at a time when they were only beginning to be utilized by banks and insurance companies and before they had been used at all in the medical or legal professions. Similarly, some industries (e.g., aluminum and ferro-alloys) are now almost fully electrified, with no further increases anticipated in electric power input per unit output, whereas certain other industries are likely to undergo significant increases in electric power use. Data to quantify these differential rates are scarce, however, and in practice, it may be necessary to assume a uniform proportional change in the substitution rate for all affected industries, except for specific cases where this would be obviously wide of the mark. This approach has been developed previously by Stone [17] and Bacharach [18].

Only substitutions crossing industrial sector boundaries affect the I–O matrix. Thus, the substitution of detergent for soap, for instance, would not affect the output coefficients in any of the major I–O models currently in use, simply because products are included in the same sector (SIC 2841), at the 4-digit level of disaggregation. Yet, the manufacture of detergents requires inputs from a different set of industries than does that of soap (viz, soap requires animal or vegetable fats from the food processing or agriculture sectors; detergents require inputs from petrochemical and mining industries). Failure of the model to distinguish between the production of soap and that of detergents will inevitably result in error as the relative outputs of these two products change with time.

In general terms, our approach to forecasting the effect of technological changes is, first, to identify the more important technological trends that can be expressed as substitutions. These are summarized in Appendix II. The next logical step is to translate this information into terms used in the I–O model, i.e., sector inputs and outputs, and ratios, and to estimate historical rates of change of the variables. This is straightforward in principle, but complicated and tricky in practice. The third logical step is to project the future time-path of the substitution process. This forecasting step involves a separate model of technological diffusion that requires explicitly detailed justification. The fourth and final step is to modify the existing input-output coefficients in conformity with the projections. This is essentially a matter of bookkeeping and computer programming and need not be discussed here.

The model of technological diffusion used for these forecasts assumes, for convenience, that the course of the substitution, defined as the percentage of the total market held by the new technology, generally takes the form of an S-shaped, or "logistic" curve.

The reasoning behind this assumption is as follows: A new technology develops slowly during the first few years after its introduction because various problems of implementation have not been fully worked out, the process or product is still being improved, its potential users have not yet learned how to use it, it has not yet achieved the economies of scale of its established competitor, it may not have commanded adequate capital for full development, and there has not been time for familiarity with the new technology to diffuse through the industry. Once this initial "learning period" is over, however, if the new technology proves to be technically and economically competitive, it is adopted much more rapidly and gradually displaces its competitor. It continues growing rapidly until it reaches the point when it is no longer improving significantly (i.e., it is a mature technology) and it begins to approach saturation of its natural market–that is, those applications for which the technology is particularly well suited. In some cases the new technology may expand into new markets. In other cases, it may never capture more than a fraction of the original market, thus segmenting it. This has happened in a number of cases, two of which are the automotive diesel engine and household laundry detergents. Neither of these has been able to displace its competitor completely; each retains a "niche". This is discussed more fully in the recent paper by Stern, Ayres and Shapanka, referred to earlier [19].

Figure 2 illustrates the form of the S-shaped curve, showing two curves of somewhat varying shape, and indicates the parameters that can be used to determine the curve. If the mathematical form of the curve is assumed to be that of the "logistic" type, and if we establish the upper and lower asymptotes, plus two other points on the curve, we can calculate the path of the entire curve, provided the relationships between these four quantitites are possible (that is, compatible) ones in terms of the curve shape. For example, it can be seen from Fig. 2 that not every value of Y_1 and $Y(u)$, respectively, would fit an S-shaped curve.

In Fig. 2, each curve starts from a lower asymptotic value Y_A (i.e., the share of the market that the new technology had when the substitution can be said to have begun). The curves rise at varying rates until they flatten out at the top at a presumed asymptotic value Y_B(i.e., the maximum share of the market that the new technology can ultimately attain). $Y(u)$ represents the mid-point between the lower and upper asymptote, reached

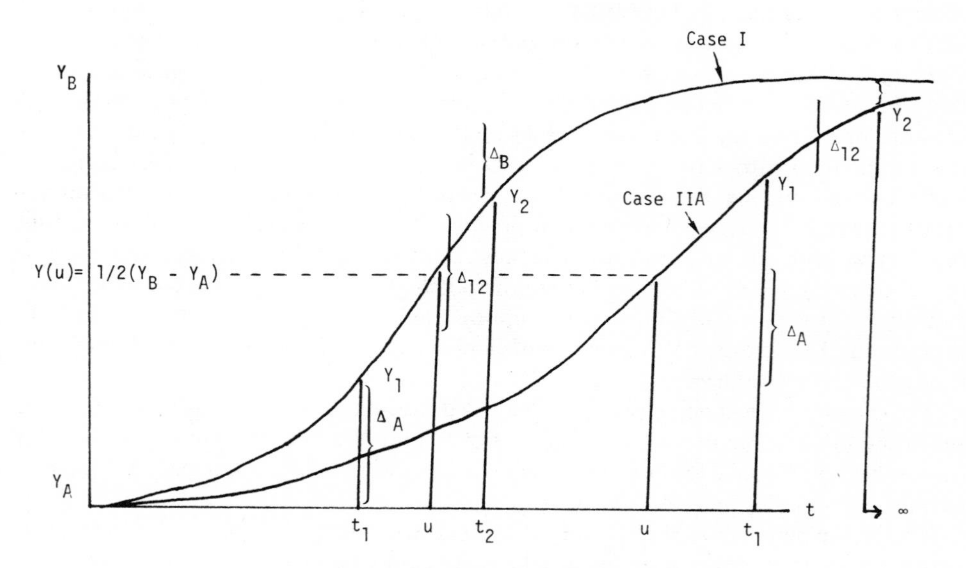

Fig. 2. Form of the Logistic Curve.

at time u (in other words, the market penetration has proceeded halfway to its ultimate point. Y_1 and Y_2 represent the share of the market attained at two selected points of time: t_1 and t_2. If the asymptotes, plus any two of the three quantitites Y_1, Y_2 and u are known, the curve can be determined. Usually, the parameters Y_1 and Y_2 (shares of the market at specific dates) are easier to estimate than u, and are therefore more commonly used. One of these dates may be in the past, making it possible to base the curve in part on actual data. The rather complex mathematics are not detailed in this paper.

The technological forecasts carried out in this project were limited to discrete, identifiable technological developments. Incremental improvements in the efficiency of various processes or products were not considered. The forecasts were based on an analysis of changes in input and output coefficients expected to result from current, known R&D efforts. In this sense, it is a rather conservative forecast.

The steps taken in making these forecasts were, in general, as follows:

After having identified the cells in the INFORUM I–O table that would be affected by each technological development (Appendix II), we made estimated projections of five quantities, of which four comprised the parameters required to calculate the S-shaped curve. Using information obtained from consultants in materials engineering and energy systems, from contacts with industry representatives, and from trade and technical journals, we estimated the following for each technology:

1. The expected extent of substitution at some specified year in the future, usually either 2000 or 2025 (expressed in fractional terms, from 0 to 1). (This is quantity Y_2 in Fig. 2.)
2. The upper asymptote of the substitution, i.e., the maximum share of the market that the new technology can ultimately attain (Y_B).
3. The lower asymptote, i.e., the share of the market that the new technology had when

the substitution can be said to have begun (Y_A). This would be zero if each technological change applied to an entire sector, but it may be non-zero because of aggregation (i.e., some parts of the sector are unaffected by the technology).
4. The fractional extent of the substitution as of 1971 (Y_1). This could usually be deduced from the Almon input-output table or from other sources.
5. What is called a "*functional equivalency ratio* (*FER*)", i.e., the value of transactions under the new technology that are functionally equivalent to a dollar of sales or purchases under the old technology. This quantity takes into account the fact that few, if any, new products or processes are exact equivalents of the ones they displace. For example, when aluminum replaces copper in electrical applications, the two substances are not pound-for-pound or dollar-for-dollar equivalents. Far from it. Aluminum is not as efficient an electrical conductor as copper; more power is lost through heat. But aluminum is lighter in weight and much cheaper to produce. These differences are discussed more fully later in this paper, and the method of calculating the FER is described in Appendix III.

With the information provided by the five quantities listed above, the INFORUM inter-industry coefficients were then projected to the year 2025 at the 185-sector level of aggregation used in Prof. Almon's study.

5. Some Types of Problems Encountered

Each technological substitution presents its own problems of estimation, and it may be of interest in the context of this paper to look at a few examples.

5.1. INPUTS TO RESIDENTIAL CONSTRUCTION

By "inputs to residential construction" we mean the new materials and methods that are now being substituted, or will eventually be substituted, for the traditional ones in construction of homes. This forecast illustrates the kind of assumptions one must make, sometimes about broad, complex areas, in order to determine the probable future of a technological trend. In this case we were trying to forecast the substitution rates between various construction materials such as lumber and other wood products, plastics, concrete, steel, and aluminum. It soon became apparent that the primary determinant of these rates is the mix of various types of dwelling units, i.e., single-family detached, low-rise multifamily, and high-rise multi-family. This, in turn, depends on what assumptions are made regarding future urban patterns and land-use policies. Policy assumptions also play a role in the development of housing technology, for example, off-site (modular factory) construction must overcome the obstacles of construction union resistance and outdated building codes. Finally, the rate of materials substitution depends upon assumptions concerning the supply and cost of lumber in the future.

These questions were resolved as follows: urban patterns were seen as developing in the direction of increased density in the central city and near suburbs, with high-rise construction predominating. Farther out from the central city we expect to see much more medium-density housing, with emphasis on integrated office-retail-residential complexes. Mobile homes, which currently have about 20% of the housing market, will probably decline in important as new types of housing become available, and as the population becomes progressively less rural and more affluent. Table 2 summarizes our assumptions about the housing mix.

As far as housing technology is concerned, we made the optimistic assumption that

TABLE 2

Dwelling	Percentage of New Units 1970	2025
Single family	45	25
Multifamily (medium-density)	29	36
Multifamily (high-density)	5	24
Mobile homes	21	15

institutional obstacles would gradually be overcome, so that by the year 2025 the majority of single-family homes (60%) would be built using factory methods. Only about 10% of total new dwelling units will be of the conventional single-family detached wood frame type.

The implications for materials are clear. Since virtually all high-rise apartment buildings are of reinforced concrete construction, and since precast concrete is a leading contender in modular systems, we forecast great growth in cement and concrete products. Plastics also look promising for factory-built units. Conversely, the fewer conventional single family homes are built, the less timber will be needed. This forecast was reinforced by a study done by the former chief of the U.S. Forest Service for the National Materials Advisory Board, which concluded that, if demand for timber continues to grow at present rates, shortages would begin to show up by the year 2000 [20].

5.2. INPUTS TO MOTOR VEHICLES

By "inputs to motor vehicles" we mean the changes in materials, equipment, and power plant that are likely to take place in the future manufacture of automobiles, trucks, etc. This forecast illustrates the difficulty of dealing with two areas of great uncertainty at the same time: automotive power plant development, particularly battery R&D, and future availability and price of various forms of energy.

The two central questions were: (1) What will be the capabilities of electric cars in the future? (2) What will be the availability and price of liquid fuel? The first question was resolved by assuming only a modest improvement in performance over today's rather limited electric vehicles. This seemed justified because of the limited R&D on batteries planned for the next few years and the fact that all current experimental batteries have serious problems. The second question above was dealt with by referring to our petroleum forecast which indicates that by the year 2000 a substantial portion (about 20%) of the liquid fuel used in the United States will have to come from either oil shale or coal, both at greater expense than petroleum extracted by conventional methods in the past. These two assumptions combine to yield a picture of electric vehicles used extensively as urban commuter cars, but probably lacking the capability for long-distance highway travel.

Another serious difficulty encountered in this forecast was that of establishing the dollar-for-dollar trade-off (or functional equivalency ratio) between the components of an electric car and those of a conventional internal combustion engine vehicle. This was accomplished by a laborious process of determining the proportion of the weight of each major metal in a car that is accounted for by the engine block and by other components that would be eliminated in an electric auto. Then the approximate value of the electric motor and the batteries had to be crudely estimated by using the specifications of an electric car now on the market and referring to the Census of Manufactures for unit costs.

Finally, in order to determine the impact on fuel consumption, it was necessary to estimate the percentage of the total *stock* of vehicles each year that would be electric (our original forecast aimed at the percent of *new* cars each year). Since this problem can only really be solved by a rather complex algorithm involving the age distribution of vehicles, average lifetime, number of cars purchased each year, etc., it was simply assumed that, with a lag of 10 years, the percentage of stock that is electric would be the same as the percentage of new cars.

5.3. SUBSTITUTION OF TELECOMMUNICATIONS FOR TRANSPORTATION

This substitution illustrates the difficulty of making a forecast that includes a large behavioral component. Development of practical video communications devices over the next twenty years will make it technically possible to eliminate a large number of business and personal trips; but whether this will actually occur will depend upon some rather subtle aspects of the communications process. In the case of business trips, the suitability of telecommunications depends upon the purpose of the trip and the relationship among the participants. If it is a meeting of new business associates, it is unlikely that any video conferencing system can provide sufficient intimacy to foster feelings of ease and informality necessary to establishing a good working relationship. Similar problems are encountered with a trip made for the purpose of selling a product or a policy. Salesmen often find that physical presence adds a large measure of persuasiveness; it facilitates extended contact and discussion and makes it more difficult for the other party to terminate the contact. Then, of course, it is simple to hand over helpful documents, whether business memoranda or product literature, in person. An improved facsimile transmission device may, however, adequately perform this function.

The categories of business travel most amenable to replacement by telecommunications are first, those for making routine communications or decisions that occur frequently, and second, meetings among individuals who know each other fairly well, such as gatherings for the purpose of exploring policy alternatives, communicating information, achieving a consensus, etc. The final difficulty with this kind of forecast is that there is a dearth of information as to what proportion of business travel is of this type. Moreover, since the cost of the telecommunications service may, in some cases be greater than the direct cost of travel, the comparative economics of the substitution will depend on how business people value their time.

6. Two Illustrations of the Forecasting Techniques

6.1 SUBSTITUTION OF ALUMINUM FOR COPPER IN ELECTRICAL APPLICATIONS

Aluminum and copper are both good electrical conductors and, therefore, compete in certain markets. The absolute conductivity (per unit volume) of aluminum is about 63% that of copper and is capable of substituting for copper in any application where volume is not a crucial limitation. Rising copper prices over the past few years have caused many users of high voltage transmission wire and other electrical goods to turn to aluminum [21]. Since aluminum is 30% as dense as copper, one pound of aluminum is the functional equivalent (for electrical purposes) of about two pounds of copper. The average price of U.S. copper in 1971 was 60¢/lb., while the average price of U.S.

aluminum in 1971 was about 30¢/lb. Thus, it costs $1.20 to obtain the same amount of carrying capacity in copper as can be had in aluminum for 30¢. The chief disadvantages of aluminum as a current carrier is that, due to its higher electrical resistivity, it wastes approximately 19% more energy as heat than does copper. Widespread utilization of aluminum for household wiring, for instance, had to await the development of better, more heat resistant insulating materials. Aluminum is also somewhat less ductile than copper; and as noted above, it requires more volume. Over 90% of the electrical transmission lines in the U.S. now employ either aluminum cable or steel-reinforced aluminum cable. About 30% to 40% of all electrical distribution lines also utilize aluminum conductors. Of the total market for electrical conductors in 1971, copper's share was 61%; aluminum had 39% of the market. We can now proceed to the forecast.

First, we must estimate what the upper limit or asymptote of the market share for aluminum might be. In other words, are there electrical applications for which aluminum could not conceivably be appropriate, given infinite time for technological improvement? Next, aluminum's market share at some specific future data must be estimated on the basis of informed opinion and expert advice. The market share forecast is shown below.

Market Share Forecast	1971	2025	Ultimate Limit (Asymptote)
Copper	61%	14%	10%
Aluminum	39%	86%	90%

We have identified the sectors relevant to this substitution in the INFORUM input-output model to be sector 84–Primary Copper, sector 87–Primary Aluminum, and sector 90-Non-Ferrous Wire Drawing and Insulating. The first two sectors are the selling factors; the third is the buying sector.

The 1971 coefficients of copper sales to wire drawing and aluminum sales to wire drawing were .31 and .05, respectively. We can now specify the asymptotic value (Y_B) for the 84–90 coefficient according to the market share forecast above, and a value for the year 2025 as well. The lower asymptote (initial value, Y_A) of the coefficient is estimated by assuming that the coefficients of 87 to 90 once had a value of zero, and using the dollar equivalency rate to calculate what the original coefficient value of 84–90 was before the substitution of aluminum for copper began.

The dollar equivalency ratio is obtained by multiplying the functional equivalency ratio of 1:2 (one pound of aluminum for two pounds of copper) by the 1971 prices of the two materials (60¢/lb for copper; 30¢/lb for aluminum), yielding a ratio of $30 Al:$1.20 Cu, or 1:4.

The asymptotic and 2025 values of the 87–90 coefficient are obtained by multiplying the change in the 84–90 coefficient by the dollar equivalency ratio and adding the result to the 1971 value of the aluminum to wire drawing coefficient. The result of all these operations are shown below.

Coefficient	Y_A	Y_{1971}	Y_{2025}	Y_B
84–90	.51	.31	.10	.06
87–90	.0	.05	.102	.15

These four data points: Y_A, Y_{1971}, Y_{2025}, and Y_B, are sufficient to estimate the values for the intervening years using an equation for a logistic curve as described in Section 4. These values are plotted in Fig. 3.

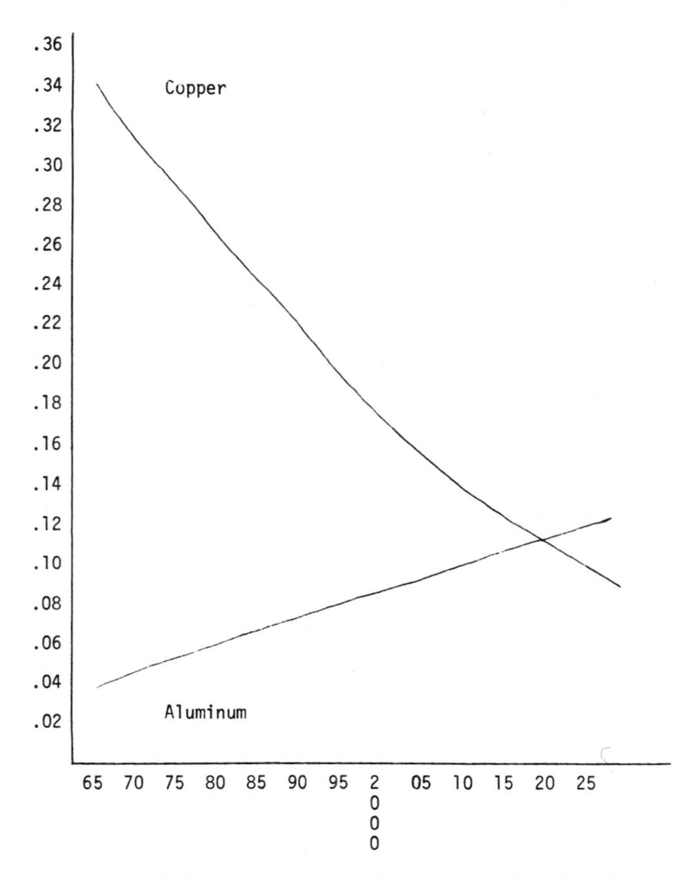

Fig. 3. Logistic Curve Estimated for Copper Vs. Aluminum (based on above parameters).

6.2. INPUTS TO THE STEEL INDUSTRY

Changes in input coefficients for the steel sector are forecast for the following reasons: The open hearth process will decrease and will become practically nonexistent by 1985. The estimated proportion of steel production in the United States derived from each of the four principal processes through 2025 is presented in Table 3.

These process changes will require more electrical energy due to the expected increase in the fraction of steel produced in electrical steel-making processes. Due to the obsolescence of open hearth, there will be reductions in natural gas consumption. The oxygen consumption in steel-making has gone up from 90,026 million cubic feet in 1964 to 186,169 million cubic feet in 1970. The natural gas consumption in steel-making has gone down from 108,540 million cubic feet in 1964 to 57,597 million cubic feet in 1970.

The data on changes in total consumption of various inputs and production of steel in the U.S. in the past are given in several tables of a U.S. Bureau of Mines circular [22]. The obsolescence of open hearth furnaces will reduce the purchase of chromite ore for refractor purposes.

6.2.1. Direct Reduction

Direct reduction is a development involving the reduction of iron oxide ore using methane, natural gas or hydrogen, and then charging the prereduced ore either in the

TABLE 3

Technology	1970	1980	1990	2000	2010	2025	Ultimate Limit (Asymptote)
Electric furnace	14%	20%	30%	40%	49%	49%	50%
Open hearth	30%	15%	19%	0	0	0	0
Basic oxygen furnace	56%	65%	69%	60%	51%	51%	50%
Direct reduction	2.5	6.3	19	30	54%	54%	55%

blast furnace or in the electric steel-making furnace. Use of direct-reduced ore in the blast furnace reduces the consumption of coke per ton of pig iron from 1100 lbs. to 500 lbs. as the percentage of direct-reduced ore increases from 0% to 100%. The advantage of using direct reduction is that it increases the blast furnace capacity and reduces the requirement of coke production that is pollution generating. The supplies of coking-grade coal are also decreasing, which increases the pressure toward reducing the coke requirement. Production of direct-reduced ore, of course, requires a reducing gas which could be natural gas, methane, carbon monoxide, hydrogen, or mixtures of these gases. One could also use coke breeze or gasified coal for the reduction.

Direct-reduced ore can also be used in electric steel-making furnaces instead of scrap. The charge can consist of a maximum of about 70% direct-reduced ore.

The ultimate in the direct reduction process will be production of steel by a single continuous process from ore to finished product. This however will just begin to become important by 2025.

6.2.2. Coking of Coal

Coke is required for blast furnace reduction of iron ore; its production by present coke-oven battery processes is a pollution generating process. Increased use of direct-reduced ore will cut down the consumption of coke to half its present quantity per ton of steel by the year 2025.

There are, however, changes forthcoming in processes for producing coke that are nonpolluting. The most promising is the form-coke process in which pulverized coal is heated and subsequently pressed to form coke briquettes. This reaction is contained and the gaseous by-products can be collected very efficiently.

6.2.3. Steel Industry

The steel industry will be much more automated and computer controlled than it is now. The industry will buy and install many on-line process control computers.

The use of nuclear power will become a reality by 1990. The nuclear energy will either be used as electrical power to run the electric furnaces or the hot gases from the reactors will be used to gasify coal or produce hydrogen from water. The gasified coal or hydrogen will be used for reduction purposes. In view of the investments ahead, it is doubtful that steel companies will build their own nuclear plants. Most probably they will have the utilities company build and maintain the reactors at the steel plants.

More and more high-strength low-alloy steel will be used in the future requiring increased quantities of ferrovanadium, ferrocolumbium, ferromolybdenum, and ferrotitanium or the elemental forms of these alloying elements. The total use of these elements per ton of steel could increase by 2025 to 15 times what it is today [23–26].

The use of aluminum in steel as an alloying element and as coatings for corrosion and oxidation resistance will increase at the expense of nickel. At present aluminum is used mainly for deoxidation purposes, and a very small amount for alloying. The total aluminum input to steel could increase by 2025 to 10 times its present value per ton of steel.

The use of tin in steels is decreasing. The techniques of electrotinning have decreased the amount of tin per square foot of steelplate. In addition, the development of electroplated chromium will further decrease the demand for tin and, to a certain extent, zinc. Tin and zinc will also be partly replaced by polymers and aluminum as coating materials [27].

The amount of tin per ton of steel could become one tenth of its present value by 2025 and the amount of zinc per ton of steel could go down to one fifth of its present value.

The use of nickel per ton of steel is likely to go to one-half of its present value due to increased use as coatings instead of tin and zinc, and more use of nickel-free chromium containing ferritic stainless steels.

The yield of steel shipments per ton of steel melted will increase from about 60% to 80% by 2025 due to increased use of high-yield processes like continuous casting. This will, to a certain extent, decrease the consumption of raw materials and energy used to produce a ton of steel. However the increased use of alloying elements and special coatings will increase the overall cost of steel to 1.5 times its present value, in 1970 prices, by 2025.

Our estimates of input coefficients for the steel industry through 2025 are as present in Table 4.

7. Final Comments

Obviously, the forecasting of the effects of technological change on trade patterns is not, and can never be, an exact science. But the application of systematic techniques raises the reliability of such forecasts far above that of the "seat-of-the-pants" methods on which businesses and governments have relied in the past. The future, while subject to forces that no one can accurately forecast, is not wholly obscure. Looking at the inputs to the steel industry just discussed, the technological developments on which these forecasts are based are quite clearly delineated. The processes expected to supplant open hearth production are not unknown or untried, although the direct reduction process is fairly new. There is no magic in these forecasts, but simply an informed view of what is happening now and of what the future is deemed likely to bring.

The approach to forecasting that this paper outlines is essentially a two-pronged

TABLE 4

Seller (Sector)	1971	1980	1990	2000	2010	2025	Ultimate Limit (Asymptote)
Coal mining	.0260	.0234	.0164	.0091	.0060	.0051	.0050
Gas utilities	.0110	.0168	.0209	.0248	.0272	.0290	.0300
Electric utilities	.0190	.0251	.0309	.0350	.0373	.0390	.0400
Primary non-ferrous metals, nec	.0100	.0113	.0149	.0230	.0384	.0700	.1000
Computers & related machines	.00001	.00004	.00018	.0008	.0029	.0080	.0100

endeavor. First, we evaluate current and anticipated technological developments, drawing upon the best knowledge available. This involves a large element of informed judgment. Second, we apply this information systematically in accordance with a defensible theoretical structure. This step is statistical and mathematical. At the present stage of the forecasting art, this seems to us a reasonable approach, and we put it forward in the hope that further developments in both the statistical and theoretical areas will lead to improvements in practicability and accuracy.

References

1. Chicago Area Transportation Study, Final Reports, Vol. 3, April 1962, p. 78.
2. Clopper Almon, Jr., *Interindustry Forecasting Project*, University of Maryland, Bureau of Business and Economic Research Monograph, Feb. 1973.
3. Based on work by Prof. W. W. Leontief and others. See the following: (1) W. W. Leontief, *The Structure of the American Economy, 1919–19329*, Oxford University Press, N.Y., 1951, and *Input-Output Economics*, Oxford University Press, N.Y., 1966. (b) Robert Dorfman, Paul A. Samuelson, and Robert M. Solow, *Linear Programming and Economic Analysis*, McGraw-Hill, N.Y., 1958. (c) Hollis B. Chenery and Paul G. Clark, *Interindustry Economics*, John Wiley & Sons, N.Y., 1959.
4. C. Almon, *op. cit.*
5. Benton F. Massell, Capital formation and technological change in United States manufacturing, *Rev. Econ. Statist.* **XLII** (May 1960).
6. E. Mansfield, Technical change and the rate of imitation, *Econometrica* **29** (4) (1961).
7. Richard A. Stone, A computable model of economic growth in *A Program for Growth* (Chapman and Hall, Eds.), 1962.
8. Michale D. McCarthy, Embodied and Disembodied-Technical Progress in the Constant Elasticity of Substitution Production Function, *Rev. Econ. Statist.* **XLVII** (February 1965).
9. *Technological Trends in Major American Industries*, Bulletin 1474, U.S. Department of Labor, U.S. Government Printing Office, Washington, D. C., 1966.
10. M. Boretsky, Comparative progress in technology, productivity and economic efficiency: USSR vs USA, *New Directions in the Soviet Economy*, Studies prepared for the Subcommittee on Foreign Economic Policy of the Joint Economic Committee, Part II–A, Economic Performance, U.S. Government Printing Office, Washington, D. C., 1966.
11. G. Chow, Technological progress and the demand for computers, *Amer. Econ. Rev.* **57**, 1117 (December 1967).
12. A. D. Little, Competition of materials, *Scientific Amer.*, Interindustry Study (1968).
13. *Indexes of Output Per Man-Hour, Selected Industries, 1939 and 1947–69*, Bulletin 1680, U.S. Department of Labor, U.S. Government Printing Office, Washington, D. C., 1970.
14. Anne P. Carter, *Structural Change in the American Economy*, Harvard University Press, Cambridge, 1970.
15. Bacharach, *Biproportional Matrices and Input-Output Change*, Cambridge University Press, 1970.
16. *Patterns of Economic Growth*, Bulletin 1672, U.S. Department of Labor, 1970.
17. R. A. Stone, 1962, *op. cit.*
18. Bacharach, *op. cit.*
19. M. O. Stern, R. U. Ayres, A. Shapanka, A model for forecasting the substitution of one technology for another, *Technol. Forecast. Soc. Change* **7** (1), 57 (1975).
20. E. P. Cliff, Timber: The Renewable Material, prepared for the National Commission on Materials Policy, August 1973.
21. *Mutual Substitutability of Aluminum and Copper*, National Materials Advisory Board Report 286, April 1972.
22. *Effects of Direct Reduction Upon Mineral Supply Requirements for Iron and Steel Production*, U.S. Bureau of Mines Circular 8583, 1973.
23. *Trends in Usage of Columbium*, NMAB Report 264, March 1970.
24. *Trends in the Use of Vanadium*, NMAB Report 267, March 1970.
25. *Trends in the Use of Ferroalloys By the Steel Industry of the US*, NMAB Report 276, July 1971.
26. *A Delphi Exploration of the U.S. Ferroalloy and Steel Industries*, NMAB Report 277, July 1971.
27. *Trends in the Use of Tin*, NMAB Report 265, March 1970.

Appendix I

185–SECTOR CLASSIFICATION SCHEME (INFORUM)

1 Dairy farm products
2 Poultry & Eggs
3 Meat, animals & misc. livestock products
4 Cotton
5 Grains
6 Tobacco
7 Fruits, vegetables & other crops
8 Forestry & fishery products
9 *** Not used ***
10 Agricultural, forestry & fisher services
11 Iron ore mining
12 Copper ore mining
13 Other non-ferrous metal ore mining
14 Coal mining
15 Crude petroleum & natural gas
16 Stone & clay mining & quarrying
17 Chemical & fertilizer mining
18 New construction
19 Maintenance & repair construction
20 Complete guided missiles & space vehicles
21 Ammunition
22 Other ordnance
23 Meat products
24 Dairy products
25 Canned & frozen foods
26 Grain mill products
27 Bakery products
28 Sugar
29 Confectionery products
30 Alcoholic beverages
31 Soft drinks & flavorings
32 Fats & oils
33 Misc food products
34 Tobacco products
35 Broad+narrow fabrics, yarns & thread mills
36 Floor coverings
37 Misc textiles
38 Hosiery & knit goods
39 Apparel
40 Household textiles
41 Lumber & wood products
42 Veneer & plywood
43 Millword & wood products
44 Wooden containers
45 Household furniture
46 Other furniture
47 Pulp mills
48 Paper & paperboard mills
49 Paper products mec
50 Wall & building paper
51 Paperboard containers
52 Newspapers
53 Periodicals
54 Books
55 Industrial chemicals
56 Business forms & blank books
57 Commercial printing
58 Other printing & publishing
59 Fertilizers
60 Pesticides + other agricultural chemicals
61 Misc chemical products
62 Plastic materials & resins
63 Synthetic rubber
64 Cellulosic fibers
65 Noncellulosic fibers
66 Drugs
67 Cleaning & toilet preparations
68 Paints
69 Petroleum refining & related products
70 Fuel oil
71 Paving & asphalt
72 Tires & inner tubes
73 Rubber products
74 Misc plastic products
75 Leather tanning + industrial leather prods
76 Leather footwear (exc. rubber)
77 Other leather products
78 Glass
79 Structural clay products
80 Pottery
81 Cement, concrete & gypsum
82 Other stone & clay products
83 Steel
84 Copper
85 Lead
86 Zinc
87 Aluminum
88 Primary non-ferrous metals, nec
89 Non-ferrous rolling & drawing, nec
90 Non-ferrous wire drawing & insulating
91 Non-ferrous castings & forgings
92 Metal cans
93 Metal barrels, drums & pails
94 Plumbing & heating equipment
95 Structural metal products
96 Screw machine products
97 Metal stampings
98 Cutlery, hand tools & hardware
99 Misc fabricated wire products
100 Valves, pipe fittings & fabricated pipes
101 Other fabricated metal products, nec
102 Engines, turbines & generators
103 Farm machinery
104 Construction, mining+oil field machinery
105 Materials handling machinery
106 Machine tools, metal cutting
107 Machine tools, metal forming
108 Other metal working machinery

109 Special industrial machinery
110 Pumps, compressors, blowers & fans
111 Ball & roller bearings
112 Power transmission equipment
113 Industrial patterns
114 Computers & related machines
115 Other office machinery
116 Service industry machinery
117 Machine shop products
118 Electrical measuring instruments
119 Transformers & switchgear
120 Motors & generators
121 Industrial controls
122 Welding apparatus & graphite products
123 Household appliances
124 Electric lighting & wiring equipment
125 Radio & tv receiving
126 Phonograph records
127 Communications equipment
128 Electronic components
129 Batteries
130 Engine electrical equipment
131 X-ray & electrical equipment
132 Truck, bus & trailer bodies
133 Motor vehicles & parts
134 Aircraft
135 Aircraft engines & parts
136 Aircraft equipment, nec
137 Ship & boat building & repair
138 Railroad equipment
139 Cycles & parts, transportation equipment
140 Trailer coaches
141 Engineering & scientific instruments
142 Mechanical measuring devices
143 Optical & ophthalmic equipment
144 Medical, surgical & dental instruments
145 Photographic equipment & supplies
146 Watches, clocks & parts
147 Jewelry & silverware
148 Toys, sporting goods+musical instruments
149 Office supplies
150 Misc manufacturing, nec
151 Railroad transportation
152 Local, surburban+interurban highway trans
153 Truck transportation
154 Water transportation
155 Air transportation
156 Pipelines
157 Freight forwarding
158 Telephone & telegraph
159 Radio & tv broadcasting
160 Electric utilities
161 Gas utilities
162 Water & sanitary services
163 Wholesale trade
164 Retail trade
165 Banks, credit agencies & brokers
166 Insurance & broker's agents
167 Owner-occupied dwellings
168 Other real estate
169 Hotels & lodging places
170 Personal & repair services
171 Business services
172 Advertising
173 Auto repair
174 Motion pictures & other amusements
175 Medical services
176 Private schools+nonprofit organizations
177 Post office
178 Federal government enterprises
179 Local government passenger transit
180 State & local electric utilities
181 Directly allocated imports
182 Business travel
183 Office supplies
184 Unimportant industry
185 Computer rental

Appendix II

TECHNOLOGICAL SUBSTITUTIONS AND THE PRINCIPAL INDUSTRIES AFFECTED

This table lists a number of explicit technological changes expected to take place over the next twenty-five years or more that would be likely to alter inter-industry trade patterns. The technological changes, called "substitutions", involve the substitution of one product or process for another, thus altering the pattern of who sells to whom. The column "Seller" refers to the industry sectors for which the substitution affects the market for their products. The column "Buyer" refers to the industry sectors for which the substitution affects their purchases. The numbers placed ahead of each sector description refer to the INFORUM classification shown in Appendix I. The expression "Inputs to" such-and-such an industry (e.g., Inputs to Motor Vehicles) refers to the changes in the products that will be used in that industry as technologies change.

Substitutions in the energy field (such as the substitution of nuclear energy for petroleum) are excluded from this table because work was not completed at the time of its preparation.

TABLE TO APPENDIX II

Technological Substitutions and the Principal Industry Sectors Affected

	Substitution		Seller		Buyer
1.	Telecommunications substituting for transportation	182	Business travel		ALL
		83	Airlines		ALL
2.	Inputs to motor vehicles	83	Steel		
		62	Plastic materials & resins		
		74	Misc. plastic products		
		87	Aluminum		
		84	Copper		
		91	Non-ferrous casting & forging	133	Motor vehicles & parts
		78	Glass		
		97	Metal stampings		
		128	Electronic components		
		120	Motors & generators		
		129	Batteries		
3.	Inputs to steel	14	Coal mining		
		161	Gas utilities		
		160	Electric utilities	83	Steel
		88	Primary non-ferrous metals, nec		
		114	Computers & related machines		
4.	Inputs to office construction	41	Lumber & wood products		
		42	Veneer & plywood		
		43	Millwork & wood products		
		46	Other furniture		
		74	Misc. plastic products		Office construction
		81	Cement, concrete & gympsum		
		82	Other stone & clay products		
		87	Steel		
		87	Aluminum		
5.	Inputs to hotel, motel & dormitory construction	41	Lumber & wood products		
		42	Veneer & plywood		
		43	Millwork & wood products		
		46	Other furniture		
		74	Misc. plastic products		Hotel, motel & dormitory construction
		81	Cement, concrete & gypsum		
		82	Other stone & clay products		
		83	Steel		
		87	Aluminum		

Continued

Table to Appendix II (continued)

Substitution	Seller		Buyer	
6. Inputs to residential construction	41	Lumber & wood		Residential construction
	42	Veneer & plywood		
	43	Millwork & wood products		
	81	Cement, concrete & gympsum		
	82	Other stone & clay products		
	83	Steel		
	87	Aluminum		
7. Substitution of plastics for paper in containers	49	Paper products nec	33	Misc. food products
	51	Paperboard containers		
	62	Plastic materials & resins		
	74	Misc. plastic products		
	49	Paper products nec	25	Canned & frozen foods (see bottle forecast)
	51	Paperboard containers		
	74	Misc. plastic products		
	49	Paper products nec	23	Meat products
	74	Misc. plastic products		
	49	Paper products nec	29	Confectionery products
	51	Paperboard containers		
	62	Plastic materials & resins		
	49	Paper products nec	24	Dairy products
	51	Paperboard containers		
	74	Misc. plastic products		
	49	Paper products nec	27	Bakery products
	62	Plastic materials & resins		
	74	Misc. plastic products		
8. Plastic bottles for glass bottles	78	Glass	67	Cleaning & toilet preparations
	74	Misc. plastic products		
	78	Glass	25	Canned & frozen foods
	74	Misc. plastic products		
	78	Glass	32	Fats & oils
	74	Misc. plastic products		
	78	Glass	66	Drugs
	74	Misc. plastic products		
	78	Glass	31	Soft drinks & flavorings
	74	Misc. plastic products		
9. Inputs to household appliance	83	Steel	123	Household appliances
	62	Plastic materials & resins		
	87	Aluminum		
	91	Non-ferrous castings & forgings		
10. Inputs to household furniture	41	Lumber & wood products	45	Household furniture
	42	Veneer & plywood		
	43	Millwork & wood products		
	62	Plastic materials & resins		
	74	Misc. plastic products		
	87	Aluminum		
	83	Steel		

Continued

Table to Appendix II (continued)

	Substitution		Seller		Buyer
11.	Inputs to plumbing & heating equipment	74	Misc. plastic products	94	Plumbing & heating equipment
		84	Copper		
		73	Rubber products		
		87	Aluminum		
12.	Inputs to structural metals	83	Steel	95	Structural metal products
		87	Aluminum		
		84	Copper		
		89	Non-ferrous rolling & drawing, nec		
13.	Inputs to aircraft turbines	83	Steel	135	Aircraft engines & parts
		88	Primary non-ferrous metals, etc.		
		89	Non-ferrous rolling & drawing, etc.		
		91	Non-ferrous castings & forgings		
		82	Other stone & clay products		
14.	Inputs to aircraft	74	Misc. plastic products	134	Aircraft
		83	Steel		
		87	Aluminum		
		89	Non-ferrous rolling & drawing, nec		
		91	Non-ferrous castings & forgings		
		128	Electronic computers		
		127	Communications equipment		
15.	Superconductors in electrical transmission	84	Copper	124	Electric lighting & wiring equipment
		87	Aluminum		
		88	Primary non-ferrous metals, etc.		
		55	Industrial chemicals		
		116	Service industry machinery		
		69	Petroleum refining & related products		
		124	Electric lighting & wiring equipment	160	Electric utilities
		90	Non-ferrous wire drawing & insulating	124	Electric lighting & wiring equipment
16.	Inputs to primary aluminum	13	Other non-ferrous metal ore mining	87	Aluminum
		16	Stone & clay mining & quarrying		
		83	Steel		
		160	Electric utilities		

Continued

Table to Appendix II (continued)

	Substitution	Seller		Buyer	
17.	Inputs to valves & pipe fittings	83 87 74	Steel Aluminum Misc. plastic products	100	Valves, pipe fittings, fabricated pipes
18.	Inputs to non-ferrous castings & forgings	84 86 87 88	Copper Zinc Aluminum Primary non-ferrous metals, nec	91	Non-ferrous castings & forgings
19.	Inputs to non-ferrous wire drawing	84 85 87	Copper Lead Aluminum	90	Non-ferrous wire drawing & insulating
20.	Fuel use by electric utilities	14 70 161	Coal mining Fuel oil Gas utilities	160	Electric utilities
21.	Fuels consumed by gas utilities	14 15	Coal mining Crude petroleum & natural gas	161	Gas utilities
22.	Fuels used by petroleum refineries	14 15	Coal mining Crude petroleum & natural gas	69	Petroleum refining & related products
23.	Industrial fuel consumption	161 70 14 150	Gas utilities Fuel oil Coal mining Electric utilities		All manufacturing sectors
24.	Commercial fuel consumption	161 70 160	Gas utilities Fuel oil Electric utilities		All service sectors

Appendix III

ESTIMATING FUNCTIONAL EQUIVALENCY RATIOS (FER)

If the new product or process were the exact equivalent of the old, dollar for dollar, the functional equivalency ratio would be one: that is, $1 of the new product or process could be substituted for $1 of the old. In practice, this is almost never the case. There are qualitative differences between materials; thus, synthetic fabrics have qualities of their own that are not exact equivalents of cotton or wool. Manufacturing costs vary and so do many other variables.

A familiar property of the input-output model is that it is based upon a balanced table and that the existence of a solution depends upon the accounting balances in the table. A balanced table has the property that sales of a sector equals costs of the sector.

Sales are given by the expression:

$$p_i x_i = \sum_j p_i q_{ij} + p_i y_i, \tag{6}$$

where p_i is the unit price of the product in sector i, x_i is the physical output of the sector, q_{ij} is the physical flow from sector i to sector j and y_i is the final demand for the product in terms of physical quantities.

Costs are similarly given by

$$p_i x_i = \sum_j p_i q_{ji} + v_i x_i, \tag{7}$$

where v_i is the value-added per unit in the ith sector.

The assumption of proportionality,

$$q_{ij}/X_j = a_{ij}, \tag{8}$$

gives us the familiar equations of output and prices of the input-output model:

$$p_i x_i - \sum_j p_i a_{ij} x_j = p_i y_i, \tag{9}$$

$$p_i x_i - \sum_j p_j a_{ji} x_i = v_i x_i. \tag{10}$$

In Eq. (9), p_i can be eliminated and in Eq. (10), x_i can be eliminated, leading to the simplified equations

$$x_i - \sum_j a_{ij} x_j = y_i, \tag{9a}$$

$$p_i - \sum_j p_j a_{ji} = v_i, \tag{10a}$$

which are equivalent to expressions (1) and (4) respectively.

There are three different possible treatments of price in a model based on physical quantities. Before the three possibilities are enumerated, it is important to consider the conditions for the existence of an inverse of an $(I-A)$ matrix. It is generally known that an inverse of $(I-A)$ will exist if every column sum of A has a sum less than one. (This is a sufficient condition but not a necessary one. Necessary conditions are much more difficult to define and they must take account of the connectivities in the economy.)

When a model is defined using physical units, then generally it would not satisfy the condition that every column sum be less than one. However, a more general existence theorem is known, that the matrix $(I - A)$ will have an inverse *if there exists* a vector Z such that

$$\sum_j Z_j a_{ji} < Z_i, \tag{11}$$

for every column of $(I - A)$. This says that if a set of prices were such that the matrix could be balanced (and every industry had a nonzero value added), then the matrix would have an inverse even though the prices were not incorporated into the matrix.

It is conceivable that the introduction of substitutions into a model might result in a matrix that could not be inverted. It seems very unlikely that this would occur, however, because the existing column sums are around 1/2, so major changes would not affect any single column sum sufficiently. Furthermore, the theorem of (11) does not state that every column sum be less than one under all price conditions, but says that there must exist *some* set of shadow prices such that every column sum would be less than unity.

Now let us turn to the three possible treatments of price in an input-output model used to predict future industry output levels. The first option is that all or some sectors can be treated using physical units as shown in (9a) rather than values as shown in (9).

The second option is that prices can be used as an aggregator in the base year. According to this formulation, a change in final demand y'_j will change output levels x'_i and x'_j but will not change prices. Hence we obtain

$$p_i x'_i - \sum_j p_i a_{ij} x'_j - p_i y'_i. \tag{9b}$$

The prices are not affected by the change in final demand, so they continue to reflect base-year ratios.

The third option is to compute a balanced table for a future date. But this requires that future prices be explicitly computed as well as future values of industry levels. The equations are

$$x_i - \sum_j a_{ij} x'_j = y'_i \tag{12}$$

$$p'_i - \sum_j p'_j a_{ji} = v'_i. \tag{13}$$

IR&T is now using the second version, which is the one most commonly used for input-output. In version two, prices are used only as conventions, and the relevant prices are the base-year prices for the year when the table was balanced. Predicted tables are not balanced unless the double prediction of version three is utilized.

Therefore, it becomes necessary to estimate how many physical units of the new material can replace a physical unit of the old material. We call this the "functional ratio of inputs". The most salient physical, mechanical or electrical properties, such as density, electrical resistivity, tensile strength, BTUs, etc. are compared to determine the functional ratio. This ratio, in physical units, is then converted to dollar purchases using base year prices.

NOTE: The material contained in Appendix III was supplied by Stedman B. Noble.

Substitution of Mechanical Corn Pickers by Field Shelling Technology— An Econometric Analysis

DEVENDRA SAHAL

ABSTRACT

Existing models of technological substitution, including those of the Blackman-Mansfield type, are the point of departure of this study. An illustrative case of substitution of mechanical corn pickers by field shelling technology (currently in progress) is studied. The results of this study suggest that technological substitution is best considered as an economic phenomenon. The assumption that the diffusion of innovation follows an S-shaped curve does not seem to be crucial. The production scale (e.g., farm size) is concluded to be the most important determinant of technological substitution.

Introduction

If an illustration of the importance of studying technological substitution is indeed required, the following is one such example even though it is unrelated to the specific case studied in this paper. According to Geo H. Seferovich [20], seven manufacturers, in 1945, enjoyed a prosperous market with about 100,000 units of cream separators worth about $4.5 million. Between 1950 and 1955, however, the technology of dairy farming began to show numerous signs of change. The advent of milking parlor technology introduced a series of new or improved devices such as pipeline milkers, liquid manure handling and bulk milk tanks. By 1955, the "diffusion" of the new dairy farming technology had proceeded so far that the cream separator sales declined from 100,000 units to slightly more than 10,000 units worth only about $1.2 million. In 1960, the U.S. Census Bureau stopped collecting data on cream separators. The transition was complete. Less than half of the original seven manufacturers of cream separators survived; one of them had to liquidate his business.

By way of conclusion it would suffice to quote Seferovich [20] himself: "If good technological forecasting had been focussed on the probable trajectory of the cream separator as early as 1945 or as late as 1960, at least one corporate life could have been saved and much economic waste prevented".

The present study is concerned with a case of technological substitution currently in progress in corn harvesting. An analysis is made of the challenge posed to mechanical corn pickers by methods and equipment called field shelling. As shown in Fig. 1, the sales of pickers have sharply declined since the mid-1950's while the manufacturers' shipments of

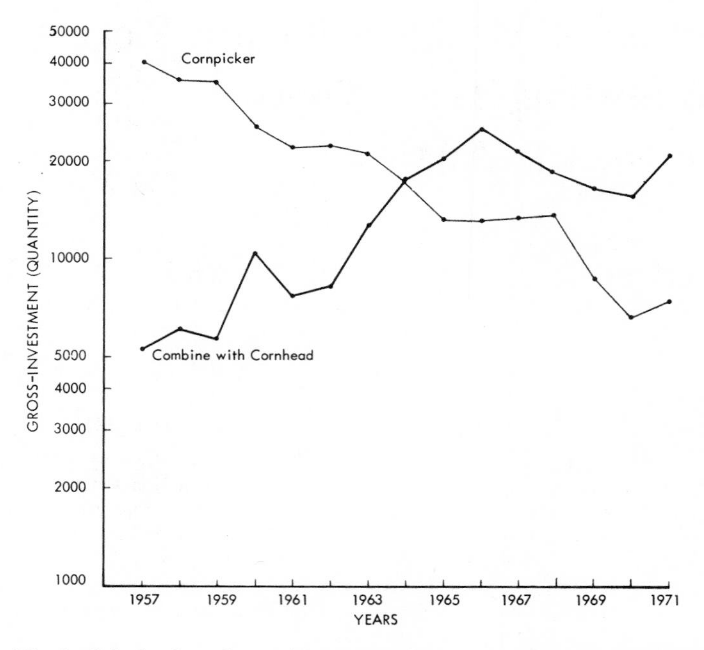

Fig. 1. Substitution of cornpicker technology by combine with cornhead.

combine with corn head have steadily increased. These opposed technologies may soon level off at something like 33% of all corn acres harvested with mechanical pickers and 65% with combines [20].

Zvi Griliches has repeatedly emphasized the importance of studying the features underlying the "diffusion" of inputs in farm production as a prerequisite "to an understanding of the behavior of the supply of the agricultural products and to an explanation of technological changes that have occurred in agriculture". [6, p. 591]. Others note with considerable justification that "one of the neglected areas . . . has been demand by farmers for inputs produced by nonfarmers". [2]. All of this is sufficient justification for the present study.

In recent years a number of "models" of the substitution phenomena have been proposed in the literature on technological forecasting. These models are based on what has long been known in sociology and economics, namely that diffusion of new technology follows an S-shaped curve [5, 16]. *While the descriptive power of these technological forecasting models is, in some ways, often good, their explanatory power is close to zero.* These models provide neither any justification of the functional form employed nor any information on the determinants of technological change–let alone on the relative importance of various determinants. Very often such models have been justified on the grounds of empirical neccessity. While there is some merit in this argument, it should be noted that it is at least partly based on circular logic. Indeed, unless the "trend" is well-established, the results based on trend extrapolation are likely to be even more unreliable. However, to say that an extrapolation is in fact based on an established trend is to imply that the data are available! There is even less truth in the claim often put forth in the literature on technological forecasting that it is possible to obtain a good forecast by means of trend extrapolation despite its neglect of other causal variables. To put it mildly, the models based on trend extrapolation are inherently more adaptable to retrospective analysis than to *ex ante* projections [21]. The chief attraction of some such models lies in the simplicity of victorian Physics. However, the cost paid is particularly heavy: the forecasted trend is either right or wrong but one cannot tell why! For the forecaster there is little to learn and much to lose from trend extrapolation. Unless, of course, the underlying causal framework is first developed.

In what follows the existing models based on trend-extrapolation are therefore taken as a point of departure. The present study, however, also differs from such welcome exceptions as the models of the type developed by Mansfield, Blackman, Stern et al. [1, 5, 13, 21], etc. The reasoning is made clear in the following section.

1. Theoretical Framework

An appropriate terminology for this study is provided by Jorgenson [11, 12], who in a series of comparatively recent studies presents a theory of investment behavior based on neoclassical theory of optimal capital accumulation. The thesis is advanced here that it is the framework provided by the theory of investment behavior that is relevant to modeling of technological substitution.[1]

In the first formulation proposed here the basic hypothesis is that the optimal or the desired level of services provided by the field shelling technology at a point in time is a function of the price of the services from field shelling technology, the price of the product and the prices of the services from other production inputs. Insofar as the services are proportional to stock, then the desired level of stock of combines will be determined by the same variables. Assuming the functional relationship to be log-linear we have

$$\log S_{1t}^{*} = a_0 + a_1 \log(P_1/P_0) + a_2 \log(P_2/P_0), \tag{1}$$

where P_0, P_1, P_2 are the (expected) prices of the (crop) output and the desired services of the combines and cornpickers and S_{1t}^{*} is the desired stock of combines.

As an alternative formulation consistent with many accelerator studies of investment (e.g., [9]) we assume that the farm firms treat output as exogenous.[2] This formulation may be written as

$$\log S_{1t}^{*} = a_3 + a_4 \log Q_t + a_5 \log(P_2/P_1), \tag{2}$$

where Q is the (crop) output.

It is important to note that Eqs. (1) and (2) are used in a static sense to mean that the farm firm achieves its desired stock, instantaneously, for any given variation in the explanatory variables. This is fictional for anything but long-run analysis. In fact, the actual stock, S, may never reach the perpetually moving desired level, S^*, since the variables that determine the latter are constantly changing over time. In short, farm entrepreneurs continually move toward, but never actually reach the desired level.

[1] The formulations developed in this study are not, however, based on explicit consideration of any underlying production function. It is the *general* framework of the investment theory [11], and not any specific formulation thereof, that is regarded as relevant here.

[2] The inclusion of the output variable in the models of investment has been a subject of considerable controversy in the econometric literature. See, however, Jorgenson's survey article [12].

Gross investment, Y in any technique of production may be considered as a means by which entrepreneurs move towards the desired level, S^*. In any given period, t,

$$Y_t = S_{3+1} - S_t + \delta S_t. \quad (3)$$

This is a general definition, e.g., in case of combines

$$Y_{1t} = S_{1(t+1)} - S_{1t} + \delta_1 S_{1t}, \quad (4)$$

and in the case of cornpickers

$$Y_{2t} = S_{2(t+1)} - S_{2t} + \delta_2 S_{2t} \quad (5)$$

where δ is the rate of replacement, treated as a constant here over the period under consideration. Further gross investment is constrained by various technical and financial considerations. These are identified here as existing services, S_{t-1}, the ratio of liabilities to assets, $(L/A)_{t-1}$, and the scale of utilization or the farm size, Z_t. Thus

$$Y_{1t} = F[S^*_{1t}, S_{1(t-1)}, S_{2(t-1)}, (L/A)_{t-1}, Z_t], \quad (6)$$

which is assumed log-linear. Incorporation of the lagged ratio of debts to assets in the formulation has its justification in Duesenberry's theory of cost of capital [4, pp. 93–99]. Thus the model takes into account the fact that financial capital market facing farmers is not necessarily perfect. The farm size variable is regarded here as an important non-neutral technical bias that may considerably affect the proportion of the inputs in the farm production, e.g., by increasing the marginal productivity of the combines with cornhead relative to cornpicker. This is in accord with a recent, and certainly one of the very best, theoretical contributions to the literature by Paul David [3] on the diffusion of innovations. It follows from David's path-breaking work that the production scale (e.g., farm size) is a crucial variable in determining the diffusion of an important class of process innovations. Furthermore, as demonstrated by the empirical evidence [e.g., 18, 19], the production scale itself is determined by a phenomenon of a predominantly stochastic nature, i.e., the determinants of the scale variable are very many and the effect of each is quite small. Thus one-way causality is assured. Put another way, the scale variable is an important determinant *of* diffusion of innovations but it is not, in turn, significantly affected *by* the process of diffusion (or for that matter, by any single variable).

In summary, Eqs. (1) and (6), and, Eqs. (2) and (6) represent the two principal models of the phenomenon under consideration. These formulations differ from those of Griliches [5–7] and Mansfield [1] in a number of ways. Unlike Mansfield's formulation, the models in this study are based on the relatively straightforward assumption of log-linearity; they do not tack on any special extraneous assumptions concerning the functional form. Also, we have chosen to explain the gross-investment in the innovation rather than the stock in the use. Explanation of the former is a fairer test of one's model [8]. In his formulation, Mansfield emphasizes profit motive only; we include technical motive as well in our explanation of technological change via the crop output and (especially) the farm size variable.[3] Both Griliches and Mansfield treat the desired stock of services as the long run equilibrium of capital. Here, instead, it is regarded as a moving target. As Jorgenson has pointed out [12], this is an important distinction for the latter interpretatioprovides an *a priori* justification to "desired" capital being conditional on the current output. In [7] Griliches assumes that investment in innovation is determined by the intersection of the marginal efficiency schedule with marginal cost of fund schedule. Here that is assumed to shift in response to change in the degree of financial risk. Thus the justification of our formulation would seem to lie in the intertemporal maximation of the utility of profit function.

2. Material for Analysis

Crop output as well as the gross investment and the stock of services have been measured intermsof quantity not value. In a production function framework, the relevant measure is in terms of physical units [see 8 and the references therein]. The stock of combines with cornhead was constructed on the basis of a 15 year "one horse shay" assumption implying that the machine has a fixed n-year life, there is no deterioration with age and that it just falls apart on its nth birthday with no scrap value. The justification of this assumption can be found in Griliches [8] and an earlier work of the author [19].

Data on cornhead and cornpicker shipments and prices have been obtained from "Current Industrial Reports", series M35A and M35S, Bureau of Census and Bulletin 419, USDA, March 1968. The index number of the prices received by the farmers (p_0) can be found in the annual price summary of the crop reporting board, SRS, USDA, June 1972. For the data on the ratio of liabilities to assets and crop output, see "Balance Sheet of Farming Sector 1971" and "Economic Tables", ERS, January 1972, respectively. Farm size data have been obtained from Bulletin 316 of USDA, June 1962 and SpSy 3(1–72) of USDA, January 1972. Due to unavailability of data, it has unfortunately not been possible to take into account the quality change in cornpicker and combines with cornhead. During the period 1957–71 the gross investment in he Combines with Cornhead (unadjusted for the quality change) increased by roughly 284%. In the same period, the relative price ratio (ratio of cornpicker price to that of Combine with Cornhead) declined by roughly 32%. It is assumed that the implicit rental on the capital services would be proportional to the stock price. While this seems justified, the exclusion of the quality change is not. However, the neglect of the "diffusion of the qualitative aspects of the technology" would be defended on the grounds of empirical necessity.

3. Results

The parametric estimates of the models formulated in this study are presented in the form of the standardized β coefficients in Table 1. To alleviate the problem due to multi-collinearity, the price ratios (p_1/p_0) and (p_2/p_0) in the first formulation (Eq. (1)) were combined into a single variable (p_2/p_1). Equation (1) corresponds to the first model. Its performance is excellent for it leaves unexplained a mere 5% variance in the observations on the gross-investment in combines (see also Fig. 2). Except for the relative price ratio, all the variables are highly significant. The price ratio variable is, however, significantly correlated with the included variables. Thus when farm size variable and the

[3] On a definition of technical motive, see, Meyer and Kuh [14].

TABLE 1

Regression Equations Explaining Gross Investment in Combines during the years 1957–71

NO.	$S_{1(t-1)}$	$(P_2/P_1)_t$	$S_{2(t-1)}$	Z_t	Q_t	$(L/A)_{t-1}$	R^2	D.W.
1.	2.19	−0.166	1.465	−6.061		6.077	0.95	2.56
	(2.19)	(1.00)	(3.43)	(2.37)		(3.21)		
2.	0.83	−0.49	0.35				0.87	2.14
	(3.73)	(2.56)	(1.39)					
3.	2.26	−0.156	1.48	−6.36	0.05	6.29	0.95	2.56
	(2.02)	(0.84)	(3.24)	(2.05)	(0.20)	(2.78)		
4.	1.0	−0.49	0.34		−0.21		0.87	2.31
	(3.00)	(2.52)	(1.29)		(0.71)			

Definitions: S_1 is the stock of combine with cornhead; S_2 is cornpicker stock; P_1 is price of combine with cornhead; P_2 is price of cornpicker, Z is average farm size; Q is crop output and (L/A) is the ratio of farm liabilities to assets. Results of t-test are presented in parentheses. R^2 is the coefficient of multiple correlation squared and D.W. is the Durbin-Watson test value.

ratio of liabilities to assets are omitted (Eq. (2)), the coefficient of the price-ratio variable becomes significant as indicated by the t-test value. It displays the sign theoretically anticipated.[4]

The second principal formulation of the study is represented by Eq. (3) and (4). The problem due to multi-collinearity is more evident here with the result that coefficients of both the crop output and the price-ratio variables are insignificant at the conventional risk level. Despite an additional variable (i.e., crop output), the overall performance of Eq. (3), as indicated by the R^2- and the Durbin-Watson test values, is not superior to that of Eq. (1). Moreover, in Eq. (4), the sign of the entry of the output variable is incorrect, possibly due to the simultaneous-equation bias. The limited evidence presented here therefore does not substantiate the Jorgensonian notions of desired capital being conditional on current output [12].

In both the formulations, Z_t and $(L/A)_{t-1}$ enter with (standardized) coefficients much larger than those of other variables. Farm size and the ratio of liabilities to assets may therefore be concluded as two most important determinants of technological substitution. The variable representing the ratio of liabilities to assets enters with a positive sign suggesting that farmers hold a preference for risk. This seems to be a surprising result. However, it is in accord with the evidence presented by Johnson [10] in an unrelated study.

The direction of the effect of the farm size variable is not entirely clear, however. In Eq. (1), for example, if the ratio of liabilities to assets is excluded, the farm size variable enters with a positive and (presumably due to multi-collinearity) insignificant coefficient.

---

[4] Recall that during the period studied the relative price ratio declined (cf. Section 2). Thus the equation implies that the more expensive are combines relative to cornpicker, the less attractive the combines should be.

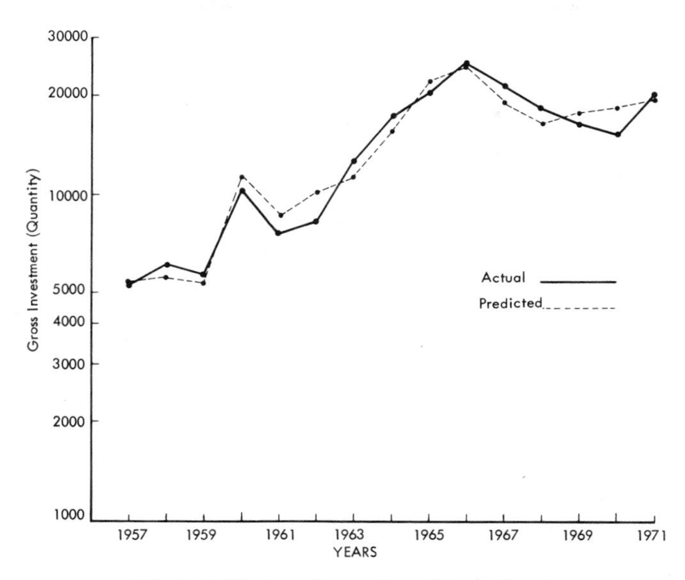

Fig. 2. Diffusion of combine with cornhead.

A priori, the sign of entry of the farm site variable is indeterminate. A positive sign is expected insofar as increase in farm size may lead to more mechanization and greater demand for the services of the combines. However, a negative sign is justified if increase in farm size is likely to lead to greater efficiency in the use of combines and therefore to a depressing influence on the demand for them.

The sign of the coefficients of lagged stock variables ($S_{1(t-1)}$ and $S_{2(t-1)}$) depends on the relative magnitude of the rates of replacement and adjustment of the actual stock to the desired level. If replacement requirements predominate, the coefficient of the stock variable has a positive sign. Alternatively, if the depressing effect of the existing stock on the adjustment process predominate, the stock exerts a negative influence on investment and consequently enters with a negative sign.[5] The positive coefficients in the estimated relationships here therefore point to a rate of replacement higher than the rate of adjustment to the desired level. Following Theil's work on specification errors [22], it is perhaps safe to conjecture that the positive coefficients of the stock variables reflecting a utility function also point to the importance of omitted variables such as desire for leisure, non-farm employment, etc., in the relationships presented here.

4. Concluding Remarks

The results of this study suggest that technological substitution is best considered as an economic phenomenon. The assumption that the diffusion of innovation follows an S-shaped curve does not seem to be crucial. Thus without any such postulate, the performance of the principal model presented in this study is found not to be lacking.

---

[5] In the literature on investment theory, replacement has sometimes been regarded as proportional to capital stock. We would, however, refrain from making any such hypothesis open to question. The purpose here is served by estimating the best statistical gross investment equation with an optimizing basis and the recognition of a replacemen related adjustment problem. We, therefore, see no great need to make a particular *a priori* hypothesis about the relationship of replacement to adjustment.

Furthermore, we have not succeeded in incorporating a pure acceleration principle in the explanation of the technological substitution for we find that the data do not convincingly support the Jorgensonian notion of desired capital being conditional on the current output. The importance of the technological background is, however, emphasized as seen from the fact that the farm size variable enters with one of the largest (standardized) coefficients, its size being anywhere twelve times larger than that of the price ratio variable. In short, the production scale has been identified as the most important determinant of the process of technological substitution.

I am indebted to Mr. Geo H. Seferovich for kindly providing me with a part of the data employed in this study and to Dr. Joseph Martino for his helpful comments on an earlier draft of this paper. To both of the my warm thanks and the usual absolution.

References

1. A. W. Blackman, The rate of innovation in the commercial jet engine market, *Technol. Forecast. Soc. Change* **2**, 268–276 (1971).
2. W. A. Cromarty, The demand for tractors, machinery and trucks, *J. Farm Econ.* **41**, 323–31 (1959).
3. P. A. David, *A Contribution to the Theory of Diffusion*, Memorandum 71, June 1969, Stanford University.
4. I. S. Duesenberry, *Business Cycles and Economic Growth*, McGraw-Hill, New York, 1958.
5. Z. Griliches, Hybrid corn: an exploration in the economics of technological change, *Econometrica* (Oct. 1957).
6. Z. Griliches, The demand for fertilizers: an econometric interpretation of technical change, *J. Farm Econ.* **40**(3), 591–606 (1958).
7. Z. Griliches, The demand for a durable input: farm tractors in the U.S., 1921–57, in *The Demand for Durable Goods* (A. C. Harberger, Ed.), University of Chicago Press, 1960, pp. 181–207.
8. Z. Griliches, Capital stock in investment functions: some problems of concept and measurement, in *Studies in Memory of Y. Grunfeld*, Stanford Univ. Press, 1963, pp. 115–137.
9. B. G. Hickman, *Investment Demand and the U.S. Economic Growth*, The Brookings Institution, Washington, 1965.
10. P. R. Johnson, Do farmers hold a preference for risk?, *J. Farm Econ.* 208 (Feb. 1962).
11. D. W. Jorgenson, Capital theory and investment behavior, *Am. Econ. Assoc. Papers and Proc.* **53**, 247–259 (May 1963).
12. D. W. Jorgenson, Investment behaviour and the production function, *Bell J. Econ. and Management Sci.* **3**, 220–51 (1970).
13. E. Mansfield, *Industrial Research and Technological Innovation*, W. W. Norton, New York, 1968.
14. J. R. Meyer and E. Kuh, *The Investment Decision*, Harvard University Press, Cambridge, Mass., 1957.
15. M. Nerlove, *Estimation and Identification of C-D Production Functions*, Chicago U. Press, Chicago, 1965.
16. E. M. Rogers, *The Diffusion of Innovations,* New York, 1962. Free Press
17. D. Sahal, Long range forecasting: A case study of demand for farm tractors in Finland, Res. Publ. 1, 1973, Inst. of Ind. Engineering, Helsinki University of Technology.
18. D. Sahal, On the dynamics of farm size distribution in Finland, *J. Sci. Agricult. Soc. of Finland* (1976).
19. D. Sahal, Stochastic models of nuclear power plant size distribution, submitted for publication.
20. G. H. Seferovich, Trajectory of a technology, *Implement & Tractor*, 28–30 (Janu. 7, 1969).
21. M. O. Stern, R. V. Ayers, and A. Shapanka, A Model for forecasting the substitution of one technology for the other," *Technol. Forecast. Soc. Change* **7**, 57–79 (1975).
22. H. Theil, Specification errors and the estimation of economic relationships, *Rev. Internat. Stat. Inst.* **25**, 41–51 (1957).

Part 4.
Applications to Energy Production

Part 4. Applications to Energy Production

Introduction

There are two important commonalities among the three articles in this chapter. First, they are all concerned with "externalities" or with the role of the technological background of the firms employing the innovation in the process of diffusion—hitherto neglected factors in the literature on the subject. Second, all three articles consider case studies with an important bearing on the production of energy.

In the first article on the diffusion of steam and hydroelectric turbines, Martino and Conver find that the pattern is best characterized by a step-wise growth rather than a smooth S-shaped curve. Diffusion of the highest performance turbines is found to be related to the growth of the total plant capacity. Specifically, the relationship indicates that the former varies in a constant proportion to the latter. Insofar as both point to the importance of the role of the scale of production in the process of diffusion, there would seem to be a resemblance between the findings of this study and those reported by Sahal in the preceding chapter.

In the second article on the diffusion of catalytic reforming processes into the petroleum refining industry in the U.S., Bundgaard-Nielsen and Fiehn hypothesize that the rate of adoption of the innovation depends on the technological background of a group of similarly situated firms (i.e., a group of firms employing almost identical production technologies and with identical production goals, as expressed by the product mix marketed by the firm at any given point in time). Their findings support the hypothesis that, provided the new technology satisfies the investment criteria used by a group of similarly situated firms, we can expect a larger rate of diffusion if the group of firms has a strong technological background relative to the diffusing technology than if the group consists of firms with a weak technological background. The results of this study also point to the importance of the refinery size in the substitution of new catalytic refining processes for the old. Thus the relevance of the plant size to the decisions to adopt a new technique is emphasized, a finding that seems to be in accord with the work of Sahal and Martino and Conver.

In the last article, Lakhani is concerned with a topic previously subject to gross neglect in the literature[1]: the social benefit-cost aspects of technological substitution. The improvement in environmental quality is studied in terms of adoption of environment-saving processes in the petroleum refining industry. Lakhani concludes in his analysis of the substitution that a diffusion rate of the hydrocracking process at the same rate as the catalytic cracking process would have been socially desirable because its marginal social benefits exceeded marginal social costs. Two important policy conclusions emerge from his analysis indicating the relatively slower growth rates of diffusion of the

---

[1] By way of exception, however, see Zvi Griliches, Research costs and social returns: Hybrid corn and related innovation, *J. Politic. Econ.*, October 1958, pp. 419–31; and K. Arrow, Economic welfare and the allocation of resources for invention, *The Rate and Direction of Inventive Activity,* Princeton University Press, NJ, 1962, pp. 609–25.

most environment-saving technique—hydrocracking. First, the overall vintage of capital equipment embodying the technique is relatively new, so that incentives or "carrots" are called for. Second, there is considerable variation in the growth rates of adoption across states in the U.S. so that the substitution could be accelerated by means of a "stick" in the form of better enforcement of pollution control legislation, say, by increasing per capita state expenditure on pollution controls. In short, both the carrot and the stick are required.

The Step-Wise Growth of Electric Generators Size

JOSEPH P. MARTINO and STEPHEN K. CONVER

Introduction

Simmonds [1] has reported a phenomenon best described as step-wise growth in a number of technologies. This phenomenon was first observed in the chemical industry. Plants built to produce a specific chemical all tend to have about the same capacity at any point in time. Since there are significant economies of scale in chemical plants, it is advantageous to have a single plant as large as possible. When a new plant is built, which is larger than previously existing plants, its product can be sold at a lower price than the product of the older and smaller plants. The companies owning these plants then close them down and replace them with new plants of the largest size that can be built, usually near-duplicates of the new pace-setting low-cost plant. This process, when plotted graphically, shows a sequence of stair-steps in plant size. Figure 1 shows one of Simmonds' plots, this one of ammonia plant capacity (million short tons per day).

One of the interesting features of this step-wise growth, not explained solely by the economics of new plant construction, is that the size of a single plant tends to be a rather constant proportion of the total plant capacity. That is, the number of plants remains about the same. As the total production grows, the size of single plants grows at about the same rate. This is clearly shown by Fig. 1, where the size of ammonia plants follows the growth of annual ammonia production.

If this principle turned out to be fairly general, that the maximum capacity of the largest single unit of some technology represented a constant fraction of the total existing capacity, it would be a useful tool for forecasting the technology involved. The total capacity would be more easily forecast on the basis of economic and demographic factors, allowing the forecast of maximum single unit capacity to be derived from this forecast of total capacity. Simmonds has found that this principle does in fact apply to a wide variety of technologies, from the size of oil tankers to the size of wine barrels. This paper reports an investigation of this relationship for maximum size of electric generators, both steam turbines and hydroelectric turbines.

In the case of electric generators, maximum size of a single unit is constrained by a factor not present in other types of technologies [2]. It has long been recognized that from the standpoint of operating economies individual generators should be as large as possible within the current state of the art. However, large generators present an unusual reliability problem. If a single unit represents a large fraction of the installed capacity of a single network, and that unit fails, then the capacity of the network is

This article appeared in *Technological Forecasting and Social Change*, Vol. 3, No. 4, 1972.

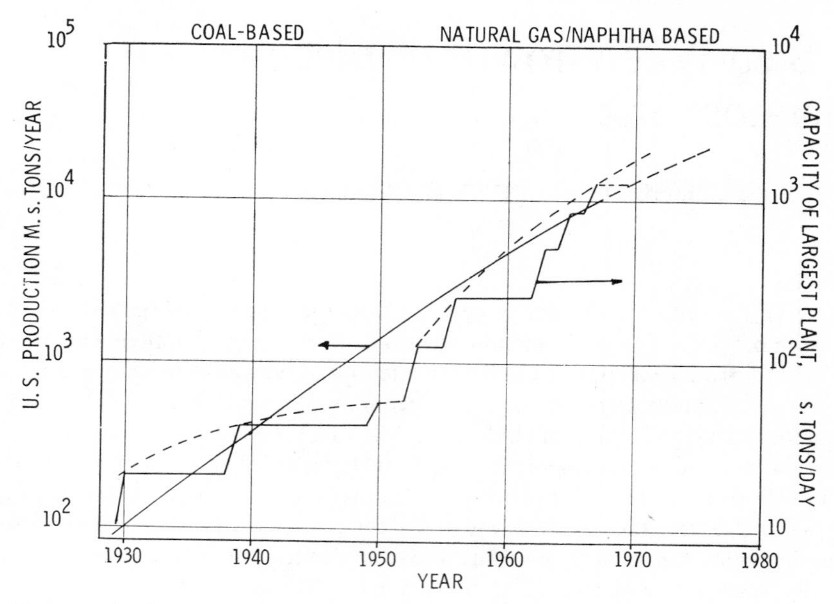

Fig. 1. Step growth of ammonia plants.

reduced significantly so long as the generator is out of service. Hence this constraint has acted to keep down the maximum size of a single generator. As the total installed capacity of a network increases, following increased consumption of electricity, then it is safe to add ever-larger units, so long as the size of a single unit does not exceed some specified fraction of total system capacity.

In the analysis reported here, however, we were concerned not with the total capacity of a single power company, but with the total installed generating capacity of the US. Since there is no single decision maker responsible for coordinating the overall growth in capacity of this aggregate electrical system, and since the activities of companies in one area would not be expected to be well coordinated with the activities of companies in geographically distant areas, there seems to be no reason to assume a consciously and deliberately chosen relationship between maximum generator size and total installed capacity. If such a relationship were found to exist, it would be indicative of the operation of factors not under the conscious control of individual decision makers. Hence the assumption that these factors will continue to operate, and will therefore be useful for forecasting size of electric generators, appears to be a reasonable one.

In the analysis, steam generators and hydroelectric generators were treated separately. This was justified partly on the pragmatic grounds that it seemed to work, and partly on the grounds that two different technologies were involved, hence separate treatment seemed justified.

Steam Turbines

Table 1 provides data on the maximum size of steam turbine available at any time in US electric power plants, as well as the total installed steam turbine generator capacity in US industrial and electrical utility plants. This data is also plotted in Fig. 2.

The plot of Fig. 2 shows that total installed capacity grew quite steadily from the beginning of the century to 1930. As a result of both the Depression and World War II,

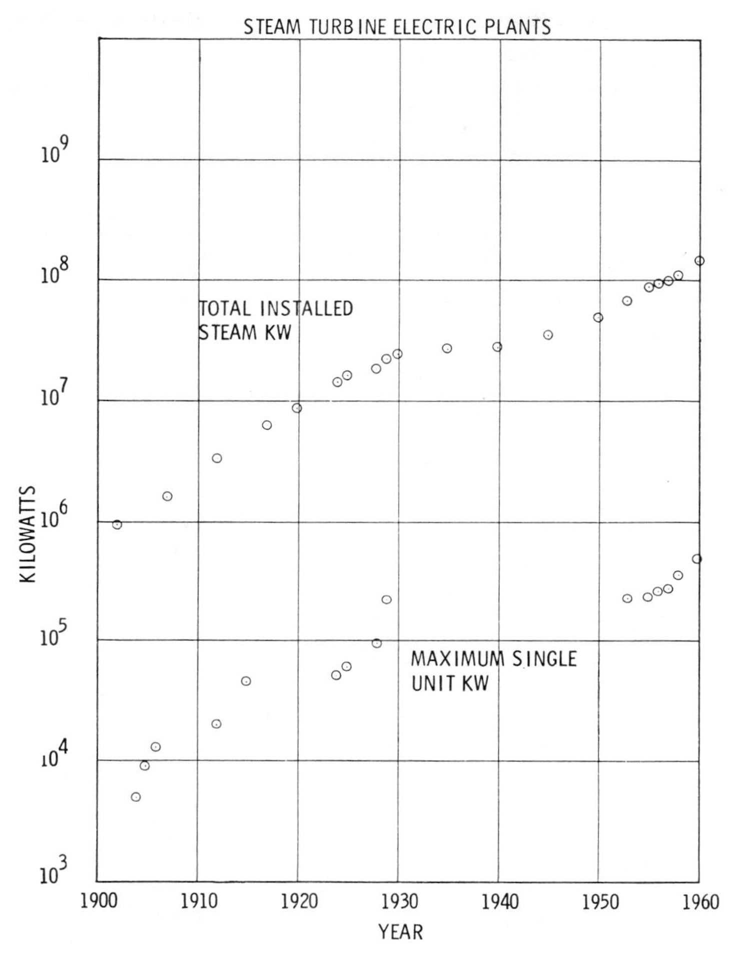

Fig. 2. Step growth of electric generators.

growth in installed capacity virtually came to a halt. After World War II, growth in installed capacity resumed at virtually the same rate as that prior to the Depression.

The growth in size of the largest single unit available, as shown in Fig. 2, roughly parallels the growth in total capacity. This parallellism is disturbed by rapid initial growth in the period prior to 1910, as well as the appearance of two "freak" units, one in 1915 and one in 1929. These were significantly larger than units introduced in the immediately previous years. Furthermore, they were not exceeded by other units for many years. In the case of the generator introduced in 1915, there was a lag of 8 years before a slightly larger generator was introduced. In the case of the generator installed in 1929, there was a lag of 24 years before a slightly larger one was introduced. If the 15-year period of stagnation in the electrical industry, from 1930 to 1945, is subtracted, this lag becomes 9 years, not much different from the lag following the 1915 generator. That is not too surprising, since both units exceeded the prior state of the art by about the same percentage. This designation as a "freak" can be justified on grounds beyond the departure from the previous general trend, at least in the case of the 1929 generator.

According to [**2**], as late as 1932 it was still the only one of its size in use, although a total of six generators of size in excess of 100,000 KW had been installed. Hence it appears that this unit was "ahead of its time." Information regarding the 1915 unit could not be obtained, hence the same argument cannot be applied.

Table 1

Steam Turbine Generators

Year	Installed Steam Turbine Capacity (1000's of KW)	Largest Steam Turbine (1000's of KW)	*Single Turbine* Total Capacity
1902	914		
1904	*	5	
1905	*	9	
1906	*	12.5	
1907	1765		
1912	3395	20	.0058
1915	*	45	
1917	6128		
1920	8920		
1924	12535	50	.0039
1925	15368	60	.0039
1928	19790	94	.0047
1929	21704	208	.0095
1930	23385		
1935	24471		
1940	27775		
1945	34112		
1950	49333		
1953	67235	217	.0032
1955	87112	217.26	.0024
1956	92591	260	.0028
1957	99542	275	.0027
1958	110633	350	.0031
1960	133282	500	.0037

Data Sources: Edison Electric Institute and Department of Commerce.

* Data not available.

The third column of Table 1 shows the size of the largest available steam turbine as a proportion of the total installed capacity. The numerical values support the conclusion drawn directly from the graph in Fig. 1, that the size of the largest available unit is a fairly constant proportion of the total installed capacity. For the eleven units shown for which it is possible to calculate this proportion, the mean value of the proportions is 0.00415454, and the standard deviation of the proportions is 0.00192873. Note that the mean value exceeds all but three of the values entering into the calculation. This represents the effect of the 1929 unit, which was a significant departure from the previous general trend of generator sizes. If this unit is omitted from the calculation, the mean value becomes 0.00362, and the standard deviation becomes 0.00097467. The ratio of standard deviation to the mean is known as the Coefficient of Variation, and is often used as a measure of the degree of variability of some parameter. The Coefficient of Variation for the ten cases not including the 1929 unit is 0.2693. This is a small value for a Coefficient of Variation, indicating that the values of the proportion are clustered about the mean.

It can be concluded that the maximum size of steam turbines for electric generators does represent a fairly constant proportion of the total installed capacity of steam electric plants in the US. This proportion has remained constant for a period of forty years.

Hydroelectric Turbines

The analysis of hydroelectric turbines parallels that of steam turbines. Table 2 provides data on the maximum size of hydroelectric turbines available at any time in US generating stations, as well as the total installed hydroelectric turbine generator capacity in US electric utility plants. This data is plotted in Fig. 3.

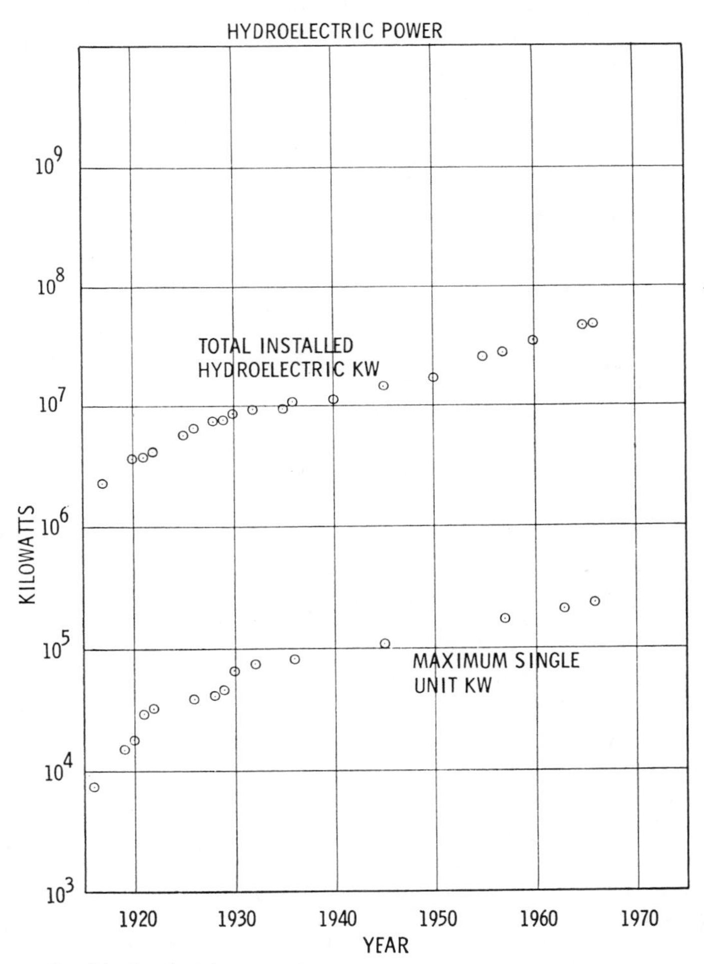

Fig. 3. Step growth of hydroelectric generators.

The plot of Fig. 3 shows a less marked effect on total capacity arising from the Depression and World War II. This is most likely because of the emphasis on building government-owned hydroelectric stations during the 1930s.

The growth in size of the largest single unit available, as shown in Fig. 3, roughly parallels the growth in total capacity. As with the steam turbines, there is a rapid growth in the early years, at a rate faster than growth of total capacity. Following this, however, the growth seems to follow that of total capacity. There apparently were no "freak" hydroelectric units, representing a radical jump above the previous general trend. At least Fig. 3 does not show any discrepancies such as the two in Fig. 2.

Table 2

Hydroelectric Turbines

Year	Installed Hydroelectric Capacity (1000's of KW)	Largest Hydroelectric Turbine (1000's of KW)	*Single Turbine* Total Capacity
1916	*	7.5	
1917	2786		
1919	*	13.5	
1920	8920	17.5	.00472
1921	3900	30	.00769
1922	4130	31.25	.00757
1925	5920		
1926	6410	40	.00624
1928	7700	40.6	.00527
1929	7810	45	.00576
1930	8590	66.7	.00776
1932	9260	77.5	.00837
1935	9400		
1936	10040	82.5	.00822
1940	11220		
1945	14910	108	.00724
1950	17680		
1955	25010		
1957	27040	167	
1960	32420		
1963	40200	204	.00508
1965	44000		
1966	45000	220	.00488

Data Sources: Harza Engineering Company and Department of Commerce.

* Data not available.

The third column of Table 2 shows the size of the largest available hydroelectric turbine as a proportion of the total installed capacity. Again, the numerical values seem to support the conclusion drawn from the graph of Fig. 2, that the size of the largest available unit is a fairly constant proportion of the total installed capacity. The proportions seem slightly higher than for the steam turbine case, however. For the 13 cases after 1920, the mean value is 0.0065369, the standard deviation is 0.0012762, and the coefficient of variation is 0.1952. Thus the variability in the proportion is even smaller than in the case of steam turbines. Another feature of interest appears in the table. The last two units listed have proportions not only smaller than the mean value, but significantly smaller than the four immediately previous units. One possible reason for this arises from an external factor which limits the size of hydroelectric turbines in US generating stations. This is the size of the rivers available at suitable hydroelectric sites. In other countries, where larger rivers are available at suitable hydroelectric sites, much larger turbines are in use. A generator with a capacity of 336,800 KW was installed in Australia in 1958, when the largest unit available in the US had a capacity of 167,000

KW. In the same year, a 450,000 KW capacity unit was installed in Sweden, and a 590,000 KW capacity unit installed at Krasnoyarsk, in the Soviet Union. Hence it is quite possible that there is not much more potential for growth in capacity of single units in US hydroelectric stations. During the period when river size was not a limiting factor, however, there was a relatively constant ratio between maximum turbine size and total installed capacity. If the last two units in Table 2 are deleted from the calculation, the mean proportion becomes 0.00682, the standard deviation becomes 0.0011841, and the coefficient of variation becomes 0.1736.

From this it can be concluded that the maximum size of hydroelectric turbines did represent a fairly constant proportion of the total installed capacity of hydroelectric plants in the US, at least for the period of time when river size was not a limiting factor. This proportion remained fairly constant for a period of about forty years. However, if river size does in fact remain a limiting factor, this proportion may not continue to remain constant but will probably decrease in coming years.

Conclusions

The analysis reported here presents additional support for the idea that certain technologies grow in step-wise fashion, with the performance of the highest-performance unit available bearing a relatively constant proportion to some measure of total need or capacity. This forecasting technique appears to be valid, and its use can be justified in those cases where it appears to be appropriate. Additional studies are needed to determine the entire range of technologies for which this technique is usable. However, at present it appears to be suitable for those technologies where an increase in scale or size brings economic benefits in terms of reduced operating costs.

Acknowledgments

We gratefully acknowledge the assistance of the Edison Electric Institute, which provided data on maximum steam turbine sizes, and of the Harza Engineering Company of Chicago, which provided data on maximum hydroelectric turbine sizes.

References

1 Simonds, W. H. C., "The Analysis of Industrial Behavior and Its Use in Forecasting," *Technological Forecasting and Social Change*, this issue.
2 Potter, J. H., The rise of the large steam turbines, *Combustion*, May, 1965.

The Diffusion of New Technology in the U.S. Petroleum Refining Industry

M. BUNDGAARD-NIELSEN and PETER FIEHN

ABSTRACT

The diffusion of new technology is described by a simple mathematical model. The parameters of the model are shown to be related to the technological background and to the prevailing market conditions of the population of industrial firms into which diffusion occurs. The results of a comprehensive case study of the diffusion of new technology into the U. S. petroleum refining industry based upon the model presented in the article are reported.

Introduction

A thorough understanding of the basic mechanisms involved in the diffusion of new technology is in essence a prerequisite to any reliable assessment of the possible impact of technical developments upon society in general. Empirical observations of the number of firms who adopt a new technology versus time tend to display the characteristics of an *S*-shaped curve. This observation implies that we may describe the diffusion mathematically as a function of time. However, time is clearly not the only variable influencing the decision to adopt a new technology. It is indeed quite conceivable that the market structure under which the firm operates as well as the technological "status" of the firm, perhaps in combination with other factors, exerts a great deal of influence upon what we actually observe, i.e., the timepath of the diffusion of a particular technology.

The purpose of this article is twofold:

(1) To present a simple mathematical model of the diffusion process in which the model parameters are related to the technological background of the firm and to the market conditions under which the firm operates, and
(2) to demonstrate how the model presented may provide us with new insight into the complex mechanisms of the diffusion of new processes into the American petroleum refining industry.

The Diffusion Process

Based upon the assumption that firms seek long-run profit maximization Alchian (1950) reached the conclusion that, since firms do not in advance know how to pursue this goal, they will imitate the behavior of other similarly situated firms who appear to be successful, i.e., information about successful implementation of a new technology will increase the probability of successful adoption, as conceived by the potential adopters.

This article appeared in *Technological Forecasting and Social Change*, Vol. 6, No. 1, 1974.

The conceived probability, P_t, is proportional to the number n_t of successful adoptions at a given point in time, t. Thus we get the relation

$$P_t = C'n_t \tag{1}$$

where C' is a constant.

Let the rate of diffusion of the new technology be defined as the increase in the number of adopters over a relatively small increment in time. According to Mansfield (1961) the rate of diffusion is proportional to P_t and to the number of firms in the group who have not yet adopted the new technology.

This in turn yields the relation

$$\frac{dn_t}{dt} = C''(N - n_t)P_t, \tag{2}$$

where the differential is the rate of diffusion, C'' is a constant, and N is the total number of similarly situated firms. Inserting (1) into (2) yields

$$\frac{dn_t}{dt} = C(N - n_t)\, n_t, \tag{3}$$

where C is the rate constant of the diffusion.

Integrating (3), assuming that n approaches O as t approaches $-\infty$ yields

$$n_t = N/(1 + \exp(-2C(t - t_0))), \tag{4}$$

where t_0 is equal to the time when half of the final number of adoptions have occurred. Equation (4) is identical to the well-known Pearl Reed Curve and thus displays the empirically observed S-shaped relationship between n_t and t.

With respect to the definition of a group of similarly situated firms, we assume that a group of firms employing almost identical production technologies and with identical production goals, as expressed by the product mix-marketed by the firm at any given point in time, constitute such a group.

The rate of diffusion is clearly influenced by the technological background of the firm. If, as an example, the firm is currently applying production methods technically similar to those of the diffusing technology, then the firm can assess the technical advantages of the new technology with greater certainty than if the production technology used differs from the diffusing technology. This, in turn, suggests the hypothesis that, provided the new technology satisfies the investment criteria used by a group of similarly situated firms, we can expect a larger rate of diffusion and consequently also a larger rate constant C in (3), if the group of firms has a strong technological background relative to the diffusing technology than if the group consists of firms with a weak technological background.

If the product mix—with respect to quality and quantity—tends to become obsolete due to market developments, then one may expect the firm to adopt a new and adequate technology at an earlier point in time than if such strategic pressures do not prevail.

Consequently we may summarize and propose the following hypothesis Bundgaard-Nielsen (1973):

The rate constant C in Eq. (3) is positively influenced by the technological background of a group of similarly situated firms as defined above, while the parameter t_0 is positively influenced by strategic pressures.

A Case Study

Over the years various processes have been used in the petroleum industry to upgrade the octane number of petroleum fractions to produce gasoline components satisfying the requirements of modern high compression engines. In the mid-forties the following processes were in use

(1) *Fractionation of crude oil* to produce high octane petroleum fractions.
(2) *The thermal cracking process,* where larger hydrocarbon molecules are cracked to smaller hydrocarbons with higher octane numbers.
(3) *The visbreaking process,* where viscous crude oil fractions are decomposed to lighter hydrocarbons with a higher octane number.
(4) *The catalytic cracking process,* where larger hydrocarbon molecules are cracked under the influence of a catalyst to yield hydrocarbons with high octane ratings.
(5) *The thermal reforming process,* where straight hydrocarbons react to yield benzene and branched hydrocarbons with high octane numbers.

In the fifties motor gasoline octane levels in U. S. A. increased significantly, and by 1960 octane ratings had reached the present day level of approximately 94 for regular gasoline and 100 for premium gasoline. The corresponding numbers in 1955 were 87 and 95, respectively. The petroleum refiners producing gasoline were, in the late forties, unable to supply the high quality gasoline in the desired quantities unless they applied a catalytic cracking process. Then, in 1949, the catalytic reforming process was introduced. The catalytic reforming process is very selective with respect to gasoline quality and quantity and was able to supply the high quality gasoline in demand. Compared with the catalytic cracking process, catalytic reforming has the advantage that it can be operated profitably on a smaller scale.

In order to test the hypothesis developed above, the diffusion of catalytic reforming processes into the petroleum refining industry in the U. S. A. was tested based upon the simple model of Eq. (4), using a non-linear regression technique to estimate the parameters C and t_0. The refineries were divided into three groups, A, B, and C, based upon technological background and the quality of the product mix marketed in 1950.

(A) Refineries who had adopted catalytic cracking before 1950 were considered to have a *strong technological background,* and these refineries were at the same time subject to *weak strategic pressure*, since they—at least—in the early fifties were able to supply the high quality gasoline in demand.
(B) Refineries who had not adopted catalytic cracking in 1950, but did so before adopting catalytic reforming, had a *rather strong technological background,* since they, by the time they adopted catalytic reforming, possessed some experience with operating a catalytic process. These refineries did, however, enjoy the advantage over the refineries in the former group that, since they adopted catalytic cracking after 1950, they were presumably operating more advanced cracking technologies which enabled them to meet the octane requirements more readily than with the processes adopted prior to 1950. Consequently, this group of potential adopters was subject to *very weak strategic pressures.*
(C) Refineries who had not adopted catalytic cracking in 1950 and who did not adopt catalytic cracking before adopting catalytic reforming. This group of refineries was subject to *strong strategic pressures,* but had a *weak technological background* due to the lack of catalytic experience prior to the adoption of a catalytic technique.

The necessary data on technological background and the time of adoption of either catalytic reforming or catalytic cracking were obtained from the Oil and Gas Journal (1948–1972).

The results are shown in Tables 1 and 2 and graphically in Fig. 1.

TABLE 1

Effect of Technological Background

Group	Number of potential adopters in 1950	Number of final adopters of cat. ref.	Technological background at time of adoption	Rate constant of diffusion C
A	81	75	strong	0.493
B	126	37	rather strong	0.425
C	126	41	weak	0.397

TABLE 2

Effect of Strategic Pressure

Group	Number of potential adopters	Number of final adopters of cat. ref.	Strategic pressure at time of adoption	Year when half of the final adoptions had occured
C	126	41	strong	1953.8
A	81	75	weak	1955.1
B	126	37	very weak	1956.5

The estimated parameters C and t_0 in the table confirm quite clearly with the hypothesis, i.e., technological background and strategic pressures influence the rate constant of diffusion as well as the t_0 parameter as predicted in the hypothesis.

Apart from technological background and strategic pressures, it is conceivable that the decision to adopt a new technology is influenced by the very size of the refinery. Schumpeter (1942) advocated that technological progress somehow presupposes protection and concentration in industry. The US refining industry consists of a number of small independent refineries, while the larger refineries are operated by the large oil companies. Accordingly, we should expect a more technically progressive trend to prevail for larger refineries with capacities of approximately 20.000 bbl/sd. than for refineries with capacities of less than approximately 10.000 bbl/sd. The diffusion of catalytic reforming into the following four groups D, E, F, and G was investigated:

(D) Refineries with capacities of 10.000 bbl/sd. or less in 1950 who had not adopted catalytic cracking in 1950 and did not do so before adopting catalytic reforming.

(E) Refineries with capacities of 10.000 bbl/sd. or less in 1950 who had adopted catalytic cracking in 1950.

(F) Refineries with capacities of 20.000 bbl/sd. or more in 1950 who had not adopted catalytic cracking in 1950 and did not do so before adopting catalytic reforming.

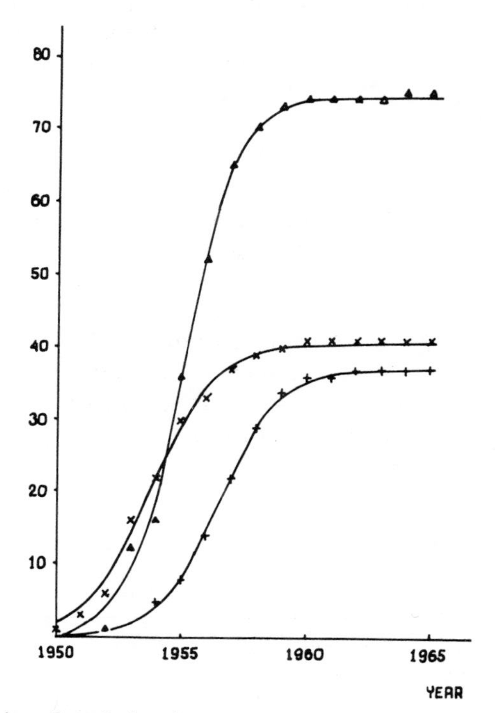

FIG. 1. Diffusion of catalytic reforming: Δ, Group A; +, Group B; ×, Group C.

(G) Refineries with capacities of 20.000 bbl/sd. or more in 1950 who had adopted catalytic cracking in 1950.

The parameters C and t_0 of Eq. (4) were estimated statistically for these four groups and the results are shown in Table 3.

TABLE 3

Effect of Size

Group	Size	Technological background	Strategic pressure	Rate constant of diffusion C	Year when half of the final adoptions had occured t_o
D	less than 10.000 bbl/sd.	weak	strong	0.37	1953.8
E	"	strong	rather weak	0.76	1956.6
F	more than 20.000 bbl/sd.	weak	strong	0.40	1953.8
G	"	strong	rather weak	0.52	1955.0

The results confirm the hypothesis of this article. Whether they also offer support to Professor Schumpeter's hypothesis is perhaps doubtful. It is probably more relevant to compare the results shown in Table 3 with some of the more recent work by Mansfield (1969). Mansfield argues that the rate constant C is linearly related to the relative profitability of the innovation and—with a negative coefficient—to the relative investment of implementing the innovation. Table 3 shows that the larger refineries, which would be relatively more capable of making a fixed-size investment, had higher rate constants than did the small refineries with a weak technology and which were subject to strong strategic pressure. And even the smaller firms with a strong technology and with a higher rate constant adopted the innovation at a later point in time than did the larger refineries. This tends to confirm Mansfield's theory concerning the effect of relative implementation investment.

We may obtain a somewhat different measure of the effect of size by considering the ratio n_{final}/N, where n_{final} is the final number of adoptions in a group of similarly situated firms. The results of this investigation are shown in Table 4 and again the hypothesis advocated in this article is verified, i.e., that technological background and strategic pressure influences the adoption decision as predicted.

TABLE 4

Effect of Size upon Final Adoptions

Size bbl/sd.	Technological background	Strategic pressure	n_{final}/N
0–10000	strong	rather weak	0.70
10000–30000	– " –	– " –	0.93
30000–50000	– " –	– " –	1.00
50000–	– " –	– " –	0.97
0–1000	weak	strong	0.36
10000–30000	– " –	– " –	0.26
30000–	– " –	– " –	0.33

Concluding Remarks

The hypothesis stated in this article is confirmed by the results reported. Clearly, the proper definition of the term "similarly situated firms" plays a crucial role in obtaining meaningful results, but, carefully applied, the approach outlined does reveal important information on inherent system characteristics as being the controlling factors in the diffusion of new technology.

With respect to the specific case study dealt with in the article, the mechanisms of the diffusion of catalytic reforming processes may receive renewed interest in the future, when restrictions upon the lead content in gasoline presumably will lead to increased interest in adoption of new gasoline quality improving techniques. The diffusion patterns uncovered in this paper may serve as useful guidelines in this context.

The authors gratefully acknowledge the support of the IBM Research Foundation in Denmark, which made this research project possible.

References

A. A. Alchian, Uncertainty, evolution and economic theory, *J. Polit. Eco.* **58**, 211–221 (1950).
E. Mansfield, Technical change and the rate of imitation, *Econometrica* **29**, 741–766 (1961).
M. Bundgaard-Nielsen and P. Fiehn, The diffusion of new technology in the chemical industry. Paper presented at *The First International Congress on Technology Assessment,* Haag, Holland, May 27 th–June 2 nd 1973.
Oil and Gas Journal, Annual Refining Section (1948–1972).
J. A. Schumpeter, *Capitalism, Socialism and Democracy, 3rd Edition,* Harper, New York (1942).
E. Mansfield, *The Economics of Technological Change,* Longmans, 120 ff (1969).

Diffusion of Environment-Saving Technological Change: A Petroleum Refining Case Study*

HYDER LAKHANI

ABSTRACT

This paper attempts a theoretical model (Section I) of a dynamic relationship, brought about by environment-saving technological changes over time, between output and water pollution. The improvement in environmental quality is studied in terms of adoption of relatively environment-saving processes in petroleum refining industry. This is done (Section II) by empirically fitting the Gompertz function or the growth rates of diffusion of the processes by the use of multiple regression equations. Social desirability of substituting the relative environment-saving process is examined in terms of social benefit-cost analysis (Section III). Thereafter, (Section IV) an attempt is made to find social policy variables which could be used in order to accelerate diffusion of environment-saving processes. The analysis concludes that both "carrot" (incentives, say, in the form of accelerated depreciation allowances for purchase of environment-saving *process* equipment) as well as "stick" (in the form of stricter enforcement of water pollution control laws) are necessary for inducing adoption of the desired technological changes over time.

I. A Theory of Diffusion of an Environment-Saving Process

THE INTRA-INDUSTRY DIFFUSION PROCESS

After a relatively new environment-saving process of production has been innovated, its acceptance or use by the producers begins to grow towards a ceiling or an ultimate equilibrium level. The ultimate equilibrium is not a static long run equilibrium, but one that shifts over time. It is ultimate only in the sense that it is a ceiling toward which the level of use of the new process is growing at any particular time. The time required to reach ultimate equilibrium, or the ceiling rate of diffusion, from the date of innovation is called the diffusion period.

This article appeared in *Technological Forecasting and Social Change*, Vol. 7, No. 1, 1975.

* This paper is taken from the author's Ph.D., Economics, dissertation completed at the University of Maryland in 1972. I gratefully acknowledge many helpful suggestions from Professors John H. Cumberland, Christopher K. Clague, Roger R. Betancourt and John E. Tilton. Many thanks are also due to two anonymous referees who recommended certain invaluable revisions in the first draft. The errors are, of course, entirely my responsibility. I must add that the paper under reference is written strictly in my personal capacity and neither this Department nor the State of Maryland are in any way responsible for the views expressed in that paper.

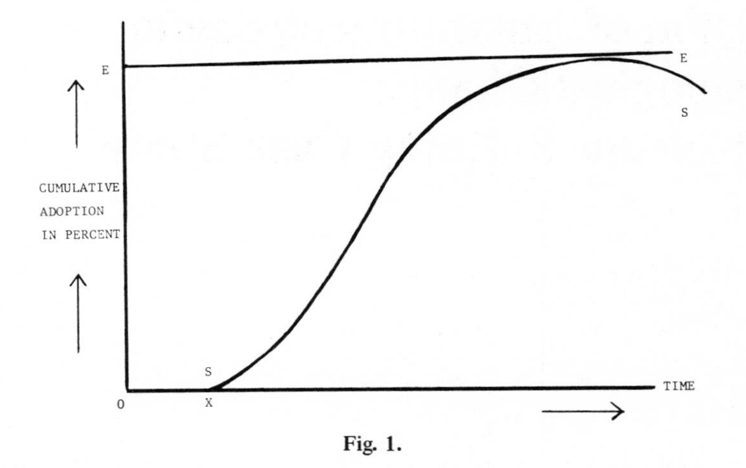

Fig. 1.

THE GROWTH CURVE

Figure 1 depicts a simplified representation of the process of diffusion. Time is indicated on the horizontal axis and the growth rate of diffusion on the vertical axis. The ceiling, or equilibrium level, is shown by the growth curve SS which depicts the diffusion rate from the time of innovation, i.e., X or zero rate of diffusion, towards the equilibrium level EE. The reasons for assuming the S-shaped pattern for the growth curve will be examined below.

REASONS FOR THE S-SHAPED PATTERN OF THE GROWTH CURVE

The S-shaped pattern of the growth curve (i.e., slow in the initial stages, acceleration in the intermediate stage and deceleration in the mature stage) could be explained in terms of three phenomena: (A) the adjustment process, (B) the ultimate equilibrium level of the new process, and (C) the changes or shifts in the equilibrium level that may take place during the diffusion period.

A. *The Adjustment Process*

There are several reasons why the acceptance of a successful innovation rarely, if ever, jumps immediately to its ceiling:

1. *Information gap.* The producer who is likely to use the new process must be informed about the innovation. Even after he is informed, uncertainty about its benefits and costs contribute to the initial slow growth rate of acceptance. With time, the producer accumulates more information, observes competitors using it, and follows the "band-wagon", contributing thereby to the accelerating intermediate phase of the growth curve.

2. *The age distribution of existing capital stock.* Many new processes require new capital equipment. But the age distribution of capital equipment of the firms varies widely. The firms whose existing capital equipment is old relative to the average age of capital in the industry would be prompt to accept the new process while other firms that have replaced their capital equipment only recently will be slow in their innovative response. Salter [1] hypothesizes the co-existence of a spectrum of techniques in terms of just such an age distribution of capital stock. He assumed that the industry price is determined by the operating plus fixed costs of the newest plants and that the older plants that continue to operate with obsolete technique do so because their capital costs

are considered sunk and their higher operating costs equal the operating plus capital costs of the firms using newer techniques.

Empirical verification of the relationship between rate of diffusion and age distribution of capital stock is found in Mansfield [2]. He analyzes the substitution of diesel locomotive for the steam locomotive by major firms in the railroad industry in the U.S. during 1925–59 by fitting a logistic function of time. The intrafirm rate of diffusion of the number of diesel locomotives is initially made a linear function of profit, *age* of the steam engine, firm size and liquidity of firms. The multiple regression equation brings out, *inter alia,* that higher the age of the then existing steam locomotive, the faster was the adoption of the diesel locomotive. The coefficients for the other independent variables were also economically and statistically significant, except that for profits. As regards the elasticitiy of adoption with respect to changes in age, size, and liquidity, the highest was that for age and the lowest was for firm size. His second version of the model related the same set of exogenous variables to *size* of freight carried by diesel locomotives instead of the *number* of locos. The conclusion, once again, confirms the relative importance of age distribution of existing capital stock and the nonsignificance of profitability.

3. *Postponement of acceptance in anticipation of changes in quality and prices.* The slow initial growth rate of diffusion could also be attributed to some potential users who anticipate an improvement in the quality of the newer process or a decline in its price. Usually such expectations are well founded. New processes are often plagued with problems which require time and production experience to overcome.

4. *Supply bottlenecks.* The diffusion period might also be prolonged by the low elasticity of supply of the new required process equipment. While the demand for the new process equipment could rise rapidly, production bottlenecks often restrict supply because the suppliers must raise capital, train labor and procure additional raw materials. The process equipment supplying firms might not wish to expand rapidly because there are generally additional costs associated with rapid expansion of capacity. This deferment is particularly likely to be true for firms which, due to a strong patent position or monopolistic market structure, have few competitors, and consequently can defer delivery of an order without much fear of losing it altogether.

According to Nelson, Peck and Kalachek, however, producers generally expand capacity fast enough to keep pace with growth in demand [3]. This suggests that growth in demand–and thus learning on the part of potential users, the age distribution of capital goods, and expectations with respect to price and quality changes–rather than supply bottlenecks constrain adjustment toward the ultimate equilibrium and determine the diffusion period.

B. *The Ultimate Equilibrium*

The ultimate equilibrium or the ceiling rate of utilization of a new producer good depends on (1) the extent to which the new process proves more effective than the process it can replace in reducing production and environmental costs and in improving the quality of output, and (2) the elasticity of output with respect to price, quality and environmental changes.

C. *The Shifts in Ultimate Equilibrium*

The upward or downward shift of the ultimate equilibrium can be explained in terms of a number of factors:

1. *Shift in the derived demand.* The demand for a producer good is a derived demand, i.e., it depends upon the demand for the outputs produced by the process. For example, the demand for a process producing low lead gasoline would be governed by the demand for the gasoline itself. A shift in demand for the gasoline would tend, therefore, to shift the ultimate equilibrium level for the process.

2. *The introduction of a substitute.* The introduction of a superior substitute process in the market would tend to depress the demand for the original process and thus lower its ultimate equilibrium level.

3. *Improvement in the quality of the process via producer learning.* Follow-on innovations and producer learning which improve the quality and expand the range of practical applications for the new process can increase the number of potential users and the amount each uses and thus shift up the ultimate equilibrium. Such new customers differ from those already described who defer purchases in *anticipation* of quality improvements or price cuts. The anticipators eventually install the new process even if their expectations fail to materialize. Consequently, their delay does not increase ultimate demands, but merely contributes to the shape of the growth curve.

4. *The band-wagon effect.* Just as the "demonstration effect" described by Duesenberry increases the long-run equilibrium of consumer demand [4], consequent upon the interdependence of consumer preferences, similarly the "band-wagon effect" can induce producers to install a new process which would not have been installed in the absence of competitive conditions. This would impart an upward shift to the ultimate equilibrium in competitive markets or a downward shift in non-competitive markets.

5. *Lower price due to economies of scale in the production of the process equipment.* Independent of the production costs referred to above, price may decline as the number of firms supplying the new process equipment increases and the market structure becomes more competitive. This often forces the initial manufacturers of the new process to reduce their profit margins by reducing prices. This, in turn, generates new demand for the process and shifts up the ultimate equilibrium.

THE ECONOMICS OF GROWTH CURVE

The S-shaped growth curve reflects not only the growth rate of diffusion of the new process but also the growth rate of supply and demand. It represents the growth rate of supply of the new equipment if we assume a closed economy (with absence of imports and exports of the equipment) and a stationary state (absence of changes in the inventories). In this case, it also represents demand growth since all that is produced is being consumed domestically. Since we noted above that during the early period of diffusion there might be supply bottlenecks, the growth curve might be construed as the growth of demand or supply, whichever is smaller at a particular time in question. This is, of course, an *ex post* construct, with no theoretical significance.

The economic reasons for the S-shaped growth pattern of this curve were noted in the discussion of the adjustment process toward an ultimate equilibrium. In short, it might be recalled that at first growth is slow because, on the demand side, potential users of the process lack information, are uncertain about its costs and benefits, or delay its use in anticipation of an improvement in its quality or a reduction in its price; and, on the supply side, bottlenecks may limit expansion of capacity. With time, the growth rate accelerates, demand picks up as more information is available to potential users and uncertainty diminishes as the process filters through the economy. Simultaneously, supply bottlenecks are overcome as more time is available to expand capacity and technological advances improve the process and reduce the price. These provide a loop for an acceleration of demand and the generation of the phase of rapid growth. This stage is followed by the phase of deceleration of the growth rate towards the ultimate equilibrium as that equilibrium is approached. This retardation is largely due to the closing of the gap between the equilibrium and the actual level of utilization resulting in a virtual saturation of the market. This factor is reinforced by the special characteristics of late adopters: an unusual degree of inertia, a protected market position, a product for which the innovation is less useful, or a special situation in the factor market. Above all, in cases where a new process replaces an existing process, for example, hydrocracking replacing the catalytic cracking process in the petroleum refining industry, the ultimate equilibrium level of the existing process is reduced to zero, its diffusion period is terminated and there is a negative growth rate of diffusion of this process. An ultimate equilibrium level is then established for the new process and a new S-shaped growth curve may be used to analyze the growth rate of diffusion of this new process.

The S-shaped growth pattern has long been familiar to economists. Nobel Laureate Simon Kuznets described it in 1929, and others have since shown that it accurately depicts growth in the use of machine tool, television, hybrid corn, coal, iron and steel, brewing, railroad, computer innovations and semiconductors [5].

THE ALGEBRA OF GROWTH CURVE–THE GOMPERTZ CURVE

An algebraic function of a few parameters could be used to statistically estimate the above-referred economic growth curve with its innumerable parameters. There are three such algebraic functions–the logistic, lognormal and Gompertz–used by economists to analyze the growth of new processes. All of these functions specify an ultimate equilibrium and an S-shaped curve which asympotically approaches one over time from zero and presents the trend of adjustment toward the equilibrium. All of these curves could be simplified and approximately represented as Fig. 1. The parameters of the functions which best fit the actual growth curve can be estimated with econometric techniques using the actual growth figures. In particular, since the functions treat separately the ultimate equilibrium and the trend toward that equilibrium, they could accommodate changes in equilibrium occurring during the diffusion period as long as these changes are known or can be estimated. This is an important attribute, for it permits the functions to describe even those occasional growth curves which deviate from the S-shaped pattern.

While it is possible to estimate statistically any of the three types of growth curves to analyze the growth rate of diffusion, we shall attempt to fit only the Gompertz curve. This choice is based on its simplicity and the fact that this curve permits a simultaneous estimation of both the growth rate of diffusion and the adjustment of the actual level of acceptance of the process towards the ultimate equilibrium level by linear estimation technique [6]. This is illustrated in the following section on the derivation of the growth rate of diffusion.

DERIVATION OF GROWTH RATE OF DIFFUSION IN THE GOMPERTZ FUNCTION

The cumulative distribution function which reflects the level of growth is:

$$Y_t = Y^* \, a^{\beta^t} \tag{1}$$

where Y_t is the cumulative level of diffusion of the new process at time t, and Y^* is the ultimate equilibrium level of the new process. This could be changing with time, as would be indicated below. a, β are parameters that are positive but less than one. The rate of growth given by the derivative of Eq. (1) with respect to time can be derived as follows:

$$\frac{dY_t}{dt} = Y^* a\beta^t(\beta^t \log a)\log \beta \tag{2}$$

But $(\beta^t \log a) = -(\log Y^* - \log Y_t)$ because since $Y_t = Y^* a^{\beta^t}$, therefore $\log Y_t = \log Y^* + \beta^t \log a$ hence

$$\log Y_t - \log Y^* = \beta^t \log a$$

or

$$\beta^t \log a = -(\log Y^* - \log Y_t). \tag{3}$$

Substituting Eq. (3) into (2);

$$\frac{d^{Y_t}}{dt} = Y^* a^{\beta t} - (\log Y^* - \log Y_t)\log \beta. \tag{4}$$

Substituting $Y^* a^{\beta t} = Y_t$ from Eq. (1) into (4):

$$\frac{d^{Y_t}}{dt} = Y_t (\log \beta) \quad [-(\log Y^* - \log Y_t)]. \tag{5}$$

Dividing Eq. (5) by Y_t and transposing the negative sign to $\log \beta$:

$$\frac{d^{Y_t}}{dt} \cdot \frac{1}{Y_t} = -\log \beta(\log Y^* - \log Y_t) \tag{6}$$

Substituting parameter γ for $-(\log \beta)$:

$$\frac{d^{Y_t}}{dt} \cdot \frac{1}{y_t} = \gamma(\log Y^* - \log Y_t). \tag{7}$$

Equation (7) reveals that the growth rate of diffusion of a new process depends directly and proportionately on the parameter γ—the greater the parameter γ, the greater is the growth rate of diffusion. Moreover, since $\gamma = -\log \beta$, the growth rate is also a negative function of β—the smaller the parameter β, the greater is the growth rate. Above all, as pointed out by Chow, it is possible to simultaneously estimate the parameter γ and the parameters of the ultimate equilibrium level Y^* if we assume that the equilibrium level Y^* is a function of certain variables with constant elasticities [7]. The ease of econometric estimation of Eq. (7) is observed in the fact that it can be rewritten as:

$$\log Y_t - \log Y_{t-1} = \gamma(\log Y^* - \log Y_t), \tag{8}$$

so that a regression of the left hand side of Eq. (8) on the right hand side (i.e., the difference between the ultimate equilibrium level and the actual level of diffusion at time t) would provide the growth rate of diffusion of the new process.

Equation (8) can also accommodate the shifts or changes in the ultimate equilibrium level, Y^*, over time, during the diffusion period. For instance, we might hypothesize that the ultimate equilibrium level is not fixed, but represents, say, 50% of the total refining capacity every year. As the total refining capacity increases consequent upon increases in the derived demand for gasoline (because of an increase in income as well as population), there would be an increase in the ceiling, Y^*_t, if it is observed to represent 50% of total refining capacity of the respective years. The estimation of Eq. (7) would then be of the form:

$$(\log Y_t - \log Y_{t-1}) = \gamma(Y^*_t - \log Y_t). \tag{8}$$

When the equilibrium Y^*_t changes, as in this instance from year to year, the resulting growth curve remains S-shaped, but the ceiling line EE, shown in Fig. 2, would be upward sloping, with time, instead of horizontal. The estimation of Eq. (8) could be for a cross section of countries, over time, or for a cross section of states within a country, over time. The empirical estimation that would be undertaken in this chapter would be for a cross section of states in the U.S. because no cross country data, over time, are available.

In short, Eq. (8) suggests that three economic phenomena could be estimated simultaneously. First, the regression coefficients γ's would provide the growth rates of diffusion. Second, the shift in the ceiling is accommodated by increasing Y^*_t from one year to another in proportion to the increase in capacity which might be construed as the demand or supply of gasoline in the state, whichever is smaller, at the specific time in question. Third, the "band-wagon" effect is also accommodated because the difference between Y^*_t and Y_t suggests the gap between the ultimate and actual level of diffusion. The smaller this gap, the greater the adoption of the new technique by the rival firms, and hence the greater the pressure on the non-adopting firms to follow the "band-wagon" and adopt the new technique.

The theoretical foundations of the growth rates (γ's) of diffusion could be further explored by relating them to economic and environmental variables. For instance, if a new process is known to be both economically profitable as well as environment-saving,

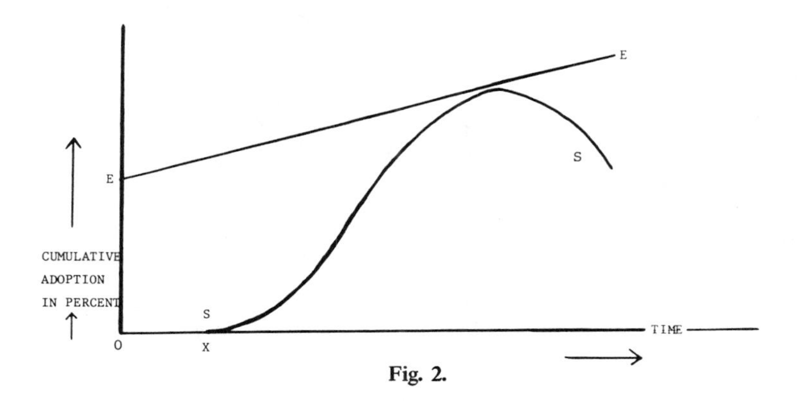

Fig. 2.

relative to the present process, one could formulate a hypothesis to explain the growth rate of diffusion in terms of certain economic and environmental parameters so as to provide certain policy conclusions for accelerating the growth rates of diffusion.

Apart from establishing the above-referred relationships between the growth rate of diffusion and the equilibrium and the actual level of diffusion on the one hand, and the growth rate of diffusion and economic and environmental parameters on the other, the Gompertz function also provides a relationship between the actual and the equilibrium level of diffusion at any point in time. This can be shown by using the parameter a in Eq. (1) because when t (time) equals zero, $Y_0 = Y^* a\beta^t$ is reduced to $Y_0 = Y^* a$, so that $a = Y_0/Y^*$, or the proportion of the equilibrium level actually achieved at time t. In terms of Fig. 1, the parameter a can be used to determine the date an innovation was available on a commercial scale in the sense that actual use reached ten or twenty percent of the equilibrium level. It must be added that a is a *constant* only at time zero which transforms to $\beta^0 = 1$. The proportion $a = Y_0/Y^*$ can be used to determine the date an innovation was available on a commercial scale and be depicted in the simplified Fig. 1. When time t is other than zero, $\beta^t \neq 1$ so that a is a *parameter.*

The last, but not the least, significant feature of the Gompertz curve which makes this curve superior to, and more realistic than the other growth curves, is that this curve is not symmetric around its inflexion point. Unlike this, the logistic curve, being symmetric around its inflexion point, suggests that the growth rate at any equidistant point above its maximum growth rate is the same as the growth rate at any equidistant point below the maximum growth rate. In case of Gompertz, it is shown below that the maximum growth rate occurs when the actual level of acceptance (y_t) equals 37% of the equilibrium level (Y^*):

$$\frac{d^{Y_t}}{dt} = Y^t (\log \beta) \;\; [-(\log Y^* - \log Y_t)] \tag{5}$$

$$\frac{d^2 Y_t}{dt^2} = -\log \beta \log Y^* + \log \beta \log Y_t + \log \beta$$

$$= \log \beta \, (\log Y_t - \log Y^* + 1);$$

this equals zero, when

$$\log Y_t / Y^* = -1;$$

$$Y_t = \frac{1}{e} Y^* = 0.37\, Y^*.$$

This is a maximum because the third derivative is negative:

$$\frac{d^3 Y_t}{dt^3} = \log \beta / Y_t < 0, \text{ since } Y_t > 0 \text{ and } \beta < 1 \text{ so } \log \beta < 0.$$

Needless to say, as the maximum growth rate occurs when the actual level of acceptance equals 37 percent of the equilibrium level, the growth rate at any point above it need not be the same as the growth rate of any equidistant point below it. This property imparts a realistic imprint to the economics of diffusion which is a multi-functional phenomena, as was referred to in the above analysis.

II. Empirical Verification of Diffusion of Environment–Saving Techniques in Petroleum Refineries

Three major processes of cracking or refining crude oil–thermal cracking, catalytic cracking and hydro cracking–have been used at various times since 1909. The distribution of crude oil inputs into these alternative processes is indicated in Appendix Table 1. Analysis of this Table suggests that catalytic cracking replaced thermal cracking in the forties and fifties and hydro cracking is in the process of replacing catalytic cracking in the sixties and seventies.

REDUCTION IN RESIDUALS FROM THE SWITCHOVER FROM THERMAL TO CATALYTIC CRACKING FROM CATALYTIC CRACKING TO HYDRO CRACKING

The changeover from thermal to catalytic cracking and from catalytic cracking to hydro cracking is associated not only with greater product yield per unit of input and better quality of the products but also with smaller emissions of residuals consequent upon the greater desulfurization of the crude oil. This observation is supported by three flow diagrams in *The Cost of Clean Water* which lists thermal cracking, catalytic cracking and hydrocracking under older, typical, and newer technology, respectively. Data on the water residual coefficients associated with each one of these technologies, after the treatment of the residuals by the most elementary primary treatment technology called American Petroleum Institute's "Separator," are also available in this publication and are reproduced below as Table 1.

It might be observed from Table 1 that, except for the average effluent level of sulfide for the hydro cracking process, there is a considerable reduction in the discharge of water residuals from a switchover from thermal to catalytic and from catalytic to hydro

TABLE 1

Total Refinery Effluent, After API Separator, per Barrel of Crude Oil Throughout, for Alternative Cracking Techniques[a]

	Waste Water		BOD		Phenol		Sulfide	
	gal/bbl		lbs/bbl		lbs/bbl		lbs/bbl	
Type of Technology	Flow avg.	range	avg.	range	avg.	range	avg.	range
Thermal	250	170–374	0.40	0.31–0.45	0.03	0.028–0.033	0.01	0.008–0.013
Catalytic	100	80–155	0.10	0.08–0.16	0.01	0.009–0.013	0.003	0.0028–0.008
Hydro	50	20–60	0.05	0.02–0.06	0.005	0.001–0.006	0.003	0.0015–0.003

[a]Source: United States, Department of the Interior, Federal Water Pollution Control Administration, *The Cost of Clean Water, Volume III, Industrial Waste Profile Number 5, Petroleum Refining,* p. 2, Table 6.

cracking processes. Consequently, the adoption of these relatively environment-saving techniques should help clean up the environment and thus reduce the social costs. But the acceptance of such techniques would tend to depend upon the reduction in private costs incurred by the entrepreneurs. If the newer techniques turn out to be more expensive, they would not be adopted by the private sector, which completely controls this industry, unless appropriate taxes were applied.

Table 2 presents engineering data to suggest that the techniques that are relatively environment-saving are also relatively less costly. From this table, it might be observed that the average variable cost is only slightly higher for catalytic cracking in the minimum range relative to that in the same range for thermal cracking. In the maximum cost range, however, the average variable cost is lower, assuming it lies at the mid-range with 56.0 cents per barrel for catalytic cracking which compares with 58.8 cents/barrel for thermal cracking. As regards the least environment-intensive hydrocracking process, it is noticed that its average variable cost of processing is not only the lowest in both ranges, but it is also as low as 50% of the cost in the corresponding ranges for catalytic cracking. This suggests the increased profitability of the hydrocracking process, assuming that both processes have equal fixed costs.

A limitation of this comparison of costs is that it ignores the fixed cost. Unfortunately, we do not have any data for fixed costs, which are considerably important in the long run analysis. Assuming, however, that fixed costs of the hydrocracking process are not so high as to swamp the fixed cost of catalytic cracking, the adoption of the hydrocracking process would be more profitable in the long run. Support for the greater profitability of the hydrocracking process as well as reduction in water and air residuals from the adoption of newer processes like hydrocracking is explicitly brought out by Dr. C. S. Russell in his linear programming simulation model of residuals management. When he simulates the substitution of the newer processes for the older ones, he finds the following effects in his basic model:

(i) An increase in profit per barrel of crude charged of about 2.5%.
(ii) A shift of the cracking function from standard catalytic crackers to hydrocrackers. This change is, in turn, responsible for most of the others observed.
(iii) An increase in production of premium gasoline of about 7 percent, because of the higher volume yield of hydrocracking products.

TABLE 2

Estimates of Average Variable Cost in Thermal, Catalytic and Hydro Cracking

Cracking Technology	(cents/barrel)	
	Minimum	Maximum
Thermal	25.0	58.8
Catalytic	28.0–32.0	51.0–61.0
Hydro	9.6–18.5	16.5–29.2

Source: W. L. Nelson, *Guide to Refining Operating Costs (Process Costimating)*, Tulsa, Oklahoma, 1971, p. 90.

(iv) Average TEL content of the gasoline falls from 1.60 to 1.08 cc/gal., largely because of increased use of the catalytic reformer.

(v) Sulfur recovery, with the same crude input, goes up by about 5 percent because of desulfurization reaction in the hydrocracker.

(vi) Heat purchase increases by almost 23 percent, while total fresh heat input goes up by only about 17 percent. The latter change is a better indicator of the overall heat intensity of the refinery. The increase in importance of purchased fuel is due to the new importance of reformer hydrogen in the hydrocracker–none is burned in the advanced refinery.

(vii) Residual heat ejected to water goes up by 14%; sulfide in water increases by 50% (again due to desulfurization of charge stocks in hydrocrackers); SO_2 emissions increase by less than 1% [8].

The increase in some of the residuals, like heat and sulfides, appears to be more than compensated by substantial reductions in the volume of some of the more harmful residuals like BOD, phenols and particulates. The reasons for and the extent of reduction in these residuals is explained by Russell as follows:

> On the other hand, phenols and BOD residuals both decrease (in agreement with the FWPCA study). These effects result partly from the tendency of the hydrocracking process to break the aromatic rings in the phenols and convert these compounds to others with lower oxygen demand, and partly from the decreased need for caustic scrubbing of gasoline because the product of the hydrocrackers is largely naptha (and is sent to the catalytic reformers), instead of cat gasoline which we have assumed is scrubbed and sent to the gasoline blending pool. The reduction in particulate emissions again reflects the shift to infrequent catalyst regeneration in the hydrocracker. If our assumption about regeneration is roughly correct, the reduction is dramatic, on the order of 50 percent [9].

THE RATE OF DIFFUSION OF NEWER PROCESSES

Having noted the importance of a faster rate of adoption of newer processes, so as to reduce the emissions of residuals, it is interesting to examine the actual rate of adoption or the diffusion of the newer processes in the industry over a period of time. Such a rate of adoption could be studied simply by noting the distribution of the percentage of plants in each of the techniques at the beginning and at the end of the period of observation. In terms of petroleum refining, however, a better approximation appears to be a study of the distribution of crude oil input into the three types of technologies, over a period of time, as shown in Appendix Table 1. In respect to this table, it might be noted that the percent crude oil input into the three processes adds up to about 50 (instead of 100) because only 50% of the total crude oil input is processed in the cracking processes which produce the gasoline, the remaining 50 percent being accounted for by non-cracking processes. This fact is also supported by Enos [10].

From the table referred to above, it seems that the adoption of newer techniques is slow. For instance, thermal cracking which, according to John L. Enos was innovated (first commercial use) in 1913 [11], took a little over a decade for its diffusion to the extent of 25%. Similarly, according to both Enos and Nelson [12], catalytic cracking was invented in 1935–36 and innovated in 1944–45 in its final phase of technological progress, but Appendix Table 1 reveals that it took 20 years before it was adopted to process 50 percent of crude oil input of the industry. History seems to repeat itself in the case of hydrocracking, whose adoption until 1971 has been less than 6% despite its invention in 1954 and innovation in 1963.

These observations with respect to the rate of diffusion are, perhaps, too crude and too aggregative to be acceptable. To be acceptable, we need fit the Gompertz function in terms of Eq. (8) derived in the preceding section, viz.,

$$(\log Y_t - \log Y_{t-1}) = \gamma(\log Y^*_t - \log Y_t), \tag{8}$$

where the left hand side represents the growth rate of catalytic cracking and the right hand side accommodates the adjustment of the actual level of adoption of catalytic cracking viz. ($\log Y_t$) to the equilibrium level of adoption of that process, as indicated by ($\log Y^*_t$). The economic rationale for such an adjustment was examined in the preceding section.

THE DATA FOR THE GROWTH RATE OF ADOPTION DURING THE DIFFUSION PERIOD, THE EQUILIBRIUM LEVEL OF ADOPTION AND THE REASONS FOR AGGREGATION BY MAJOR REFINING STATES

The estimation of Eq. (8) would require data on the level of adoption of the catalytic cracking process, over time, during the diffusion period. The growth rate of adoption could then be computed by simply converting the levels into logarithms and taking the first difference so as to yield the left hand side of Eq. (8).

The period of diffusion would logically extend from the date of innovation (first commercial use) of catalytic cracking to the date of innovation of hydrocracking which came to replace the former. Both Enos and Nelson suggest that catalytic cracking was innovated in 1944–45 so that this should be the beginning of diffusion period for this process. Unfortunately, the data on this process are available beginning with 1949 only, when, as shown in Appendix Table 1, this process had already accounted for 23% of the total refining capacity in the industry as a whole. As regards the termination date of this process, or the date of initiation of hydrocracking, we assume it to be 1963 because, as indicated in Appendix Table 1, columns (5) and (6), in that year, catalytic cracking reached its ceiling and hydrocracking accounted for 0.2% of total capacity in the industry for the first time. We would, therefore, have fifteen observations, 1949 to 1963, for the diffusion period of catalytic cracking process.

The equilibrium level of adoption of catalytic cracking towards which the growth rate would move asymptotically might be taken to be the same as that reached by thermal cracking at its peak. The equilibrium for thermal cracking was 50% of total refining capacity of the industry. This 50 percent is an observed norm, as indicated by Enos [13] as well as in Appendix Table 1, column (4), where the ceiling reached for thermal cracking was 48.6% of total capacity. The remaining 50% of crude oil input forms a throughput in processes other than the cracking processes. The assumption of 50% of total crude oil input capacity for each of the years from 1949 to 1963 and converting them into logarithms would, therefore, provide the ($\log Y^*_t$) values for Eq. (8). As the capacity would be increasing from year to year consequent upon the increased income and population and reflected in the increased demand for gasoline in the country as a whole, we would, in fact, be analyzing the shifting equilibrium level from one year to another. In this context, the ceiling or equilibrium line EE drawn in Fig. 2 is upward sloping, with time, the height at each point of time being 50% of the total refining capacity of the state at that point of time. This, however, fails to affect the growth curve in any way [14].

The second term of the right hand side of Eq. (8), viz. ($\log Y_t$), is obtained by taking the logarithm of the actual level of catalytic cracking capacity in the respective years. While ($\log Y_t$) would be acceptable in terms of the theory referred to above, in practice it would be necessary to take ($\log Y_{t-1}$) so as to eliminate the correlation of error terms in ($\log Y_t$) which is present on both sides of Eq. (8). The entire right hand side would then be obtained by simply taking the difference between ($\log Y_t^*$) and $\log Y_{t-1}$). The result would then serve as an independent variable to explain the growth rates of diffusion in consonance with the theory outlined in the preceding section.

These regressions were computed for 18 major refining states, one each for a state. A major refining state was defined as the state having three or more petroleum refineries. Such an aggregation of refineries into states was deemed necessary because the power to control water pollution has been vested in the states so that the growth rates of diffusion could be further explained by regressing the γ's for each of the states on the environmental variables like states' expenditures on water pollution controls.

In summary, it might be noted that the estimation of a single equation referred to above accommodates a number of parameters which would otherwise have required more than one equation. For instance, Eq. (8) provides for (1) the equilibrium level of the new process, (2) the changes in this equilibrium level over time, and (3) the adjustment of the actual level towards the changed equilibrium level, over time, and in tune with the "band-wagon" effect.

THE ESTIMATES OF DIFFUSION RATES FOR CATALYTIC CRACKING FOR 18 STATES AND THEIR STATISTICAL SIGNIFICANCE

Table 3 presents the estimates of regression coefficients (γ's) or the growth rates of diffusion of catalytic cracking for the eighteen states over the period 1949–63.

TABLE 3

Regression Results Showing Growth Rates of Diffusion of Catalytic Cracking Process, 1949–1963

State	Regression coefficients for growth rates (γ's)	t ratio	Durbin-Watson statistic	R^2
1. Arkansas	0.207[a]	2.938	2.507	0.293
2. California	0.068	1.099	3.119	0.014
3. Colorado	0.127[a]	1.947	2.750	0.105
4. Illinois	0.207[a]	3.712	2.186	0.293
5. Indiana	0.088[a]	2.394	3.026	0.027
6. Kansas	0.276[a]	5.429	2.658	0.591
7. Kentucky	0.218[a]	3.658	2.363	0.404
8. Louisiana	0.242	1.592	2.944	0.041
9. Michigan	0.109[a]	1.897	3.233	0.038
10. Montana	0.215[a]	2.378	2.643	0.198
11. New Jersey	0.134[a]	1.740	2.069	0.075
12. New York	0.189[a]	1.831	2.594	0.153
13. Ohio	0.353[a]	8.551	2.286	0.800
14. Oklahoma	0.241[a]	5.197	2.408	0.556
15. Pennsylvania	0.143	1.391	3.012	0.009
16. Texas	0.134[a]	3.680	1.641	0.254
17. Utah	0.326[a]	4.036	2.446	0.454
18. Wyoming	0.049	0.737	1.889	0.019

[a]Significant at 0.05 level (t ratio greater than 1.740).

From Table 3 it might be observed that, as expected, all the eighteen coefficients are positive, fourteen of which are significant at 0.05 probability level. The growth rates for the states with significant coefficients vary from a minimum of 8.8% for Indiana to 35.3% for Ohio. The coefficient of determination (R^2 corrected for degrees of freedom) is between 0.075 and 0.800 for the states with significant regression coefficients. The Durbin-Watson statistic exceeds 2 in all the states, except Texas and Wyoming, so that there is absence of any positive autocorrelation among the error terms. Table 4 presents the regression results for hydrocracking process.

From Table 4, it might be observed that, as expected, all the seven regression coefficients are positive, three of which are significant at 0.05 probability level. The growth rates for states with significant coefficients varies from 0.08% for Pennsylvania to 10.7% for Ohio. The coefficient of determination (R^2 corrected for degrees of freedom) is between 0.005 and 0.253 for states with significant regression coefficients. The Durbin-Watson statistic exceeds 2 in all the states, except Louisiana and Texas, suggesting the absence of any positive autocorrelation among the random disturbance terms in these five states.

The relatively low coefficients of determination in this and subsequent tables need an explanation. These coefficients would tend to be higher the greater the number of plausible explanatory variables. As outlined in the theoretical section, these variables could be (i) information gap, (ii) age distribution of existing capital stock, (iii) postponement by the holdouts, (iv) supply bottlenecks, (v) relative profitability of the newer process, (vi) innovation propensity of sectors, (vii) growth rates of firms, (viii) firm size, (ix) liquidity, (x) profit trend, and (xi) age or outlook of entrepreneurs/managers. Unfortuantely, none of these data are available at the plant/refinery and state level—the major thrust of inquiry of this study. It must, however, be noted that some of these variables have been incorporated in other studies of diffusion at *inter-firm or intra-firm* levels. For instance, Mansfield [15] observed diffusion to be a direct and significant function of (i) proportion of firms using it and (ii) profitability, and an indirect function of (iii) size of required investment, but additional variables like (iv) durability of the replaceable equipment, (v) growth rate of firms, (vi) time elapsed since innovation and

TABLE 4

Regression Results Showing Growth Rates of Diffusion of Hydrocracking Process, 1967–71

State	Regression coefficients for growth rates (γ's)	t ratio	Durbin-Watson statistic	R^2
1. California	0.072	1.642	3.040	0.028
2. Illinois	0.124	1.118	2.502	0.059
3. Louisiana	0.062	1.912	1.427	0.180
4. Mississippi	0.134	1.239	2.605	0.097
5. Ohio	0.107[a]	2.261	2.451	0.253
6. Pennsylvania	0.008[a]	2.457	2.436	0.005
7. Texas	0.072[a]	2.483	0.591	0.284

[a]Significant at 0.05 level (t ratio greater than 2.132).

(vii) the expansionary phase of business cycle were not significant for the overall level. At the individual firm level, the rate of substitution was significantly explained by (i) firm size, (ii) profitability of innovation, and (iii) size of the required investment, but other variables like (iv) profit of firm as a proportion of net worth, (v) profit trend, (vi) innovative propensity of sectors, as indicated by concentration of technical leadership, and (vii) liquidity were not significant. Lastly, at the interfirm level in the railroad industry, he observed dieselization significantly explained by (i) profitability, (ii) age distribution of capital stock, and (iii) liquidity; but not by (iv) firm size, (v) trend in profit, and (vi) length of haul.

COMPARISON OF GROWTH RATES OF DIFFUSION OF CATALYTIC CRACKING AND HYDROCRACKING

From Table 4, it might be observed that the average growth rates of diffusion for these seven states are only between 0.8 and 13.4% and between 0.8 and 10.7% for the states with significant regression coefficients. These rates are unduly low when compared to that between 4.9 and 35.3% for catalytic cracking for all states or to that between 8.8% and 35.3% for catalytic cracking for states that have significant regression coefficients for catalytic cracking. Moreover, most of the states have either failed to adopt the hydrocracking process, despite (1) its innovation in 1963, (2) its relatively smaller environment-intensity, and (3) greater cost-saving characteristic, or have adopted it but failed to adopt it at an increasing rate, over time.

The comparison of growth rates of diffusion of catalytic and hydrocracking, referred to above, could be analytically deficient, because while the growth rates of catalytic cracking refer to a period of fourteen years, that of hydrocracking refer to only five years. In order to make the two sets of growth rates comparable, regression coefficients were estimated for catalytic cracking for the intermediate period. In other words, growth rates of diffusion were computed for the period 1949–53, i.e., five years after catalytic cracking was innovated in 1944–45. This compares with the period 1967–71, i.e., five years after hydrocracking was innovated in 1963–64. The comparison of these growth rates for the relevant periods for the same sample of states is indicated in Table 5.

From Table 5, it might be observed that the lower diffusion rates for hydrocracking is vindicated in three out of four states—Illinois, Ohio and Texas. The only state for which the growth rate of diffusion of hydrocracking exceeds that of catalytic cracking is California, but this is not statistically significant.

The relatively lower growth rates of diffusion of the hydrocracking process could, perhaps, be due to the fact that we have used historical data instead of current data. In this context, Fein [16] has rightly pointed out that historical data tend to underestimate the *future* growth rates of diffusion relative to the forecasts made from current data. He illustrated the difference by using both the approaches and noted the relatively higher, and hence more realistic, estimates of projections of residential demand for electricity given by current data. There is, however, a crucial difference between Fein's approach and the methodology of this paper. Fein does not address the question of adjustment of actual consumption of electricity to the desired level of its consumption; he pinpoints his attention only on the *actual* levels, whereas we are directly interested in this adjustment process. Fein also leaves out economic variables like price of electricity, prices of its substitutes like gas, income, etc., some of which tend to act in opposite directions in influencing residential demand for electricity and thus give different results than that extrapolated by him from only the *observed* levels of consumption. Above all, the growth

TABLE 5

Comparison of Growth Rates of Catalytic and Hydrocracking for Their Intermediate Periods

State	Regression coefficients for growth rates	*t* ratio	Durbin-Watson statistic	R^2
1. *California*				
Cat. Cr.	0.041	0.793	2.679	0.013
Hydro. Cr.	0.072	1.642	3.040	0.028
2. *Illinois*				
Cat. Cr.	0.223[a]	2.412	1.640	0.232
Hydro Cr.	0.124	1.118	2.502	0.059
3. *Ohio*				
Cat. Cr.	0.353[a]	4.445	2.248	0.743
Hydro Cr.	0.107[a]	2.261	2.451	0.253
4. *Texas*				
Cat. Cr.	0.179[a]	2.462	1.733	0.121
Hydro Cr.	0.072[a]	2.483	0.591	0.284

[a]Significant at 0.05 level (*t* ratio greater than 2.132).

rates of the hydrocracking process estimated by us are not projections. We are simply asserting that these growth rates, if they tend to continue into the future, are, indeed, low relative to growth rates for catalytic cracking, and some policy measures need be traced to accelerate adoption of hydrocracking process.

III. Social Benefit-Cost Implications of Accelerated Diffusion of Environment-Saving Process

In order to undertake an exhaustive social benefit-cost analysis, we would need data on (1) marginal external benefits like reduction in damages to society in the form of reduction in diseases due to reduction in water residuals consequent upon hydrocracking replacing catalytic cracking, (2) marginal external cost like the amounts spent by society to restore the effluents to its pristine quality, (3) marginal internal benefits like profits and value of higher prestige earned by refineries for reducing the residuals by using the new process and (4) marginal internal cost like the expected incremental capital and operating expenses required for setting up the hydrocracking process. Given these data, we need convert them into dollar values and then add up the values of marginal external and internal benefits to get marginal social benefit and values of marginal external and internal cost to obtain marginal social cost. We could then simply divide marginal social benefit by marginal social cost. If the resulting marginal social-benefit-cost ratio is greater than one, the switchover to hydrocracking is advisable, if it is less than one, it is inadvisable and we would be indifferent if it equals one, provided that the distribution of income after the substitution is not worse than that prior to it [17].

The only information available to us in the present context pertains to part of the internal marginal cost and the marginal physical quantities of external cost. For instance, Table 2 lists the minimum and maximum ranges of operating cost of the three cracking processes. Since the operating cost of hydrocracking is lower than that of catalytic cracking in both the ranges, it appears that the substitution of hydrocracking will not raise marginal internal cost. In the absence of any knowledge of fixed cost of setting up the hydrocracking process, however, we may assume that the annualized fixed cost just equal the savings in operating cost in hydrocracking. Furthermore, we also know that marginal private benefits of accelerated adoption of hydrocracking are positive, since they emanate less residuals and generate higher profits, thereby increasing social benefits. In such a case, we could infer that the switchover is socially beneficial because it would help reduce the water residuals. Thus the marginal reduction in these residuals tantamounts to an increase in social benefits or a reduction in social cost (since it indicates the extent to which society will have reduced burden of treatment of water wastes). The physical quantites of reduction in such residual emission coefficients for waste water, B.O.D. and phenol were estimated by subtracting the relevant coefficients for hydrocracking from that for catalytic cracking. For WW, B.O.D. and phenol these amount to 50 gals/bbl, 0.05 lb/bbl and 0.995 lb/bbl, respectively. Second, we must estimate approximately the average national growth rate of catalytic cracking and hydrocracking and compute the difference between these two rates. Such a differential growth rate is estimated at about 10% from Tables 3 and 4. Assuming that diffusion of the hydrocracking process at the same rate as catalytic cracking process should have occurred in the 1967–71 period, the differential growth rate of diffusion could be applied to the total quantities of crude oil refined (50% of that in Appendix Table 1). We could, thus, estimate the marginal increase in quantities of crude oil that would have been refined by hydrocarcking during the 1967–71 period. Third, multiplication of marginal residual coefficients by the marginally expected increase in quantities of crude oil under the hydrocracking process would simply give us the physical quantities of reduction in marginal social cost, or "damages avoided," or the physical quantities of marginal social benefits. The estimates of these marginal social benefits are indicated in Table 6.

Unfortunately, the quantities of marginal social benefits in Table 6 cannot be converted into monetary units because neither the cost of treatment of these wastes nor the

TABLE 6

Marginal Social Benefits of Desired Increase in Diffusion Rate of Hydrocracking Process, 1967–71[a]

Years	Incremental quantities of crude oil refined at desired growth rate (Th. barrels)	Marginal Social Benefits		
		Waste water (Th. gal.)	B.O.D. (Th. lbs.)	Phenol (Th. lbs.)
1967	598	29,900	29.9	595.01
1968	569	28,450	28.45	566.155
1969	604	30,200	30.2	600.98
1970	632	31,600	31.6	628.84
1971	634	31,700	31.7	605.47

[a]Source: Tables 1, 3, 4, and Appendix Table 1.

shadow prices of public water bodies (into which these effluents are dumped) are available. Alternately, we could have converted them into monetary values if we had information regarding damages done by the residuals to human health, animal and plant lives, etc. It is only in terms of physical quantities of residuals, however, that we could conclude that social benefits of the substitution of hydrocracking for catalytic cracking are considerably greater than social costs, assuming once again that the assimilative capacities of public water bodies into which the residuals are dumped have been virtually exhausted and that distribution of income after the substitution was in no way worse than that prior to the switchover.

IV. Social Policy Variables for Inducing Diffusion of Relatively Environment-Saving Techniques

Having noted the unsatisfactory growth rates of diffusion of the environment-saving hydrocracking process relative to the growth rates of catalytic cracking process of production, one would seek the reasons for such a slow diffusion and would look for social policy variables that might induce accelerated adoption of the desired techniques. One of the reasons for the slower diffusion of hydrocracking could be the average vintage or age distribution of the existing capital stock in petroleum refining. While precise data for the various vintages are not available, it appears that the existing capital equipment, which always embodies technical change, is relatively new so that it is not economical to replace it. This observation is based on the assumption that the life of cracking equipment in thermal as well as catalytic cracking is the same, say about 30 years. If so, the relatively rapid growth rate of adoption of catalytic cracking in the mid-forties could be due to the fact that most of the capital stock in thermal cracking, which was in existence for about thirty years, had depreciated. Such a favorable factor did not seem to exist in the sixties because the equipment in catalytic cracking had been in existence for less than twenty years.

In order to bring about a rapid replacement of the catalytic cracking equipment, incentives in the form of accelerated depreciation allowances and/or tax holidays for purchase of the environment-saving processes is called for.

In this context, it must be noted that the incentives suggested herein refer to purchases of "process equipment" and do not refer to "treatment equipment." It is well-known that many states, including Maryland, provide incentives for the purchase of the latter but none seem to have any provision for the former. Needless to say, the former incentives are more important than the latter in the same sense as prevention is better than cure.

While the "carrots" referred to above could help induce overall acceptance of the relatively environment-saving processes, they would not explain the differences in the growth rates of diffusion across the states. To the extent that the diffusion of the newer techniques is slower, there is greater emission of water residuals into the public waterways.

EXPLAINING THE GROWTH RATES OF DIFFUSION OF CATALYTIC AND HYDROCRACKING IN TERMS OF AN ENVIRONMENTAL ECONOMIC PARAMETER

One of the ways in which the discharge of water residuals could be reduced would be for the state governments to pass legislation restricting the discharge of water residuals into public waterways. The passage of such laws would, however, be a necessary but not a sufficient condition for the reduction of water pollution in the states. The necessary and sufficient conditions would be the passage of such laws as well as their enforcement. The enforcement of a law is more important than its enactment–witness the *River and Harbor Act, 1899,* which did not help reduce water residuals for a long time. The extent of enforcement of these laws is, however, difficult to measure. One crude way to measure it would appear to be the environmental-economic variable in the form of the amount of money spent by a state on water pollution controls. This measure is too crude because the larger states would always tend to spend more than the smaller states. The difference in sizes of the states could, however, be normalized by using the measure of per capita expenditure on water pollution controls in the respective states.

In view of the above, it is hypothesized that the greater the per capita expenditure on water pollution control in a state, the greater would be the pressure on the industries in the states to reduce the quantities of water residuals discharge. One of the ways in which the industries could reduce it would be by adoption of the relatively environment-saving processes, like catalytic cracking instead of thermal cracking or hydrocracking instead of catalytic cracking, as the case may be at the time in question. In short, we can test this hypothesis by regressing the growth rates of diffusion of the process under consideration in the various states on the per capita expenditure on water pollution controls of the respective states during the diffusion period. We expect a positive and significant coefficient for the independent variable.

Regression equations were computed to explain the growth rates of diffusion of catalytic cracking and hydrocracking from Tables 3 and 4, respectively. The data for the independent variable were available from the Environmental Protection Agency in the form of total state expenditures, including federal grants, on water pollution controls only, i.e., excluding the expenditure on sewage treatment. The data for the diffusion period for catalytic cracking were available only for a part of the period, viz. from 1959 to 1963, whereas that for hydrocracking were available for the entire period. The annual expenditures were averaged for the period for each state and normalized by dividing the average by the average population of the respective states during period under consideration. The results of the regressions are presented as Table 7, with t ratios in the parentheses.

From Table 7, it might be observed that the coefficient for per capita expenditure of

TABLE 7

Regression Results Explaining the Growth Rates of Diffusion by Environment-Economic Parameter

Regression Coefficients				
Process	Constant	Per capita expenditure	Durbin-Watson statistic	R^2
Catalytic cracking	23.2674[a] (5.389)	−0.0758 (1.245)	2.554	0.031
Hydro cracking	4.2712 (1.933)	0.4913[b] (2.212)	2.603	0.394

[a]Significant at 0.05 level (t ratio greater than 2.110).
[b]Significant at 0.05 level (t ratio greater than 2.015).

states for water pollution control is negative, so that an increase in the expenditure did not accelerate the growth rate of diffusion of catalytic cracking. This might largely be due to the fact that the period covered by catalytic cracking was 1949–53, when there was hardly any concern for the protection of the environment in the states. Fortunately, the coefficient is not statistically significant and also bears a low coefficient of determination (R^2).

In sharp contrast to the coefficient for catalytic cracking referred to above, the coefficient for hydrocracking has the positive sign and is also significant. This suggests that the consciousness to protect the environment has been aroused in the last ten years. The states could pressure the refineries into accepting a relatively environment-saving process like hydrocracking by enforcing the water pollution laws, i.e., by spending more money on the enforcement of water pollution control legislation. The coefficient of determination (R^2) of this equation is also relatively high and the Durbin-Watson statistic reveals absence of positive autocorrelation.

The low absolute values of coefficients of determination (R^2) in the equations referred to above are due to the exclusion of a number of other economic variables that could also explain a great deal of variation in growth rates of diffusion. Some of these economic variables were outlined in the preceding section on the theory of diffusion. For instance, the inclusion of variables like (1) the age distribution of capital equipment, (2) the savings in the cost of cracking by the newer hydrocracking process relative to the older, (3) the increased profitability of the newer process consequent upon the increased demand for gasoline due to the improvement in its quality and/or the reduction in its price, (4) the measures of supply bottlenecks restricting the adoption of the newer process, (5) the duration of information lag between the innovators and imitators, and (6) the extent of increase in competition in the process equipment supply sector resulting in a reduction in the price of the process equipment would all tend to increase R^2.

Summary and Conclusions

In Section I, diffusion of environment-saving technical progress was theoretically analyzed in terms of a growth curve, like the Gompertz function. In Section II, the environment-saving technical progress was empirically studied in the context of growth rates of diffusion of thermal, catalytic and hydrocracking processes in the petroleum refining industries. In Section III, we attempted an approximate social benefit cost analysis of the substitution and concluded that a diffusion rate of the hydrocracking process at the same rate as the catalytic cracking process would have been socially desirable because its marginal social benefits exceeded marginal social costs. Section IV brought out two reasons (and hence policy conclusions) for the relatively slower growth rates of diffusion of the most environment-saving technique–hydrocracking. First, the overall vintage of capital embodying the technique is relatively new so that incentives or "carrots" are called for. Second, there is considerable variation in growth rates of adoption across states so that the diffusion could be accelerated by means of a "stick" in the form of better enforcement of pollution control legislation, say, by increasing per capita state expenditure on pollution controls. In short, both the carrot as well as the stick are required.

References

1. W.E.G. Salter, *Productivity and Technical Change,* Cambridge University Press, Cambridge, 1969, Chap. IV.

2. Edwin Mansfield, *Industrial Research and Technological Innovation–An Econometric Analysis,* Norton, New York, 1968, Chap. 9.
3. Richard R. Nelson, Morton J. Peck and Edward D. Kalachek, *Technology, Economic Growth and Public Policy,* The Brookings Institution, Washington, D.C., 1967, p. 105.
4. James S. Duesenberry, *Income, Saving and the Theory of Consumer Behavior,* Harvard University Press, Cambridge, 1949.
5. Simon Kuznets, "Retardation of Industrial Growth," *J. Econ. Bus. History,* (August 1929), pp. 534–560, also reprinted in Simon Kuznets, *Economic Change,* Norton, New York, 1953, pp. 253–278; Harry Jerome, *Mechanization in Industry,* National Bureau of Economic Research, New York, 1934; Thomas Dernburg, "Consumer Response to Innovation: Television," in *Studies in Household Economic Behavior,* Yale University Press, New Haven, 1958; A.D. Bain, *The Growth of Television Ownership in United Kingdom,* Cambridge University Press, Cambridge, 1964; Zvi Grilliches, "Hybrid Corn, An Exploration in the Economics of Technological Change," *Econometrica,* (October, 1957), p. 25; Edwin Mansfield, "Technical Change and the Rate of Imitation," *Econometrica* (October, 1961), pp. 741–766, reprinted in Edwin Mansfield, *Industrial Research and Technological Innovation,* Norton, New York, 1968, Chap. 7, and summarized in Edwin Mansfield, *The Economics of Technological Change,* Norton, New York, 1968, Chap. 4; Gregory C. Chow, "Technological Change and the Demand for Computers," *Amer. Econ. Rev.* **LVII** (December 1967), 1117–1130; and John E. Tilton, *International Diffusion of Technology; The Case of Semi-Conductors,* The Brookings Institution, Washington, D.C., 1971.
6. Gregory C. Chow, *op. cit.,* p. 1120.
7. *Ibid.,* p. 1120. It might be added that Chow makes Y^* a function of Gross National Product because almost all the industries demand computers so that *national demand* for computers is reflected in the G.N.P. We would refrain from making Y^* a function of G.N.P. because in our disaggregative analysis Y^* would be a function of installed capacity in the *state.*
8. C.S. Russell, *Models of Response to Residuals Management Action: A Petroleum Refining Case Study,* draft, Resources for the Future, Washington, D.C., August 1971, p. 3–7.
9. *Ibid.,* p. v-7.
10. John L. Enos, *Petroleum Progress and Profits: A History of Process Innovation,* The M.I.T. Press, Cambridge, 1962, p. vii.
11. John L. Enos, "Invention and Innovation in the Petroleum Refining Industry," in *The Rate and Direction of Inventive Activity: Economic and Social Factors,* (ed.) National Bureau of Economic Research, Princeton University Press, Princeton, 1962, p. 300.
12. W.L. Nelson, *Guide to Refining Operating Costs,* p. 98.
13. John Lawrence Enos, *Petroleum Progress and Profits–A History of Process Innovations,* The M.I.T. Press, Cambridge, Mass., 1962, p. vii.
14. R. L. Chaddha and S.S. Chitgopekar, "A Generalized Long Range Forecast (1966–1991) of Residence Telephones," *Bell J. Econ. Manag. Sci.* (Autumn, 1971), pp. 542–560. This is also supported by Chow, *op. cit.,* p. 1129.
15. Edwin Mansfield, *Industrial Research and Technological Innovation* Norton, New York, 1968, Chaps. 7–9.
16. Elihu Fein, "Projecting from Current Data," *Technol. Forecasting Soc. Change,* **5**, 91–93 (1973).
17. Such an analysis was done for another similar subject, *vide,* Hyder Lakhani, "Internalizing Some Externalities: Social Benefit-Cost Analysis of Substituting Iron for Lead Shot in Waterfowl Hunting in Maryland," working paper, Maryland Department of Economic and Community Development, Annapolis, Maryland, January 4, 1974.

APPENDIX TABLE 1

Distribution of Crude Oil Inputs by Technologies[a]

Year Jan. 1	Number of operating refineries	Crude oil capacity (Th. barrels/sd)	Process capacity as % crude oil capacity		
			Thermal cracking	Catalytic cracking	Hydro cracking
(1)	(2)	(3)	(4)	(5)	(6)
1909	–	–	–	–	–
14	–	–	2.0	–	–
18	–	–	6.5	–	–
20	373	–	13.1	–	–
24	357	–	24.1	–	–
27	327	–	33.2	–	–
30	358	3,964	37.0	–	–
31	346	4,151	40.5	–	–
32	365	4,226	40.3	–	–
33	372	4,095	42.1	–	–
34	454	4,125	43.8	–	–
35	435	4,167	47.4	–	–
36	422	4,334	48.3	–	–
37	423	4,521	46.9	–	–
38	431	4,580	48.6	0.6	–
39	431	4,746	45.0	1.5	–
40	461	4,873	44.1	3.1	–
41	420	4,967	41.5	3.4	–
42	430	5,218	39.4	4.0	–
43	386	5,160	35.8	5.0	–
44	384	5,580	33.6	10.3	–
45	380	5,580	32.7	17.5	–
46	364	5,566	32.9	19.1	–
47	361	5,683	31.7	20.3	–
48	352	6,284	30.3	21.0	–
49	336	6,859	28.6	23.0	–
50	320	6,884	26.7	24.8	–
51	325	7,402	24.3	26.8	–
52	327	7,697	22.0	29.8	–
53	315	8,094	19.4	34.2	–
54	308	8,437	16.6	33.5	–
55	296	8,639	13.9	37.8	–
56	313	8,958	11.0	41.4	–
57	304	9,569	8.7	41.6	–
58	298	9,704	7.1	43.5	–
59	291	10,154	6.5	45.0	–
60	290	10,361	6.5	46.2	–
61	298	10,500	7.4	47.1	–
62	299	10,594	6.7	48.5	–
63	293	10,456	6.4	52.0	0.2
64	288	10,725	5.8	52.2	0.7
65	275	10,762	6.4	51.7	1.0
66	265	10,721	4.8	51.2	1.9
67	261	11,952	–	51.2	2.4
68	262	11,382	–	50.0	3.7

APPENDIX TABLE 1

Year Jan. 1	Number of operating refineries	Crude oil capacity (Th. barrels/sd)	Process capacity as % crude oil capacity		
			Thermal cracking	Catalytic cracking	Hydro cracking
69	263	12,079	–	48.1	4.1
70	262	12,651	–	46.2	4.7
71	253	12,681	–	47.1	5.8

[a]Sources: Wilbur L. Nelson, *Guide to Refining Operating Costs (Process Costimating)*, (2nd ed.) The Petroleum Publishing Company, Tulsa, Oklahoma, 1970), adapted from the chapter "Processing in U.S.A.," p. 107 for 1909–1969; and computed from *Oil and Gas Journal*, April 6, 1970 and March 22, 1971 for the years 1970 and 1971, respectively.

Part 5.
Diffusion Processes

The Multidimensional Diffusion of Technology

DEVENDRA SAHAL

ABSTRACT

Much of the literature on the diffusion of innovations focuses on the rate of adoption of innovation. The question of why the diffusion should continue in the first place has been generally neglected likewise, it is one of the most commonly noted empirical regularities that the diffusion of technology evinces an S-shaped pattern. However, only meager attention has been paid to the question of why the observed pattern is what it is. In this study an attempt has been made to explore some of these hitherto neglected questions. The thesis is advanced here that a consideration of both qualitative and quantitative dimensions of the evolutionary process is a prerequisite to development of an adequate theory of diffusion and substitution of technology. A theoretical model of multidimensional diffusion of technology is presented and illustrated by means of a case study.

Introduction

The increase in the adoption of a new technology is generally accompanied by numerous qualitative changes. This is taken to imply here that the diffusion of a technology is essentially a multidimensional phenomenon. By way of illustration, we show here in Figs. 1–3, the growth in the number of tractors on farms, their average horsepower, and the total tractor horsepower on farms during the years 1940–1974. It can be seen that the combined diffusion of technology (growth in the total tractor horsepower on farms) far exceeds either the growth in the number of tractors ("quantitative diffusion") or growth in the average tractor horsepower (one aspect of the "qualitative diffusion") during the same period. Indeed, as the diffusion proceeds, the technology seldom remains the same physical entity as before. While a new technology is substituting the old, within itself, numerous major and minor innovations can be found to be replacing their functional predecessors. All this should be a commonplace observation. However, the qualitative aspects of the phenomenon of diffusion have, for long, remained neglected in the literature on the subject. This is sufficient justification for the present study in which an attempt is made to develop a framework for studying the diffusion of technology in both its qualitative and quantitative aspects.

The plan of this study is as follows: Sections I and II seek an explanation of the mechanism underlying the observed S-shaped pattern in the diffusion of technology. Indeed, while a number of studies have shown that the diffusion of technological innovation tends to conform to S-shaped time trends, an explanation of *why this is so* is generally lacking. Section III is addressed to the significance of the observed S-shaped pattern. Section IV, based on an earlier work of the author, is devoted to the development of an index of quality change. The theoretical model of the multidimensional diffusion of technology is developed in Section V and its application to the illustrative

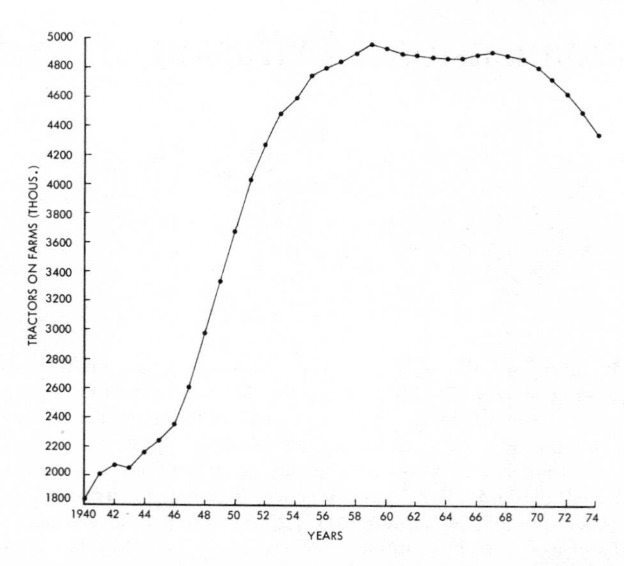

Fig. 1. Post-war growth in the use of tractors on farms. Source: File #18400 (cf. Paul E. Strickler, USDA), Farm and Industrial Equipment Institute.

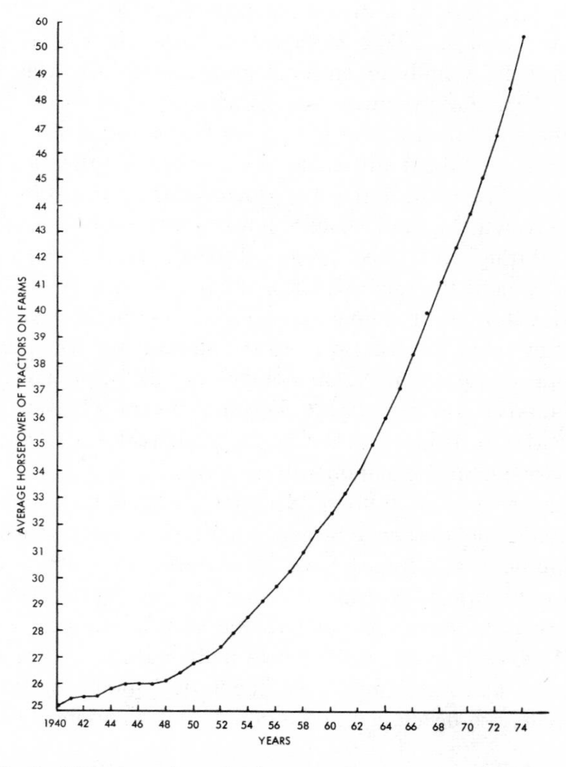

Fig. 2. Diffusion of average horse-power of tractors on farms.

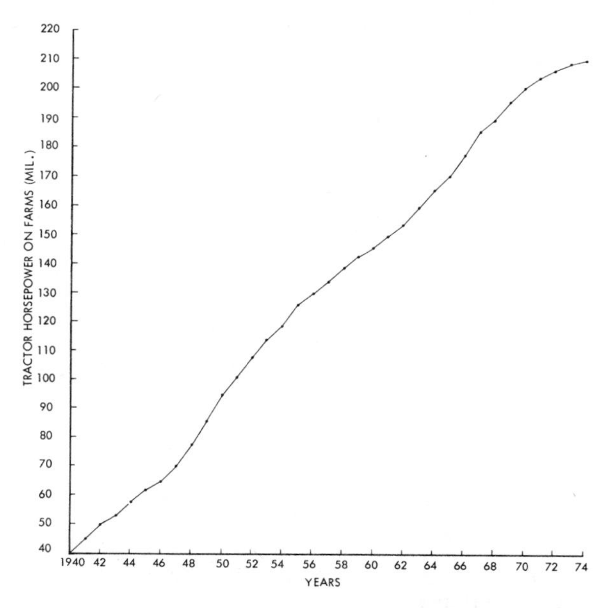

Fig. 3. Post-war diffusion of tractor horsepower. Source: File #18400 (cf. Paul E. Strickler, USDA), Farm and Industrial Equipment Institute.

case of farm tractors is presented in Section VI. The study and its implications are briefly summarized in Section VII.

I. Why Do the Quantitative Aspects of Diffusion Processes Conform to S-shaped Trends?

1. DIFFUSION AS A "DISEQUILIBRIUM PHENOMENON"

A number of economists including Griliches [17], Mansfield [28, 29] and others, regard diffusion as a disequilibrium phenomenon. According to Nelson, Peck, and Kalachek [31, pp. 97–99], "the diffusion process may be viewed as a movement from the old to the new economic equilibrium. Prior to the time when one new product or process proved itself in trial use, there was one equilibrium set of levels of use on various processes and sales of various products. The introduction of a new product or process broadens the range of choice of producers and consumers and the equilibrium is altered". In the real world, there is only gradual adjustment over the course of time to the new equilibrium level. A simple formulation of this adjustment process would be to assume that percentage adjustment in any one period is proportional to the percentage difference between the actual level of adoption of the innovation (Y) and the level corresponding to the new equilibrium (K). In symbols

$$\log Y_t - \log Y_{t-1} = b(\log K - \log Y_t), \tag{1}$$

where b is the "adjustment coefficient". One interpretation of the above expression is that the time path of the diffusion of an innovation may be approximated by means of an

S-shaped curve based on the Gompertz function

$$Y = K(a)^{\beta^t}; \tag{2}$$

since this leads to

$$\log Y = \log K + \beta^t \log a,$$

then

$$\frac{d\log Y}{dt} = \beta^t \log a \log \beta,$$

or

$$\frac{d\log Y}{dt} = -\log\beta(\log K - \log Y).$$

Upon approximating the derivative of $\log Y$ by its difference $\log Y_t - \log Y_{t-1}$ the above expression may be written as

$$\log Y_t - \log Y_{t-1} = -\log\beta(\log K - \log Y), \tag{3}$$

and provided $b = -\log\beta$ this is identical to Eq. (1).

In short, according to the first explanation advanced here, the observed S-shaped pattern may be viewed as a consequence of regarding the diffusion as a disequilibrium phenomenon.

A parameter of considerable interest is b, the rate of adoption of an innovation. In econometric studies it is generally assumed that the speed of the diffusion process depends, among other things, on the profitability of employing the innovation [17, 19, 29]. A formulation such as Eq. (1) may, therefore, be regarded as an economic analogue to the classic psychological law relating reaction time to the intensity of the stimulus [29]. It is also analogous to the principle of the sequential sampling that, in an uncertain environment, it takes a shorter time to find out that there is a difference, if that difference is large [18].

Note that there is nothing unique about the Gompertz function in this explanation. Mansfield [29], for example, implies that *lagged* response to a constant economic stimulus is responsible for the observed S-shaped curve of diffusion. However, this reasoning leads him to choose the logistic function, instead.

A more elaborate economic reasoning along these lines has been presented by Lakhani [27]. At first the rate of adoption of an innovation tends to be usually slow because, on the demand side, potential users lack information, are uncertain about its cost and benefits, or delay its use in anticipation of an improvement in its quality or a reduction in its price; and, on the supply side, bottlenecks may limit the expansion of capacity. With time, the growth rate accelerates. This is followed by retardation that is largely due to the closing of the gap between the equilibrium and the actual level of utilization of the innovation.

2. DIFFUSION AS AN "INTERACTION PHENOMENON"

In contrast to the preceding econometric explanation, the sociological viewpoint regards the variables, such as the nature of interaction between adapters and non-adapters, to be responsible for the observed pattern in the process of diffusion. Casetti [9] presents a particularly illustrative model of this view based on the following postulates: (1) that the adoption of technological innovation by potential users results primarily from "messages" emitted by adapters; (2) that potential users have different degrees of resistance to change; (3) that within any region there are potential users with different degrees of "resistance"; (4) that resistance is overcome by a sufficiently large repetition of messages. It can be shown that interaction of these postulates causes the proportion of adapters to increase slowly at first; then rapidly; then slowly again, until the saturation is reached.

According to the sociological view the observed S-shaped trends are generated by such postulates [36, 37] rather than those based on economic theory. Babcock [2] emphasizes the complementarity of the sociological and the economic variables. He shows that Mansfield's derivation of the logistic function as a summary description of the diffusion process is, in fact, based on this complementarity. Indeed, Mansfield [29] starts his theory by positing a functional relationship between the rate of adoption on one hand, and a group of variables on the other, among which are *both* economic quantities and the interactive effect. Babcock further shows that in a certain Taylor's expansion, the constant term of the function relating the slope parameter of the logistic function and the economic variables may be interpreted as a sort of coefficient of "pure interaction".

3. THE NORMAL DISTRIBUTION OF THE RELEVANT VARIABLES

According to this explanation the S-shaped diffusion curve would be generated by the aging of the capital if the ages of the component elements of stock were normally distributed [3; 31, p. 99]. An implicit assumption underlying this proposition is that unit operating costs vary in direct ratio to the age of the equipment being used [11, ch. 2, p. 24]. This seems reasonable. Furthermore, the significance of the relationship between rate of diffusion and age distribution of capital stock is strongly supported by the empirical evidence [29, ch. 9].

This explanation can be generalized. Thus simulation models developed in Hägestrand's path-breaking work assumed that resistance to adoption of an innovation is normally distributed throughout the population and generated respectable S-shaped curves of diffusion over time [25]. Stapleton has presented relatively direct evidence of the normal distribution of data on substitution of technology [47]. It seems reasonable to conclude, therefore, that sociological studies point to normal distribution of resistance, innovativeness, and time of adoption; and economic studies suggest a normal distribution of the elements constituting the profitability and marginal return from the adoption of an innovation.

4. THE LOG-NORMAL DISTRIBUTION OF FIRM-SIZE

This explanation, perhaps most satisfactory of all, has been recently presented by Paul David in an ingenious contribution to the theory of diffusion [11]. According to this thesis, it is the distribution of actual firm size (gauged in terms of production scale) that is critical to the distribution of decisions within the industry as to the adoption of a new technique. If, as found in a number of studies [23, 41, 46], the distribution of the size of the firms is log-normal, it can be shown that the path of diffusion of any labor-saving, fixed-capital–using production technique will be a standard cumulative normal curve, when plotted against a positive linear transform of the time variable. Because of the limitation of space, however, the proof is omitted here. Suffice here to say that the critical importance of production scale in determining the diffusion of process innovations is strongly supported by empirical evidence [30, 40]. Thus, a study of the substitution of mechanical cornpickers by field-shelling technology concludes that farm size is the most important determinant of the diffusion of the combines [44].

It should be noted that both this as well as the preceding explanation ("The Normal Distribution of Relevant Variables") emphasize that there are a host of factors at work in the diffusion and substitution of technology. That there is *a* strategic determinant (e.g., production scale) is, however, best brought out by the present explanation. Given that distribution of personal income tends to be approximately log-normal, it should be possible to show, via not very dissimilar reasoning, that diffusion curves of consumer durables also tend to conform to S-shaped time trends.

Last, but not least, this explanation need not be restricted to the diffusion of a class of process innovation. It would seem to be also applicable to product innovations, i.e., to the qualitative diffusion of technology [45]—the theme of the following section.

II. Why do the Qualitative Aspects of the Diffusion Processes Conform to S-shaped Trends?

It is interesting to note that when the data are smoothed out, the qualitative diffusion, like the quantitative increase in the adoption of a technology, evinces an S-shaped pattern.[1]

In an attempt to study the nature of the underlying mechanism, the following explanations are advanced. These should not be regarded as mutually exclusive.

1. AN ECONOMIC EXPLANATION

One important determinant of the level of technology is the price elasticity of demand for the innovation, i.e., the sensitivity of buyers to the price of the product. In that the first stage of innovation is characterized by the existence of monopoly and a captive market unresponsive to price change, advance in technology is expected to be minimal due to low price elasticity of demand. When this first stage is past, however, rapid technological change and an approaching "limit" is expected as a consequence of the manufacturers' anticipations of an eventual decline in the price elasticity of demand with the reduction in the price of the product [26, 42]. A second explanation is that, with the approaching "saturation" level of quantitative adoption of an innovation, the qualitative diffusion slows down because concern about production costs begins to take the place of the concern about the product characteristics.

However, what guarantees that "saturation" level would be approached in the first place, i.e., the diffusion would tend to completion? One plausible hypothesis, utilized in the development of the model of multidimensional diffusion of technology in a later

[1] More accurately the pattern is best described in terms of successively smaller S-shaped curves. The question of what if any significance may be assigned to the fact that the successive S-shaped curves usually tend to be smaller and why such a pattern is what it is, has been considered elsewhere by the author [42].

section of this paper, is that the saturation level of the quantitative diffusion of technology is, *inter alia*, determined by the qualitative changes in the innovation. Thus, the quantitative and qualitative aspects of diffusion would seem to be interdependent and consequently consideration of *both* would seem to be essential for adequate modelling of the phenomenon.

There is some evidence that the S-curves of qualitative diffusion of technology over time may as well be a consequence of the condition of the ultimate use of the innovation, i.e., various factors related to the users' rather than the manufacturers' side [45].

2. THE DIFFUSION AS A "LEARNING PROCESS"

There is some merit in the parallel that is occasionally drawn between learning and evolution [34, 38]. The choice of S-shaped functions to describe the qualitative diffusion of technology is not inconsistent with this parallel. Thus, the differential equation leading to the S-shaped function of the logistic form is the very same as the one leading to the well-known learning model of Type B. The S-shaped (Gompertz) function is particularly representative of a selective learning process [38]. The latter may be interpreted as implying that easier problems are attacked first and the more complex, costly problems afterwards.

3. THE DIFFUSION AS A "SELECTION PROCESS"

Consider the case of a product that is described by a sales coefficient, a. If an entrepreneur develops a number of different products, it will market only the best, that is the product with the highest a coefficient. So long as the mean of the underlying distribution exists (i.e., it is not an arc-sine distribution) the expected value of this largest coefficient will increase as a function of sample n, from a given population but with diminishing returns. The proof is based on the statistical theory of extreme values [22, ch. 4]. It is, however, omitted here because of limitation of space.

This proposition is similar to one examined by Stigler [48] and Nordhouse [32]. The essence of it is as follows: An economic agent, engaged in a search process tries a sample of n from a given population (of bid prices, wages, etc.). His only interest is in the maximum of this sample that may account for the process of diminishing returns to set in the search process.

It is easy to see that this reasoning applies to both qualitative and quantitative diffusion of technology. In the latter case, an entrepreneur would choose only the maxima or minima of the available sample, e.g., only the technique with the lowest labor coefficient, or the highest labor-saving coefficient. Again, the same reasoning applies to the behavior of the expected value of this largest or smallest coefficient with increase in the sample size.

III. On the Significance of the S-shaped Pattern of Diffusion

That the time path of diffusion of technology tends to take the form of an S-shaped curve is the most commonly noted empirical regularity in the literature on the subject. However, a "description" does not necessarily constitute an "explanation". In the two preceding sections, our attempt, therefore, has been to identify the causal factors responsible for the S-shaped curves of diffusion over time.

One further point should be noted. The empirical diffusion curves show a large *variety* of S-shaped trends. Furthermore, data presented by Hägerstrand [25], for example, clearly indicate that detection of S-curve among the sample points may not be always easy. Rapoport [35] and Brown [6, 7] show that it is possible to obtain statistically good fit to the data by means of functions yielding a curve that increases monotonically to an upper asymptote with no inflection in modelling the diffusion phenomena. Indeed, in the literature on the subject, widely different functions have been used to produce S-shaped curves [4, 15, 17, 27, 33, 47] each of which implies a different set of assumptions concerning the underlying process. The framework postulating a causal underpinning of the phenomenon does, however, explain why, despite the fact that it is observed in most cases, the S-shaped pattern need not conform to a unique functional form.

The function used most often to generate the diffusion curves is the logistic. However, the diffusion processes implied by the logistic curve are not the most plausible ones in the light of empirical research [8, p. 552]. Its popularity, therefore, seems to stem from the fact that its parameters can be readily estimated by the least-square methods. Whatever be the case, even a close agreement of the logistic with observational data does not in itself and without further investigation imply the correctness of the assumptions underlying the mathematical deduction of the logistic. To quote William Feller from a not very dissimilar context [12, p. 53], there are infinitely many similar curves and the first question that imposes itself inevitably is: "... if instead of the logistic, we chose at random another S-shaped curve involving three arbitrary constants of the described sort, what is the chance of getting as good or better fit than the one obtained by applying the logistic?"

In answer to the question[2], results of Feller's empirical analysis are worth quoting in some detail [12, p. 52]:

"... At the present stage, the recorded agreement between the logistic and actually observed phenomenon of growth does not produce any significant new evidence in support of the logistic, beyond the great plausibility of its deduction. Nor is the close agreement as strange as it appears at first sight—since equally good results can be obtained by applying other mathematical forms corresponding to quite different ... assumptions."

To summarize the discussion thus far, it is easy to obtain good *ex post factum* fit to the empirical observations by means of one or the other form of an S-shaped growth curve. However, the value of such a model is limited because it sheds little light on the nature of the underlying mechanism. More important, such a model is likely to be of little help in *prediction* because of the difficulty in choosing (especially at an early stage in the process of diffusion) a specific form from a variety of S-shaped curves that would be appropriate.

The assumption that diffusion of technology evinces an S-shaped pattern is not, however, indispensable. If, as suggested by the two preceding sections, the observed pattern is indeed due to identifiable causal factors rather than some "law", the *raison d'être* for the use of S-shaped functions becomes largely unnecessary. It should be possible to do away with any such otherwise extraneous assumption concerning the functional form by *explicitly* incorporating the relevant causal factors in the model itself. This is supported by the empirical evidence. An analysis, for example, of the substitution of mechanical cornpickers by combines concludes that the assumption of S-shaped diffusion curve can be dispensed with, without any loss in the explanatory power of the model [44].

To conclude, if the aim is to predict, a causal framework is necessarily required. It is such a framework to which we next turn.

---

[2] See also Feller [13, p. 52].

IV. Quantification of Quality

In the formulation of a model of multidimensional diffusion of technology a question that immediately presents itself is: How does one combine the dimensions themselves of the quality to obtain an appropriately weighted index?

A number of approaches to measurement of quality change have been reported in the literature, but they all suffer from serious limitations. The most prominent work in this area is due to Professor Zvi Griliches and his co-workers [21]. In its essence, the Griliches et al. approach consists of regressing the observed price of a product on its technical characteristic on combined cross-section data and interpreting the coefficients of the dummy-time variables as a direct estimate of the pure price changes. This leads to observed–pure-price ratio as an estimate of quality change. Several variants of this technique have been developed and applied to a variety of cases [21].

However, as noted by Griliches himself, this approach suffers from many limitations. It is prone to errors that may arise from a variety of sources such as changes in production technology, a decline in the marginal utility of money as a consequence of general increase in the real income, costless improvements, and so on.

An alternative approach, presented elsewhere by the author in greater detail [40] is based on discriminant analysis.[3] Let $x_1, x_2, \ldots, x_p$ be the various design and performance characteristics of the technology under consideration. The data on these characteristics may be then grouped, e.g., by regarding the data from time periods t_1 and t_2 as constituting groups 1 and 2, respectively. The index of quality in this approach is represented by a desired discriminant function, such as,

$$Y = a_1 x_1 + a_2 x_2 + \cdots + a_p x_p, \tag{4}$$

where the weighting coefficients, $a_1, a_2, \ldots, a_p$ are determined such that the t-statistic or the F-ratio between the groups will be maximum. The function to be maximized, as first defined by R. A. Fisher, is the ratio of the between-group variance to the within-group variance

$$f(a_1, a_2, \ldots, a_p) = \frac{n_1 n_2}{n_1 + n_2}\left[\frac{(a_1 d_1 + a_2 d_2 + \cdots + a_p d_p)^2}{\Sigma\Sigma\, c_{ij} a_i a_j}\right] \tag{5}$$

$$= \frac{n_1 n_2}{n_1 + n_2}\left[\frac{a' d d' a}{a' C a}\right]$$

where $d' = [d_1, d_2, \ldots, d_p]$ is the vector of the mean differences on the p original measures, C is the within-group covariance matrix, and n_1 and n_2 are the observations in groups 1 and 2. The quality index is determined by applying the weighted coefficients to the data on the original variables

$$Y^{(i)} = a_1 x_1^{(i)} + a_2 x_2^{(i)} + \cdots + a_3 x_3^{(i)}, \quad \text{for } i = 1, 2, \ldots, n. \tag{6}$$

Grouping in this approach may be determined in several ways, e.g., by means of an appropriate clustering algorithm [24], this being a generally preferred alternative to

---

[3] The proposed approach seems to be free from the limitations of the existing methods. See also two other works of the author on this subject [39, 43].

grouping on the basis of the time variable. Also, the approach need not be restricted to the case of two groups. The discriminant analysis of the data in more than two groups is well-known and, therefore, needs no further discussion here.

In what follows we consider the diffusion of an important class of innovations, namely those used as inputs in the production process. When referring to their "quality", a multidimensional index such as the one represented by Eq. (6) is implied.

V. A Model of Multidimensional Diffusion of Technology

In the proposed formulation of the multidimensional diffusion of technology, the basic hypothesis is that the optimal level of the overall (quantitative and qualitative) adoption of an innovation is, *inter alia*, a function of the expected levels of the price of the services from it, price of the product and price of the services from other production inputs. Assuming the functional relationship to be log-linear we have

$$S_t^* = F\left[\left(\frac{P_1}{P_p}\right)_{t-1}, \left(\frac{P_f}{P_p}\right)_{t-1}, \ldots\ldots, Z\ ,\right] \tag{7}$$

where

S_t^* = optimal level of the overall adoption of the innovation

$P_1 = P_0/q_0$, i.e., the ratio of the observed price of the innovation to its quality

$P_f = P_2/q_2$, i.e., the ratio of the price of the other factors to their quality

P_p = price of the output of the firms under consideration

Z = other variables

Assuming that the expectations are based on the past experience, the price-ratio variables in the above formulation are lagged by one time period.

The distinction between the optimal level and the actual level of the overall adoption of the innovation is crucial in this model. In the real world, there are many obstacles of an institutional, technical and psychological nature to the achievement of the optimal level. In fact, the actual stock of services of the innovation, S, may never reach the perpetually-moving desired level, S_t^*, since the variables that determine the latter are constantly changing over time. It seems realistic to assume, therefore, that the entrepreneurs continually move toward, but never actually reach, the desired level.

One hypothesis of adjustment of the actual level of adoption of the innovation to the desired level, as first proposed by Zvi Griliches [19], assumes that

$$\frac{S_t}{S_t} = \left(\frac{S_t^*}{S_{t-1}}\right)^c . \tag{8}$$

Another interpretation of the above equation is (recall the derivation of Eq. (3)) that the time path of the diffusion of an innovation may be approximated by means of a Gompertz function

$$S_t = S^* a^{b^t}, \qquad \text{where} \quad b = \exp(c/c - 1). \tag{9}$$

Thus, the above formulation is not inconsistent with the hypothesis that diffusion of technology evinces an S-shaped pattern. Note that $S = q_1 S_1$ where q_1 is the quality of the observed stock S_1. The relationship between q_1 (quality of the stock in use at given time period) and q_2 (quality of the purchases) will depend on the rate of replacement.

One further point should be noted. It is true that the dependent variable is the stock adjusted by the index of quality q_1, and the right-hand side of the model contains the terms such as the price adjusted by the index of quality q_2, which is obviously related to q_1. However, the quality indices on the right-hand side enter via the price ratio variables, *not* as *independent* explanatory variables.

Combining Eqs. (7) and (8), while assuming the functional relationship in (7) to be log-linear, we obtain

$$\log S_t = ca_0 + ca_1 \log\left(\frac{P_1}{P_p}\right)_{t-1} + ca_2 \log\left(\frac{P_f}{P_p}\right)_{t-1} + \cdots + ca_n \log Z_n + (1-c)\log S_{t-1} + cu_t, \tag{10}$$

where any given coefficient ca measures the short-term effect of change in the corresponding independent variable, any given coefficient a is the measure of the long-term effect, and u_t is a disturbance term.

The preceding model (or similar reduced forms thereof) is recognizable as a well-known formulation in investment theory. Unfortunately, however, except perhaps in a few isolated cases [10, 19, 44], the relevance of this formulation has simply not been recognized in the literature on diffusion of innovations.

It seems, therefore, worthwhile to indicate that a formulation based on investment theory such as above provides a clear alternative to the Mansfield-Blackman type models [5, 28, 29] of diffusion of technology. Specifically, the former differs from the latter in many important ways. First, the Mansfield type of models assume that the parameters of the observed growth curve of diffusion remain constant throughout the observation period, i.e., they do not change *over time*. In particular, the Mansfield formulation assumes that there exists a saturation level in the adoption of an innovation. In contrast, the above formulation postulates an equilibrium that is constantly changing over time in response to changes in the relevant variables. Consequently, in contrast with the Mansfield's formulation, it is dynamic in that it permits separation of long-term and short-term effects of change in the explanatory variables on the level of adoption of an innovation. The assumption that the diffusion evinces an S-shaped pattern is at best implicit in that it is not crucial to the above formulation. Second, the Mansfield formulation (like much of the literature on the subject) focuses entirely on the rate of adoption of an innovation; it simply does not consider the question of why the diffusion should continue in the first place. The formulation here, however, postulates that, whether or not the diffusion would tend to possible completion, is *inter alia* determined by the qualitative aspects of the phenomenon.

However, what determines these qualitative aspects of the diffusion of technology? One plausible hypothesis is that the optimal level of quality is, *inter alia*, determined by the experience accumulated in the production and the utilization of the innovation [see 38 and the references therein]. Assuming the functional relationship to be log-linear, we have

$$\log Y_t{}^* = \alpha_0 + \beta_1 \log X_t + v_t, \tag{11}$$

where $Y_t{}^*$ is the optimal level of quality, X_t, the accumulated experience in the production and utilization of the innovation, and v_t is a disturbance term. Further it is assumed that

$$\log Y_t - \log Y_{t-1} = \gamma(\log Y_t{}^* - \log Y_{t-1}), \tag{12}$$

which is not inconsistent with the hypothesis (recall the derivation of Eq. (3)) that (like the quantitative growth in the adoption of an innovation) the time path of quantitative diffusion evinces an S-shaped pattern represented by

$$Y_t = Y^* \alpha \beta^t, \quad \text{where} \quad \beta = \exp(\gamma/\gamma - 1). \tag{13}$$

Combining Eqs. (11) and (12) we obtain

$$\log Y_t = \gamma\alpha_0 + \gamma B_1 \log X_t + (1 - \gamma)\log Y_{t-1} + \gamma v_t. \tag{14}$$

Following an earlier work of the author [38], it may be further hypothesized that *ceteris paribus*: (i) a lower value of parameter γ (implying a lower rate of adjustment) is expected in the case of innovations characterized by greater complexity, (ii) the greater the ease of duplication of technology and the more "science-based" the industry in question, the higher is the expected value of γ likely to be. Note, however, that rate of qualitative diffusion is a direct function of γ. It follows, therefore, that the rate of qualitative diffusion is inversely related to the complexity of the design and the use of the innovation.

It is *a priori* assumed that the disturbance terms of Eq. (10) and (14) are independent of each other. Put another way, despite the potentially simultaneous character of the system, it is possible to estimate Eqs. (10) and (14) separately as single equations. This seems justified in view of the differences in the determinants of the dependent variables of the two equations and the time lags involved. Thus, the model may be regarded in two parts, where the first part is represented by Eq. (10) and the second part by Eq. (14). The former (or a similar reduced form thereof) has been extensively tested in the context of investment theory by Fox [16] in the case of demand for tractor horsepower[4], and by Chow [10] in the case of demand for computer services. It, therefore, does not require any further verification. An application of the second part of the model is presented in the following section.

VI. Illustrative Analysis

This section is concerned with an application of the model of qualitative diffusion of technology to the case of farm tractors. In Figs. 4 to 8, provided by the University of Nebraska Tractor Testing Laboratory, the various dimensions of the qualitative diffusion of technology are shown. Following the approach presented in Section IV, it should be possible to combine them to obtain an appropriately weighted index of quality. However, since the underlying data are not readily available, the following test is restricted to an explanation of the growth of the average maximum belt horsepower of tractors purchased.

The data and their sources are presented in Table 1. "Experience" was measured both in terms of [1, 14, 38], cumulated production quantities (X_1), and the elapsed time

[4] See also Zvi Griliches [19].

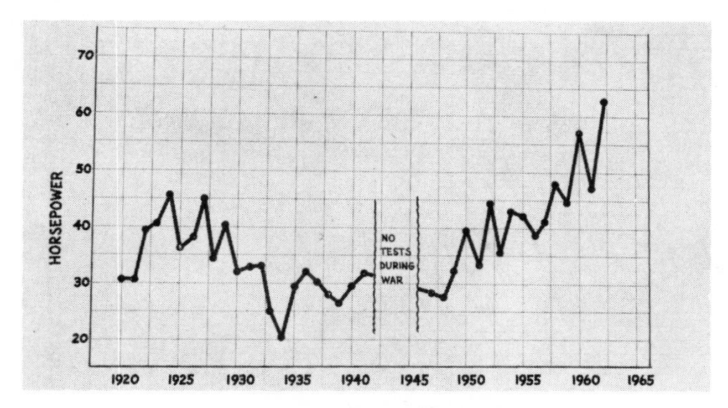

Fig. 4. Average maximum belt horsepower of wheel-type tractors.

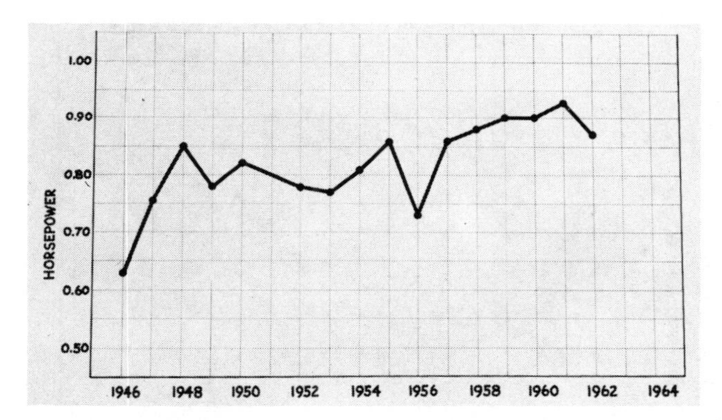

Fig. 5. Average maximum belt horsepower per 100 lbs of tractor weight.

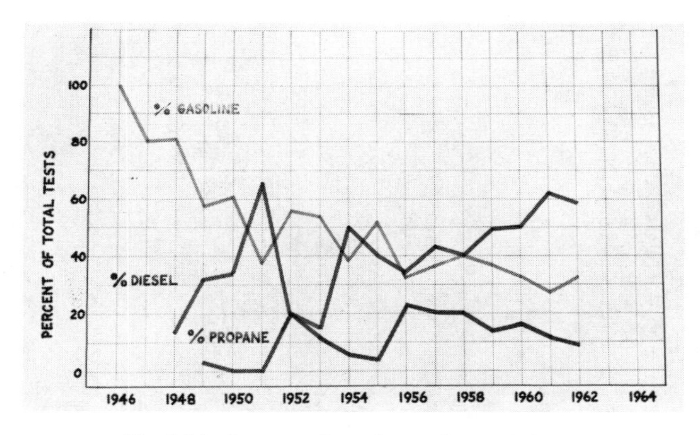

Fig. 6(a). Percent of total tests for each fuel.

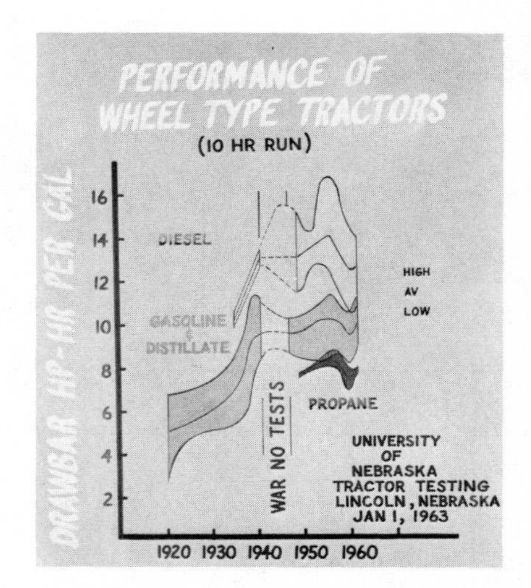

Fig. 6(b)). Trends in fuel consumption efficiency.

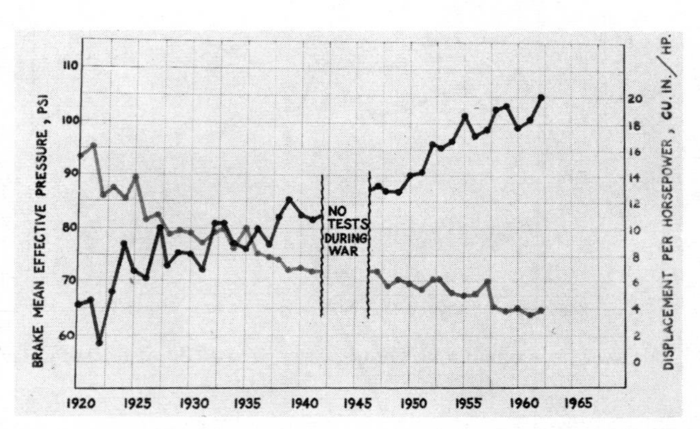

Fig. 7. Trends in brake mean effective pressure and displacement per horsepower.

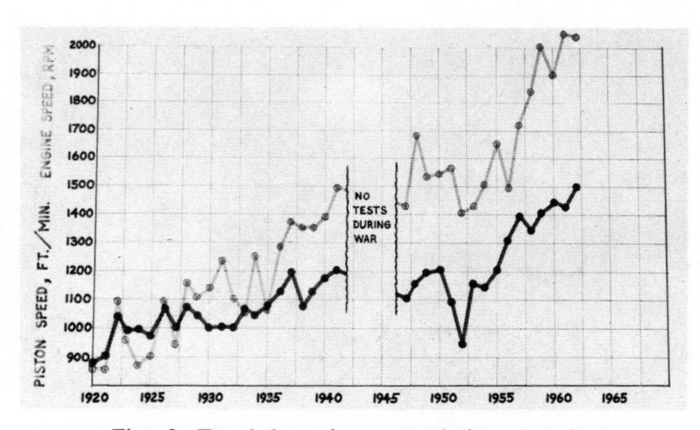

Figg. 8. Trends in engine rpm and piston speed.

TABLE 1

Data for an Explanation of the Growth of the Average Maximum Belt Housepower of Tractors Purchased

Years	Average Maximum Belt Horsepower of Tractor Purchased	Cumulated Tractor Production	Elapsed Time
1921	20.6	72.84	21
1921	100	100	21
1922	101.942	112.987	22
1923	103.398	130.409	23
1924	104.854	145.813	24
1925	106.796	167.504	25
1926	108.252	190.884	26
1927	109.709	216.227	27
1928	111.65	237.122	28
1929	114.563	264.017	29
1930	116.505	288.193	30
1931	117.961	296.691	31
1932	119.417	299.725	32
1933	120.874	302.759	33
1934	122.33	311.01	34
1935	123.786	329.97	35
1936	125.243	356.59	36
1937	126.699	389.237	37
1938	127.67	412.905	38
1939	129.126	438.385	39
1940	127.67	471.389	40
1941	129.126	514.415	41
1942	130.097	538.042	42
1943	129.126	552.485	43
1944	138.835	586.683	44
1945	131.068	620.236	45
1946	124.272	655.684	46
1947	125.243	715.17	47
1948	131.553	787.878	48
1949	137.379	864.141	49
1950	140.777	932.606	50
1951	140.777	1010.5	51
1952	151.456	1067.41	52
1953	167.961	1121.	53
1954	187.864	1154.74	54
1955	197.087	1200.05	55
1956	192.233	1229.52	56
1957	209.709	1260.96	57
1958	221.845	1294.08	58

Continued

TABLE 1–*Continued*

Years	Average Maximum Belt Horsepower of Tractor Purchased	Cumulated Tractor Production	Elapsed Time
1959	224.272	1329.76	59
1960	245.146	1350.66	60
1961	261.165	1374.19	61
1962	271.845	1400.01	62
1963	281.553	1427.94	63
1964	287.864	1457.21	64
1965	306.311	1490.71	65
1966	319.903	1527.87	66
1967	331.068	1561.12	67
1968	337.379	1590.38	68
1969	553.398	1617.24	69
1970	363.107	1640.8	70

Source of Data. Average Maximum belt horsepower purchased: (For a plot see Fig. 9. This refers to average belt horsepower of tractors *purchased,* not to average horsepower of tractors on *farms* (Fig. 2). It also differs from the average horsepower of tractors tested at Nebraska Laboratory (Fig. 4), in that (unlike the latter) it has been weighted by sales.) Data from 1920 to 1939 were obtained from Agricultural Economic Report No. 103 of USDA (by A. Fox), Nov. 1966, p. 33. Data from 1940 to 1974 were obtained from Farm and Industrial Equipment Institute, File # 18400 (by Paul E. Strickler, USDA).

Cumulated Tractor production (00): The data were cumulated from 1909. Data from 1909–1930 were obtained from USDA Report 157 (by W. M. Hurst), April 1933. The data from 1931–1956 were obtained from W. A. Cromarty, The Demand for Farm Machinery and Tractors, Michigan Agriculture Experiment Station, Bulletin 275, 1959. The data from 1957–1970 were obtained from Implement & Tractor, Nov. 21, 1971, p. 48. The resulting series was then set to 1921 = 100.

(X_2), i.e., the cumulated years of experience. Due to lack of information, the cumulated years of experience in the production and utilization of tractors were approximated by means of calendar years. The origin of the calendar-years index is arbitrary because of the difficulty in determining as to how far back of the past experience one should consider as relevant. Admittedly, somewhat better results might be obtained by shifting the origin.

The results from the analysis of data for the period 1922–1970 are shown in Table 2. Because of the presence of the lagged-dependent variable in these equations, the Durbin-Watson statistic is biased towards 2 and should not be regarded as evidence against correlation in the residuals. Its estimate is presented here as a test for the presence rather than absence of serial correlation. Indeed, both Eqs. (1) and (2) must be rejected as unsatisfactory because of their low Durbin-Watson test values despite the presence of the lagged-dependent variable on the right-hand side. Equation (3), incorporating both cumulated production and elapsed time variables seems satisfactory. Its parametric

TABLE 2

Analysis of the Diffusion of Average Maximum Belt Horsepower of Tractors Purchased, 1922–1970

$$[1]\quad \log Y = -0.07 + \underset{(1.61)}{0.018\log X_1} + \underset{(43.39)}{0.99\log Y_{t-1}}$$
$$R^2 = 0.99,\ \text{D.W.} = 1.59,\ S = 0.03$$

$$[2]\quad \log Y = -0.12 + \underset{(1.07)}{0.03\log X_2} + \underset{(39.88)}{0.99\log Y_{t-1}}$$
$$R^2 = 0.99,\ \text{D.W.} = 1.56,\ S = 0.03$$

$$[3]\quad \log Y = 0.27 + \underset{(3.10)}{0.17\log X_1} - \underset{(2.83)}{0.40\log X_2} + \underset{(41.73)}{0.99\log Y_{t-1}}$$
$$R^2 = 0.99,\ \text{D.W.} = 1.91,\ S = 0.03$$

Definitions. Y is average maximum belt horsepower of tractor purchased, X_1 is cumulated tractor production and X_2 is cumulated years. R^2 is coefficient of determination, D.W. is Durbin-Watson test value and S is the standard error-of-estimate. Values of t-test are presented in the parentheses.

Source: See Table 1.

estimates are well-determined and at a 5% level (for 45 degrees-of-freedom); its Durbin-Watson value well beyond the significance range. It leaves a mere 1% variation unexplained in the diffusion of average tractor hosepower over the time period considered. This is shown in Fig. 9.

Nevertheless, the estimated coefficient of the lagged variable is so high as to be suspicious. It is also likely that we may have estimated hybrid coefficients in view of the length of time span considered. It seems, therefore, useful to present separate estimates for the periods 1922–1947 and 1948–1970. The results for the former period are presented in Table 3 and those based on post-war data in Table 4. Due to multi-collinearity, in many cases the coefficients of the cumulated production and elapsed time variable are insignificant. For all the equations in both Tables the R^2-values are quite high and Durbin-Watson test values satisfactory at a 5% level. Based on these results the estimates of "elasticity" of qualitative diffusion with respect to explanatory variables are presented in Table 5. The "short-run elasticity" is provided by the coefficient itself of any given explanatory variable and the "long-run elasticity" is obtained from dividing this coefficient by $1 - \gamma$, where γ is the coefficient of the lagged dependent variable, Y_{t-1}.

From these results two conclusions may be drawn. First, the role of adaptive elements becomes more important late in the history of qualitative aspects of diffusion. Second, the cumulated years of production and utilization of innovation constitute a more potent variable than cumulated production quantities. Further case studies are required to determine the generality of these conclusions.

In conclusion, the results of this illustrative analysis confirm the role of adaptive elements in the qualitative diffusion of technology. Despite the good explanatory power of the model, however, there would seem to be considerable merit in developing alternative formulations of the phenomenon of qualitative diffusion.

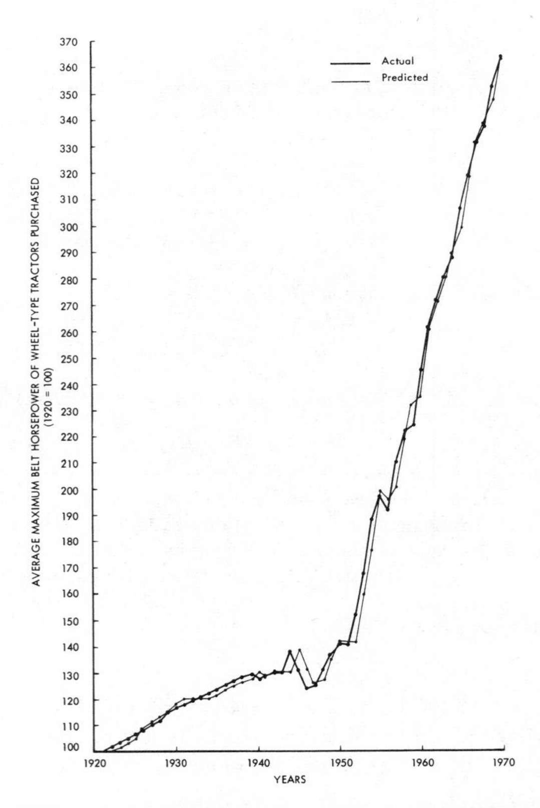

Fig. 9. Average maximum belt horsepower of wheel-type tractors purchased.

VII. Concluding Remarks and Areas of Further Research

Much of the literature on diffusion of innovations focuses on the rate of adoption of innovation. The question of why the diffusion should continue in the first place has been generally neglected. Likewise, it is one of the most commonly noted empirical regularities that the diffusion of technology evinces an S-shaped pattern. However, only meager attention has been paid to the question of why the observed pattern is what it is.

In this study an attempt was made to explore some of these hitherto neglected questions. It is suggested that the diffusion of technology constitutes a potentially simultaneous system. This is best illustrated in considering the question as to why some innovations capture the entire market (fraction f of the market captured = 100%) while others diffuse through only half of a potential market (f = 50%) and yet others simply fail to make inroads beyond their preliminary adoption. The hypothesis advanced here is that how far the diffusion would proceed is, *inter alia*, determined by the qualitative changes in the innovation. The qualitative aspects of the phenomenon, in turn, depend, among other things, on the increase in the quantitative adoption of the new technology. The "quantitative" and "qualitative" aspects of diffusion would, therefore, seem to be interdependent. The thesis is advanced here that *a consideration of both qualitative and quantitative aspects of the evolutionary process is a prerequisite to development of an adequate theory of diffusion and substitution of technology.* Put another way, the diffusion of technology is necessarily a multidimensional phenomenon.

TABLE 3

Analysis of the Diffusion of Average Maximum Belt Horsepower of Tractors Purchased, 1922–1947

$$[1]\ \log Y = 0.92 + \underset{(0.70)}{0.02\log X_1} + \underset{(4.72)}{0.78\log Y_{t-1}}$$
$$R^2 = 0.94,\ \text{D.W.} = 2.12,\ S = 0.02$$

$$[2]\ \log Y = 0.99 + \underset{(0.75)}{0.06\log X_2} + \underset{(3.79)}{0.75\log Y_{t-1}}$$
$$R^2 = 0.94,\ \text{D.W.} = 2.06,\ S = 0.02$$

$$[3]\ \log Y = 0.99 + \underset{(0.05)}{0.004\log X_1} + \underset{(0.25)}{0.05\log X_2} + \underset{(3.63)}{0.75\log Y_{t-1}}$$
$$R^2 = 0.94,\ \text{D.W.} = 2.06,\ S = 0.02$$

Definitions. See Table 2.

Based on causal considerations, an attempt has been made to develop a preliminary model of the multidimensional diffusion of technology. It is indicated that considerations of investment theory lead to a formulation that provides a clear alternative to the Mansfield-type models of diffusion of technology. Unlike the Mansfield-Blackman type of models that assume that the parameters of the observed diffusion curve remain constant throughout the observation period, the proposed alternative formulation postulates an equilibrium that is constantly changing *over time* in response to a change in the relevant variables. This is in keeping with the reality that, as new uses of an innovation are developed, the potential market for it grows with the consequence that the ceiling in the diffusion process remains not fixed, but changes with time.

It is easy to obtain good *ex post factum* fit to the data on diffusion of technology by means of one or the other form of an S-shaped growth curve. However, the value of such a model is limited. Insofar as it sheds little light on the nature of the underlying

TABLE 4

Analysis of the Diffusion of Average Maximum Belt Horsepower of Tractors Purchased, 1948–1970

$$[1]\ \log Y = -0.73 + \underset{(1.76)}{0.22\log X_1} + \underset{(11.45)}{0.85\log Y_{t-1}}$$
$$R^2 = 0.99,\ \text{D.W.} = 1.96,\ S = 0.03$$

$$[2]\ \log Y = -3.10 + \underset{(2.56)}{1.45\log X_2} + \underset{(2.48)}{0.48\log Y_{t-1}}$$
$$R^2 = 0.99,\ \text{D.W.} = 1.57,\ S = 0.03$$

$$[3]\ \log Y = -2.94 + \underset{(0.51)}{0.07\log X_1} + \underset{(1.81)}{1.26\log X_2} + \underset{(2.50)}{0.51\log Y_{t-1}}$$
$$R^2 = 0.99,\ \text{D.W.} = 1.67,\ S = 0.03$$

Definitions: See Table 2.

TABLE 5

Percent Effect on Qualitative Aspects of Diffusion by a 10% Increase in the Values of the Explanatory Variables

	Cumulated Production	Time Period
	1922–1947	
Short-run	0.04	0.5
Long-run	0.16	2
	1948–1970	
Short-run	0.7	12.6
Long-run	1.4	25.7

mechanism, it is a trivial restatement of facts. The postulate that diffusion of technology evinces an S-shaped pattern is concluded to be inherently unsuitable for *ex ante* predictions unless there is *a priori* justification for choosing a specific form from a wide variety of S-shaped curves that would be appropriate. However, a framework for choosing an appropriate functional form at an early stage in the process of diffusion is lacking.

Clearly, the appropriate form of the time path of diffusion and substitution of technology is an important area in which considerable further research is required.[5] To begin with, an explanation of why diffusion of technology evinces an S-shaped curve is still somewhat tenuous.

Further investigation of the "qualitative" aspects of the diffusion phenomenon would seem to be another fruitful area of research. In the literature on the subject it is generally assumed that the characteristics of the "innovation" do not change during the process of adoption. This is a patently unrealistic assumption because changes in the innovation itself often lead to new uses and markets for the innovation, thereby significantly affecting the course of adoption. Indeed, surveys indicate that process development efforts consume less than a fourth of the U.S. industrial R & D expenditure, the remaining three quarters going into product development and improvement.

Much remains to be learned. The sociological and the econometric explanations of the phenomenon would seem to be complementary. However, a unified explanation is still lacking. What is required is the interdisciplinary diffusion of the research itself on diffusion! The study of diffusion of non-market (e.g., organizational) innovations and assessment of the socio-political implications of diffusion is still in an early stage. There have been a flood of articles written on the subject, however, the research has not significantly advanced beyond the descriptive stage. Despite the valuable works of Torsten Hägerstrand, Zvi Griliches, Paul David, and others, it seems hardly an exaggeration to say that research in this area has merely begun.

References

1. Arrow, K. D., The economic implication of learning by doing, *Rev. Econ. Studies* **29**, 153–73 (June 1962).

[5] This problem would seem to be related to another important but little explored problem of determining the appropriate unit of adoption in diffusion.

2. Babcock, J. M., Adoption of hybrid corn: A comment, *Rural Sociology* **27**, 332–338 (1962).
3. Bain, A. D., The growth of demand for new commodities, *J. Roy. Statis. Soc.* **126**, Series A (1963).
4. Bain, A. D., *The Growth of Television Ownership in the United Kingdom Since The War. A Lognormal Model*, Cambridge University Press, Cambridge, 1964.
5. Blackman, A. W., The market dynamics of technological substitutions, *Technol. Forecast. Soc. Change* **6**, 41–63 (1974).
6. Brown, L. A., Diffusion dynamics: A review and revision of quantitative theory of the spatial diffusion of innovation, *Lund Studies in Geography*, No. B–29, Lund, Gleerup, 1968.
7. Brown, L. A., Diffusion of innovations: A macro view, *Econ. Dev. Cultural Change* **17**, 189–211 (1969).
8. Brown, L. A., and K. R. Cox, Empirical regularities in the diffusion of innovation, *Annals Assoc. Am. Geogr.* **61**, 551–559 (1971).
9. Casetti, E, Why do diffusion processes conform to logistic trends?, *Geogr. Anal.* **1**, 101–105, 1969.
10. Chow, G. C., Technological change and the demand for computers, *Am. Econ. Rev.* **57** (Dec., 1967).
11. David, P. A., A contribution to the theory of diffusion, Memorandum No. 71, Research Center in Economic Growth, Stanford University, June 1969.
12. Feller, W., On the logistic law of growth and its empirical verification in biology, *Acta Biother.* **5**, 51–66 (1940).
13. Feller, W., *An Introduction to Probability Theory and Its Application*, Wiley, New York, 1966, Vol. 2.
14. Fellner, W., Specific interpretations of learning by doing, *J. Econ. Theory* **1**, 119–40, August 1969.
15. Fourt, L. A., and Woodlock, J. W., Early prediction of market success for new grocery products, *J. Marketing* **XXV**, 31–38 (1960).
16. Fox, Austin, Demand for farm tractors in the U.S., USDA ERS, *Agr. Econ. Rep.* **103** (Nov. 1966).
17. Griliches, Z., Hybrid corn: An exploration in the economics of technological change, *Econometrica* **25** (Oct. 1957).
18. Griliches, Z., Hybrid corn and the economics of innovation, *Science* **27**, 275–80 (July 1960).
19. Griliches, Z., The demand for a durable input: Farm tractors in the U.S., 1921–57, in *The Demand for Durable Goods* (A. C. Harburger, ed.), University of Chicago Press, Chicago, 1960, pp. 181–207.
20. Griliches, Z., Profitability versus interaction: Another false dichotomy, *Rural Sociology* **27**, 327–330 (1962).
21. Griliches, Z., *Price Indexes and Quality Change*, Harvard University Press, Cambridge, Mass., 1971.
22. Gumbel, E. J., *Statistics of Extremes*, Columbia University Press, New York, 1966.
23. Hart, P. E., The size and growth of firms, *Economica*, **29**, 113 (Feb 1962).
24. Hartigan, J. A., *Clustering Algorithms*, John Wiley, New York, 1975.
25. Hägerstrand, T., *Innovation Diffusion as a Spatial Process*, University of Chicago Press, Chicago, 1967.
26. Kuznets, S., *Economic Change: Selected Essays in Business Cycles, National Income and Economic Growth*, Norton, New York, 1953, Chap. 9.
27. Lakhani, H., Diffusion of environment-saving technological change–A Petroleum Refining Case Study, *Technol. Forecast. Soc. Change* 7(1), 33–55 (1975).
28. Mansfield, E., *Economics of Technological Change*, New York, 1968. Norton
29. Mansfield, E., *Industrial Research and Technological Innovations*, New York, 1968. Norton
30. Martino, J., and S. Conver, The step-wise growth of electrical generator size, *Technol. Forecast. Soc. Change* **3**, 465–471 (1972).
31. Nelson, R. R., M. J. Peck, and E. D. Kalachek, *Technology, Economic Growth and Public Policy*, The Brookings Institution, Washington, D.C., 1967.
32. Nordhaus, W. D., *Invention, Growth and Welfare: A Theoretical Treatment of Technological Change*, The MIT Press, Cambridge, Mass., 1969.
33. Perry, A., G. A. Sullivan, R. J. Dolan, and G. P. Marsh, The adoption process: S-curve or J-curve?, *Rural Sociology* **32**, 220–222 (1967).
34. Pringle, J. W. S., On the parallel between learning and evolution, *Behavior* **3**, 174–215 (1951); reprinted in *General Systems*, Vol I, 1956, pp. 90–110.

35. Rapoport, A., Spread of information through a population with socio-structural bias, *Bull. Math. Biophys.* **15**, 523–47 (1953).
36. Rogers, E. M., and E. Havens, Adoption of hybrid corn: Profitability and the interaction effect, *Rural Sociology* **26** 409–414 (1961).
37. Rogers, E. M., Rejoinder to Griliches another false dichotomy, *Rural Sociology* **27**, 168–170 (1962).
38. Sahal, D., A reformulation of the technological progress function, *Technol. Forecast. Soc. Change* **8**(1), 75–90 (1975).
39. Sahal, D., On the conception and measurement of technology, A case study of aircraft design process, *Technol. Forecast. Soc. Change*, to appear in **8**(4) (1976).
40. Sahal, D., Another measure of technology: The generalized mahalanobis distance; A discriminant analytic framework for technology assessment, *Technol. Forecast. Soc. Change*, **to appear in 9** (1976).
41. Sahal, D., Stochastic models of nuclear power plant size distribution, submitted for publication.
42. Sahal, D., Evolving parameter models of technology assessments, *J. Int. Soc. Technol. Assessment* 11–20 (1975).
43. Sahal, D., System complexity: Its conception and measurement in the design of engineering systems, *IEEE Transaction on Systems, Man and Cybernetics,* 1976.
44. Sahal, D., The substitution of mechanical cornpickers by field shelling technology–An econometric analysis, *Technol. Forecast. Soc. Change*, to appear in **9** (1976).
45. Sahal, D., Stochastic models of evolution of technological systems (tentative title), in preparation.
46. Simon, H. A., and C. P. Bonini, The size distribution of business firms, *Am Econ. Rev.* **48** (Sept., 1958).
47. Stapleton, E., The normal distribution as a model of technological substitution, *Technol. Forecast. Soc. Change* **9** (1976).
48. Stigler, G. J., The economics of information, *J. Political Econ.* **69**, 213–225 (June, 1961).

Part 6.
The Future

Part 6. The Future

Introduction

In this final section we bring the discussion back to some fundamental questions such as those raised in the Introduction to the book. As we suggested the primary trigger for technological substitution in the future may well be the depletion of material resources. The first article, admittedly speculative, confronts this question head on. Goeller and Weinberg posit that energy, rather than mineral resources, is the key to the solution. They note that the central role of inexhaustible energy was already recognized by H. G. Wells sixty years ago, a striking confirmation of the profound perception of this foremost futurist. The article concludes that energy is the ultimate raw material, the determinant of all other resource substitutions. They echo H. G. Wells in another respect, their doubts about the ability of man's social institutions and planning to match his technological advances.

H. G. Wells: "Here were we British, 41 millions of people, in a state of almost indescribably aimless economic and moral muddle that we had neither the courage, the energy, nor the intelligence to improve."

–*In the Days of the Comet,* 1906[1]

Goeller and Weinberg: "Our *social* message is less clear and certainly less optimistic . . . Heilbroner's Wars of Redistribution . . . could collapse the society long before our carbon runs out . . . We urge attention to those institutional deficiencies that now prevent us from passing through Stage II of our human voyage without causing the boat to capsize."

The second article, by Linstone, lists some of the implications and issues raised in conjunction with the subject of technological substitution. They are largely unresolved and suggest important questions that need attention. We face the inexorable frustration that new insight creates new ignorance.

[1] Quoted from I. F. Clarke, The Pattern of Prediction: 1763–1973, *Futures* **2**, 175 (June 1970).

The Age of Substitutability* or What Do We Do When the Mercury Runs Out?

H. E. GOELLER and ALVIN M. WEINBERG

ABSTRACT

The primary resource crisis is the energy crisis. Materials other than energy resources pose relatively little problem if (1) world population can be stabilized at some reasonable value—we use 10,000 million persons—and if (2) at least one economic infinite energy resource (fission breeders, fusion, solar) can be developed and deployed before most of the fossil fuels are consumed. This rather surprising finding is made plausible by the realization that only 0.3% of the quantity and 14% of the value (excluding fossil fuels) of all non-renewable resources are derived from resources in limited supply, and that most of the uses of these limited materials are substitutable by materials in near-infinite supply (sand, stone, iron, aluminum, magnesium, etc.). Based on these facts, the seriousness of non-energy resource scarcities is greatly diminished. Of course there will be some supply-demand imbalances at certain times and places, but none of the scope and gravity of the energy crisis. In the steady-state Age of Substitutability the amounts of unit energy used, the unit costs, and the environmental impacts of mining of *all* non-renewable resources will not be more than about twice present values. We see as the major problem the development of suitable policy to get us safely from our present methods of materials management to the steady-state several centuries hence.

Two conflicting views dominate our perception of man's long-term future. The "catastrophists" believe the earth's resources will soon be exhausted and that this will lead to a collapse of society. The "cornucopians" argue that most of the *essential* raw materials are infinite: that as society exhausts one raw material, it will turn to lower grade, inexhaustible substitutes. Eventually the society will subsist on renewable resources and on those elements, such as iron and aluminum, that are practically infinite. According to this view, society will settle into a steady state of substitution and recycling. This asymptotic society we call the Age of Substitutability.

We are cornucopian even in this era of dwindling resources, despite the dire predictions of the Club of Rome, of *The Limits to Growth* [1], and of *Mankind at the Turning Point* [2]. We came to this position some 25 years ago after reading Sir Charles Darwin's *The Next Million Years* [3]. Darwin rightly insisted, as do some of the present-day forecasters and systems modelers, that short-term forecasting was impossible (as well as being disconcerting for the forecaster, who presumably would live to see his forecasts tested against fact). But *very* long-range forecasting, in which fluctuations in human experience

---

*This article appeared in *Science,* **191**, 683–689 (Feb. 20, 1976). It was originally an Eleventh Annual Foundation Lecture presented before the United Kingdom Science Policy Foundation Fifth International Symposium—"A Strategy for Resources"—Eindhoven, The Netherlands, September 18, 1975. The authors wish to thank E. R. VanArtsdalen and G. Marland for their help in preparing the manuscript.

are blips on an underlying statistical state, could be done in the same sense that statistical mechanics predicts the future of a big ensemble. Darwin asserted that over the *very* long term, man's fate, on the average, was predestined by the Malthusian disequilibrium: man procreates in a finite world. Man's continual scramble for finite resources condemns him to eternal social instability, to violence, to brutishness.

Darwin, as all Neo-Malthusians, realized that no panacea could work unless population leveled off. But even this was not enough: as resources dwindled, the pressures on what remained would grow. Darwin also recognized, as did H. G. Wells some 40 years earlier [4], that an inexhaustible source of energy would be necessary—perhaps even sufficient—to rescue man from the resource catastrophe. Most of the materials that man uses, e.g., iron and aluminum, are found in nature in an oxidized state. To reduce these oxides, i.e., to extract iron, or to change alumina to aluminum, requires energy (more accurately, free energy). Thus man's economic transaction with nature involves expenditure of free energy; and as his high-grade resources dwindle, he expends more energy. Darwin saw only one possible inexhaustible source of energy: nuclear fusion. We disagree with Darwin: there are other inexhaustible sources of energy (notably the breeder reactor, possibly solar energy) that, if developed, would allow man to avoid the hopeless future Darwin predicted.

The Principle of Infinite Substitutability

The catastrophists, when discussing a particular resource such as aluminum, usually say that when all our bauxite has been used, we shall have to live without aluminum. Thus we read in *The Limits to Growth*, "The effect of exponential growth is to reduce the probable period of availability of aluminum, for example, from 100 years to 31 years. . ." [5]. This cannot be correct. As geologist Dean F. Frasché said in 1962, "Total exhaustion of any mineral resource will never occur: Minerals and rocks that are unexploited will always remain in the earth's crust. The basic problem is how to avoid reaching a point where the cost of exploiting those mineral deposits which remain will be so costly, because of depth, size, or grade, that we cannot produce what we need without completely disrupting our social and economic structures" [6]. Will we ever completely disrupt our social and economic structures because we run our of mineral resources? In this paper we shall examine these questions; although much of what we say must be regarded as tentative and speculative, we expound these views in the hope that they will stimulate further study.

Actually, we can conceive of depletion of resources and substitution in three stages. Stage I, that almost surely will persist for the next 30 to 50 years, is a continuation of present patterns of use of our non-renewable resources. During Stage II, when society still would depend on reduced C and H found in nature, i.e., coal, there would be little oil and gas, and people would begin to turn away from widespread use of a few of the non-ferrous metals and toward much greater use of alloy steels, Al, Mg, and Ti. Stage II might last several hundred years. Finally in Stage III, the truly asymptotic Age of Substitutability, all the fossil fuel would be exhausted, and society would be based almost exclusively on materials that are virtually unlimited. It is our basic contention that, insofar as limits to mineral resources can be discerned, the condition of life in Stage III—the Age of Substitutability—would not be drastically different from our present condition of life: we have the *physical* possibility of living in the Age of Substitutability and not "completely disrupting our social and economic structures". To reach this state

Table 1

The Average Non-Renewable Resource Used by Man in 1968[a]–Demandite

United States			
$(CH_{2.14})_{.8022}$	$(SiO_2)_{.1115}$	$(CaCO_3)_{.0453}$	$Fe_{.0110}$
$Al_{.0011}$	$(Cu, Zn, Pb)_{.0004}$		$Mg_{.0004}$
$N_{.0076}$	$O_{.0053}$	$Na_{.0053}$	$Cl_{.0053}$
$S_{.0023}$	$P_{.0008}$	$K_{.0007}$	$X^b_{.0008}$
World			
$(CH_{1.71})_{.6660}$	$(SiO_2)_{.2117}$	$(CaCO_3)_{.0815}$	$Fe_{.0145}$
$Al_{.0007}$	$(Cu, Zn, Pb)_{.0004}$		$Mg_{.0004}$
$N_{.0068}$	$O_{.0045}$	$Na_{.0045}$	$Cl_{.0045}$
$S_{.0023}$	$P_{.0007}$	$K_{.0007}$	$X^b_{.0008}$

	United States	*World*
Total Quantity, 10^6 tons	3,360	17,300
Total Value, $\$10^6$	42,200	158,500
Average Unit Value, $/ton	12.55	9.16
Average Energy Used for Recovery, kwh/ton	1,140	800
Total Quantity, 10^6 ton moles	140.4	551.3
Average Molecular Weight	23.9	31.4
Per Capita Consumption, tons	17	5
Per Capita Energy, kwh	18,800	3,800
Per Capita Energy Rate, kw	2.14	0.43

[a]Applies only to new metal from ore. For total metal, see Tables A1 and A2, Appendix 2.

[b]*X* represents all other chemical elements; highest in order of demand are Mn, Ba, Cr, F, Ti, Ni, Ar, Sn, B, Br, Zr; all others <100,000 tons/year (world) or 30,000 tons/year (U.S.)

Source: Derived, with some modifications, from data in *Mineral Facts and Problems, 1970.*

without immense social disruption will, however, require unprecedented foresight and planning.

A convenient way to place the matter in perspective is to examine the "chemical" composition of one molecule of "Demandite", the average non-renewable resource man uses. We take the total number of moles of each element that man extracts from the earth, the sea, and the air, and from this compute the average mole percent composition of man's average non-renewable resource. We exclude renewable resources, agricultural products, wood, and water. Similarly, we can define the average metal, "Avalloy"; the average fertilizer, "Avfert"; the average industrial gas, "Avgas"; the average industrial heavy chemical, "Avchem", and the average refractory, "Avrefract". In Tables 1 and 2, and in Appendix 1, we give the chemical formulae and other properties of Demandite and Avalloy. All data are given for 1968.[1]

Now let us compare the composition of these average materials with the average

[1] These formulae are based on the highly consistent statistics given in *Mineral Facts and Problems, 1970*, U.S. Bureau of Mines Bulletin 650 (and modified as needed), U.S. Government Printing Office, Washington D.C., 1970.

composition of the earth's crust, the sea, and the air (Table 3). Actually, the resource situation is even better than this comparison implies, because the majority of most-used elements in Demandite, plus a few others, have near-infinite resources of much higher quality and concentrations in sources other than average rocks; these are summarized in Table 4.

The data in Tables 3 and 4 show that of the 13 most widely used atoms only extractable CH_x and phosphorus are not essentially infinite: whereas oxidized C and H are very common, extractable CH_x is rare—only 25 to 30 ppm in the first kilometer of the earth's crust. Yet extractable CH_x dominates Demandite: fully 80% of U.S. Demandite, and 66% of world Demandite is CH_x. Thus, when we speak of exhausting our resources, *by far the most important scarce resource is extractable* CH_x. Since energy can reduce CO_2 and H_2O to CH_x (see Appendix 3), we can paraphrase the basic point of both Wells' and Darwin's Malthusianism: the primary resource shortage is one of reduced CH_x or its equivalent, energy. Shortages of almost all other minerals are second order compared to the shortages of CH_x.

The central position of reduced carbon, or of reducing agents, is seen also when we examine the chemical composition of Avalloy. Iron atoms constitute about 86% of Avalloy, aluminum atoms, 8%: these two inexhaustible metals constitute 94% of the total. Extracting iron from its oxides by presently used processes requires reduced carbon, and this accounts for about 12% of all coal used in the United States. Although

Table 2.

The Average Virgin Metal Used by Man in 1968[a]—Avalloy

United States						
$Fe_{.8570}$	$Mn_{.0119}$	$Si_{.0105}$	$Cr_{.0050}$	$Ni_{.0015}$	$X^b{}_{.0003}$	$Al_{.0822}$
$Cu_{.0138}$	$Zn_{.0123}$	$Pb_{.0025}$	$Sn_{.0003}$	$Mg_{.0021}$	$Ti_{.0002}$	$Y^a{}_{.0004}$
World						
$Fe_{.8983}$	$Mn_{.0176}$	$Si_{.0071}$	$Cr_{.0045}$	$Ni_{.0009}$	$X^b{}_{.0002}$	$Al_{.0447}$
$Cu_{.0135}$	$Zn_{.0097}$	$Pb_{.0020}$	$Sn_{.0003}$	$Mg_{.0009}$	$Ti_{.0001}$	$Y^c{}_{.0002}$

	United States	*World*
Total Quantity, 10^6 tons	85.6	424.3
Total Value, $ $\times 10^6$	16,050	75,775
Average Unit Value, $/ton	187.50	178.60
Average Value, $/ton, of Fe, Si, Al, Ti, Mg in Avalloy	154.95	145.40
Average Energy Used for Recovery, kwh/ton	12,300	11,100
Average Molecular Weight	53.7	54.8
Per Capita Consumption, tons	0.43	0.12
Per Capita Energy, kwh	5,290	1,340
Per Capita Energy Rate, kw	0.60	0.15

[a]Applies only to new metal from ore. For total metal, see Table A1, Appendix 2.

[b]*X* includes all other ferrous metals.

[c]*Y* all other non-ferrous metals.

Source: Derived from data in *Mineral Facts and Problems, 1970.*

Table 3.

Average Mole Percent Composition of the Earth

Crust (Topmost Kilometer on Continents)					
$(CH_x)(\text{Extractable})_{.00004}$		$(CH_x)(\text{Unextractable})^a{}_{.0083}$		$C(\text{Oxidized})_{.0153}$	
$O_{.5907}$	$Si_{.1943}$	$H(\text{Other})_{.0658}$	$Al_{.0507}$	$Ca_{.0175}$	
$Na_{.0142}$	$Fe_{.0132}$	$K_{.0123}$	$Mg_{.0120}$	$Ti_{.0016}$	
$Cl_{.0014}$	$S_{.0009}$	$F_{.0007}$	$P_{.0004}$	$Mn_{.0003}$	
$X_{.0004}$					
where X is all other elements, and for which X =					
$Ba_{.000072}$	$Sr_{.000064}$	$V_{.000040}$	$Zr_{.000034}$	$(Cu, Zn, Pb)_{.000032}$	
$N_{.000027}$	$Rb_{.000027}$	$Cr_{.000026}$	$\text{Rare Earths}_{.000024}$	$(Co, Ni)_{.000022}$	
$\text{Other}_{.000021}$					
Seawater (Except for Water)					
$Cl_{.4859}$	$Na_{.4180}$	$Mg_{.0485}$	$S_{.0255}$	$Ca_{.0091}$	$K_{.0088}$
$C_{.0021}$	$Br_{.0007}$	$P_{.000002}$	$Si_{.0001}$	$(Fe, Al, Ti)_{.0000005}$	
$\text{Other}_{.0009}$					
Air (Excluding Variable Amounts of CO_2 and H_2O)					
$N_{.7805}$	$O_{.2100}$	$Ar_{.0093}$	$Ne_{.0002}$	$(He, Kr, Xe) \ll .0001$	

[a] (CH_x)(Unextractable) represents the very large amount of hydrocarbon in the topmost kilometer of shale. Almost all of this is too dilute to extract with a positive energy balance and therefore cannot be used as a source of energy. In principle, it might be used as a source of CH_x for petrochemicals.

Avalloy itself is made largely from inexhaustible Fe and Al, to produce Avalloy requires reduced carbon, which is a rare material, or some other source of energy.

Let us now state the principle of "infinite" substitutability: *with three notable exceptions–phosphorus, a few trace elements for agriculture, and energy-producing fossil fuels, i.e, CH_x–the society can subsist on infinite or near-infinite minerals with relatively little loss of living standard.* Such a civilization would be based largely on glass, plastic, wood, cement, iron, aluminum, and magnesium: whether such a civilization would be anything like our present society will depend upon how much of the ultimate raw material–energy–we can produce and how much energy will cost, both economically and environmentally.

To give substance to so broad a claim, one must examine in detail, not simply in principle, exactly how one would propose to live without a particular finite resource. This we do not claim to do here; rather we quote from previous work on the subject by H. E. Goeller [7].

Certainly there are no substitutes for the elements required to sustain life. The major life elements are: H, O, C, and N; next come Ca, P, Cl, K, S, Na, and Mg, which constitute less than 1% of the total in living things. Besides these are at least 13 required trace elements: F, Si, V, Cr, Mn, Fe, Co, Cu, Zn, Se, Mo, Sn, and I. Modern agriculture requires

Table 4.

Present or Future Near-Infinite Resources for the Most Extensively Used Elements

Element	Resource	Maximum % in Best Resource	World Resource (Tons)	R/D Ratio[a] Years
CH_x(Extractable)	Coal, Oil, Gas	<75	10^{13}	2,500
C (Oxidized)	Limestone	12	2×10^{15}	4×10^{6}
Si	Sand, Sandstone	45	1.2×10^{16}	5×10^{6}
Ca	Limestone	40	5×10^{15}	4×10^{6}
H	Water	11	1.7×10^{17}	$\sim 10^{10}$
Fe	Basalt,[b] Laterite	10	1.8×10^{15}	4.5×10^{6}
N	Air	80	4.5×10^{15}	1×10^{8}
Na	Rock Salt, Seawater	39	1.6×10^{16}	3×10^{8}
O	Air	20	1.1×10^{15}	3.5×10^{7}
S	Gypsum, Seawater	23	1.1×10^{15}	3×10^{7}
Cl	Rock Salt, Seawater	61	2.9×10^{16}	4×10^{8}
P	Phosphate Rock	14	1.6×10^{10}	1,300
K	Sylvite, Seawater	52	5.7×10^{14}	4×10^{7}
Al	Clay (Kaolin)	21	1.7×10^{15}	2×10^{8}
Mg	Seawater	.012	2×10^{15}	4×10^{8}
Mn	Seafloor Nodules	30	1×10^{11}	13,000
Ar	Air	1	5×10^{13}	2×10^{8}
Br	Seawater	–	1×10^{14}	6×10^{8}
Ni	Peridotite	0.2	6×10^{11}	1.4×10^{6}

[a]R/D ≡ resource to demand ratio in 1968. We realize that as demand grows, these values will be reduced; however, we also anticipate that population must eventually level off, followed by an ultimate leveling off of demands for energy and non-renewable resources.

[b]**It must be noted that no process now exists for obtaining iron from basalt or nickel from the** ultrabasic rock peridotite; however, given a century for development of such processes, the likelihood for success seems quite high.

large amounts of calcium, nitrogen, potassium, and phosphorus, plus relatively small quantities of some of the trace elements. Of the major life-sustaining elements, only phosphorus is not on our list of infinite elements. The average natural abundance of phosphorus in rocks is about 1000 ppm; if society eventually had to depend on average phosphorus, the costs of agriculture might become intolerably high. However, high-grade resources of phosphorus are very large: the present resource-to-demand ratio is 500 years for world reserves and an additional 800 years for potential resources. In addition, even though speculative resources are poorly known, they are regarded as very large (but considerably smaller than fixed nitrogen from air or potash from seawater). Nevertheless, phosphorus can hardly be regarded as "inexhaustible". This led H. G. Wells many years ago to imply that ultimately we shall have to recycle bones as fertilizer [8]; and we are in no position to refute this view.

With regard to the trace elements that are present in the soils and are needed only at low concentrations, modern agriculture slowly depletes soil of these elements. In the near term, shortages can undoubtedly be supplied from inorganic sources; in the long run, we shall undoubtedly be forced to return agricultural and animal wastes to the soil, particularly for the trace elements with limited resources such as copper, zinc, and cobalt.

Beyond these non-substitutable elements, we use many rather scarce metals because of their special properties: copper because it is a good electrical conductor; nickel, chromium, and niobium because they confer corrosion resistance and high-temperature strength on iron; mercury because it is a metallic liquid at room temperature. Goeller has studied in some detail the extent to which one might substitute for cadmium [9], zinc, lead, copper, tin, and mercury [10], the main conclusion being that for most of their uses, substitutes derived from infinite or near-infinite materials are available. To illustrate, we summarize some of the results in the case of mercury.

The Case of Mercury

The average annual consumption of mercury by use and form in the United States, 1964–1973, is given in Table 5. About 2250 tons of mercury were used each year during this period. Suppose we had no mercury—could we survive? Obviously, yes. To persuade ourselves of this we need only to demonstrate, for each present use of mercury, a plausible alternative that does not require mercury. These alternatives are listed in Table 5. For example, the largest use of mercury (34% or 770 tons per year) is for caustic chlorine production. The diaphragm cell is an alternative that was in wide use, before mercury cells were introduced, and still accounts for 70% of the U.S. production of caustic. These cells require relatively common materials such as concrete for the cell body, asbestos (which itself is made of calcium and magnesium silicates) for diaphragms, and copper and graphite for electrodes.

Acceptable alternatives are now known for all the major uses of mercury except possibly high-performance electric batteries (which would require other scarce materials

Table 5.

Average Annual Consumption of Mercury in the United States, 1964–1973

Uses	Substitutes	Metric Tons	% of Total
Caustic-Chlorine Production			
Inventory for New Facilities[a]	Diaphragm Cells	306	13.5
Makeup		463	20.5
Subtotal		769	34.0
Electrical (Mainly Batteries)	Zn-MnO_2-Graphite Dry Cell	598	26.4
Industrial and Control Instruments	*See* Note 1	252	11.2
Biocidal Paints	Plastic Paints, Copper Oxide Paint	299	13.2
Dental Amalgams	Metal Powders, Porcelain, Plastics	99	4.4
Agriculture[3]	Organic Biocides	88	3.9
Catalysts	Ethylene Process for PVC Mfg.	54	2.4
Laboratory Use	*See* Note 1	57	2.5
Pulp and Paper Mfg.[2]	Organic Biocides	17	0.8
Pharmaceuticals	Sulfur Drugs, Iodine, Etc.	17	0.8
Amalgamation[2]	Gold Recovery by Cyanide Process	6	0.3
Total		2255	100.00

NOTES

1. Substitution may be based on alternative physical properties and may include bimetallic couples, aneroid barometers, etc. Liquid gallium and sodium-potassium alloys may be substituted at room temperatures but not at sub-zero temperatures.
2. These uses of mercury were discontinued in the U.S. in 1971. Large former uses included mercury fulminate detonators and felt manufacture; both have been replaced decades ago.
3. Use of methyl mercury on seed grains banned by U.S. Department of Agriculture; banning of other uses may follow.

[a] Includes small quantities for other uses.

such as Cd and Ag) or electric switches.[2] For minor uses such as pharmaceuticals, or odd laboratory uses, which amount to <1%–5% each of the total, alternatives have not been sought very seriously because the amount of mercury required is small. We can hardly imagine the society collapsing, or even being seriously impeded *in the long run*, if we have to do without mercurial pharmaceuticals or mercury batteries.

Other Metals

If we played the same game of what do we do when the high-grade ores of copper, silver, zinc, tin, etc., run out, would the aggregate additional cost—both in energy and in dollars—of the substitutes or of the materials themselves, ultimately derived from "dirt", be intolerable? To be more specific, let us estimate how much the aggregate unit cost of these metals could rise without causing the cost of Avalloy to rise more than twofold.

We can look at the matter by again examining the chemical composition of Avalloy. Ninety-five percent of Avalloy consists of iron, aluminum, silicon, magnesium, and titanium. These account for 95% of the energy/ton of Avalloy, 80% of the cost/ton of Avalloy. All these materials are in infinite supply and their ultimate energy cost will not increase by more than a factor of two as shown in Table 6. The remaining 5% of Avalloy consists mainly of Cu, Zn, Mn, Cr, Pb, Ni, and Sn; these represent about 20% of the total value of Avalloy (Table 2). At present, extraction from ore represents only about one-half the total cost of these metals. The aggregate cost of this group of metals that is sensitive to the grade of ore therefore is not more than 10% of the total unit price of Avalloy. Thus, even if the price of these materials increased five to tenfold, the unit price of Avalloy would only double.

This is almost surely an overestimate since we have not allowed for substitution. For example, in 1968, copper accounted for 12% of the total cost of Avalloy in the U. S. But electrical copper is, in the long run, almost entirely replaceable by aluminum; and structural copper and brass largely by steel, aluminum, titanium, and plastic. Moreover, Mn (in sea floor manganese nodules) and Ni (in the peridotites) are nearly infinite: it is hard to see how the latter two metals can ever cost more than ten times their present price.

Some substitutes can be identified for the remaining elements—Zn, Cr, Pb, and Sn. Galvanized iron (which uses zinc) can, in good measure, be replaced by plastic-bonded steel; tin plate in cans can be largely replaced by plastic or glass. However, no very good substitute for Cr in stainless steel is presently known. Ultimately, this may force the society to use titanium (which is in near-infinite supply) for most uses that are now served by stainless steel. Nevertheless, the remaining elements—Zn, Cr, Pb, and Sn—could increase in cost by a very large factor, and the price of Avalloy would still remain within a factor of two of its present real cost. This is true also of the remaining nonmetals, the metal oxides, the refractory metals, and the non-ferrous by-product metals. These are discussed in Appendix 2. Thus we arrive at our basic observation: that *Avalloy and indeed Demandite (with the extremely important exception of* CH_x*) are so heavily dominated by elements derived from infinite sources or elements for which substitutes are available*

[2] A convenient small battery that does not use scarce materials and lasts a long time is an obvious target for research. However, if such a battery were never developed, we could revert to standard miniaturized dry cells (based on zinc and manganese) with only slight inconvenience. It must be remembered that in a 25-g cell used to power a portable computer, there are only 10 g of mercury. Even if the price of mercury rises one hundredfold, the overall cost of the computer would rise by $12 if we elected to stay with the mercury battery instead of converting to shorter-lived, cheaper Zn–Mn dry cells. In the longer term, a substitute for the lead battery also may be needed.

Table 6

Energy Requirements for the Production of Abundant Metals and Copper

Metal	Source	Gross Energy[a] (kwh/ton of Metal)	E_L/E_H
Magnesium Ingot	Seawater	100,000	1
Aluminum Ingot	Bauxite	56,000	1
	Clay	72,600	1.28
Raw Steel	Magnetic Taconites	10,100	1
	Iron Laterites	11,900	1.17 (with carbon) ~2 (with electrolytic hydrogen)
Titanium Ingot	Rutile	138,900	1
	Ilmenite	164,700	1.18
	Titanium Rich Soils	227,000	1.63
Refined Copper	Porphyry Ore, 1% Cu	14,000	1
	Porphyry Ore, 0.3% Cu	27,300	1.95

[a]At 40% thermal efficiency for generation of electricity.

that their unit price is relatively insensitive to depletion of mineral resources. The tentative conclusion to be drawn is that in principle our social and economic structures are *unlikely*, in the long run, to be disrupted because we shall have to exploit lower grade mineral resources–provided always that we find an adequate inexhaustible source of relatively cheap energy to substitute for CH_x. We put this hypothesis forward, realizing that it is in sharp opposition to the currently fashionable Neo-Malthusianism, but we believe the Neo-Malthusians have been misled by their habit of lumping all resources together without regard to their importance, ultimate abundance, or substitutability.

Though we can conclude from a global viewpoint that the cost of Avalloy in the Age of Substitutability will not be more than two to three times its present price, there will be specific sectors of the economy that are likely to be hit severely. A point requiring additional study is the possibility that some material used in rather small quantity has extremely high leverage on a strategic industry, and that a drastic increase in cost or even collapse of that industry would rock the entire economy. An example might be helium if we are forced to obtain it from air. If superconducting magnets forever required liquid helium, and if fusion absolutely required superconducting magnets, then a drastic rise in the price of helium might cause a corresponding rise in the cost of energy. However, one must realize that if the cost of fusion energy exceeds that of solar energy or energy from breeders, then the society will choose solar or breeders, not fusion, for its primary inexhaustible source of energy. Thus we cannot accept the view that helium is absolutely essential for the long-term future of our industrial society.[3]

Recycle in the Age of Substitutability

As prices for materials increase, there obviously will be stronger incentive to recycle, to bring the empty bottles back. The remelt energy required to put scrap metal back into

[3]Other examples of elements with finite resources that may have very high leverage are silver in photography; tungsten in tool-making; lead and antimony in storage batteries; beryllium in beryllium-copper alloys; and perhaps manganese in steel-making. Obviously more study is needed to identify other such critical situations.

productive use is very much less than that needed to reduce and refine ores. For Mg it is 1.5% of that required to win the metal from ore; for Al it is 3%–4%; for Ti it is 30%. One must recognize that recycle cannot provide a total answer, since the recovery in each recycle is never 100%; in fact, unless recycle is very efficient, it will not make very much difference. Thus, if the recovery of copper, say, is 0.9 per cycle, then recycling reduces the required amount of virgin copper to be added to the system each year by a factor of $(1-0.9)^{-1} = 10$. But if recovery is only 0.4 per cycle, this factor is reduced to only 1.7.

Recycle is simplified by human intervention. If materials are not recycled, they become diluted and dissipated in the environment: their entropy increases. Thus when an intelligent being sorts used material into separate bins, he diminishes the entropy of the original waste material. However, such a macroscopic Maxwell demon does not change the entropy of mixing appreciably unless, by his intervention, he prevents useful materials from becoming dissolved as individual molecules that then can be widely dispersed in soils or in the oceans.

The Energy Budget in the Age of Substitutability

As we exhaust the high-grade materials and have to use lower grade ones, the energy required to recover our needed materials will grow. Yet, because the composition of Avalloy is so dominated by essentially infinite iron and aluminum, the energy required for extracting a ton of metal will not grow nearly as much as one might think. Although Frasché was correct in stating in 1962, "The extraction of mineral raw materials from low-grade rock is a problem in the application of energy—at a price" [6], he should have added that the total mass of useful minerals that have a finite resource base is small. Therefore, the effect of their depletion on the entire energy system is less than Frasché's statement might imply.

Of course, even the infinite elements will take more energy to extract as we go from ores to more common rocks. The energy requirements for production and recycle of the abundant metals—Mg, Al, Fe, Ti, and Cu—were estimated by J. C. Bravard, H. B. Flora, II, and C. Portal in 1972 [11], and are summarized in Table 6. In this table we compare E_L, the amount of energy required to extract a ton of metal from low-grade, essentially inexhaustible ores, with E_H, the energy required to extract the metal from high-grade ores.

It is remarkable that the estimated energy required to produce these infinite metals from essentially inexhaustible sources is in every case not more than about 60% higher than the energy required to win the metals from high-grade ores. Even when all the reduced carbon is gone, the ratio for iron—by far the most important—is ~2.0.[4]

Extraction of useful metal from ore consists of two separate steps: first, mining and beneficiation of the ore; and second, reduction and refining of the metal from the beneficiated ore. Generally the second step requires considerably more energy and expense than the first, e.g., to mine and beneficiate one ton of Fe from presently used magnetic taconite ores requires about 5% as much energy as is required for the total production of steel from ore. Thus, the overall energy required is not very sensitive to the grade of ore until the ore becomes extremely dilute; and this will never happen for the infinite metals.

[4] Could we imagine charcoal from wood, used in metallurgy until 150 years ago, replacing coke today? Is this practical?

Table 7

Possible Annual Energy Budget for Demandite, Avalloy, and Agriculture in Stage III of The Age of Substitutability[a]

	United States		*World*	
	Present	Stage III	Present	Stage III[b]
Population × 10^6 persons	200	300	3600	10,000
Industry				
Avalloy[c]	0.60	1.20	0.15	0.60
Inorganic Chemicals	0.86	1.12	0.13	0.56
Cement, Stone, Clay, Glass	0.21	0.27	0.08	0.14
Petroleum Refining & Petrochemicals	0.47	–	0.07	–
Reduced C and H (Petrochemicals)	–	0.24	–	0.12
Total Demandite	2.14	7.83	0.43	1.42
Agriculture				
With Desalted Water	–	1.70	–	1.70
Without Desalted Water	0.87	0.87	0.68	0.87
Total Industry and Agriculture with Desalted Water	3.01	4.53	1.1	3.12
Total, All Other Uses	8	10.5	0.4	4.5
Total Energy Use	11[d]	15	1.5	7.6
Increase in Stage III				
Industry	–	0.69	–	0.98
Agriculture	–	0.83	–	1.02
Total	–	1.51	–	2.00

[a]All values in kw/capita unless otherwise noted.
[b]World values taken as half of U.S. values for industry and equal to U.S. values for agriculture.
[c]New metal from ore only; total metal = ~1.5 times new metal.
[d]Total in U.S., 1968, was 9.6 kw/person versus current value of 11 kw/person.

In 1968, the metal industry consumed 8.5% of the total energy used in the United States; some 90% of this was expended in the production of Fe, Al, Mg, Ti, and Cu. The per capita annual energy expenditure for metal production came to about 8,000 kwh/person or 0.91 kw/person. However, when only *new* metal from ores is considered, the primary metal industry accounts for 5.7% of total U.S. energy consumption; in this case, per capita energy use is reduced to 5,300 kwh/person. In the Age of Substitutability (assuming electrolytic H_2, not C, is used), it seems fair to double the energy expended for these metals—to 10,600 kwh/person/year or 1.2 kw/person—assuming that the amount of metal used per capita and the composition of Avalloy remain as they now are.[5] *Even in the Age of Substitutability, the amount of energy required to provide Avalloy is hardly twice (rather than ten or one hundredfold) per unit of metal used in the present era. This is because the dominant metals—Fe, Al, Mg—can be extracted from inexhaustible resources that demand relatively small additional amounts of energy for their extraction.*

We have tried to estimate how much the per capita energy budget for producing Demandite and desalted water for agriculture would increase in Stage III, though we realize that these estimates are speculative and require more study. In Table 7 we summarize these estimates. We also estimate what these values might be in Stage III,

[5] Even if Cu is replaced by Al, the energy per ton of Avalloy changes by less than 10%.

assuming the per capita use of Demandite remains constant, but that agriculture requires 100 gallons/person/day of desalted water. The details of the calculations that lead to the numbers in Table 7 are given in Appendix 3.

The most striking speculation that we make in Table 7 is that the increase in energy required in Stage III for non-renewable resources and desalted water is about 2 kw/person. Adding 2 kw for other uses, we arrive at a guess (justified in part by the considerations in Appendix 3) that the U.S. per capita energy for all uses in Stage III might increase from its present 11 kw to 15 kw, and the world per capita use from 1.5 kw to 7.5 kw. If the world's population grows to 10×10^9, and the entire world reaches half the projected U.S. per capita energy expenditure, the world's total production of energy would amount to about 75×10^9 kw. This is 12 times the present world-wide man-made energy, but still represents only 0.1 percent of the solar energy absorbed and reradiated by the earth.

We shall not dwell on the environmental, technical, and institutional implications of so large a production of energy, since these matters have been discussed elsewhere by many authors [12]. The main conclusions that we can draw are:

1. All fossil fuels, at this rate of energy expenditure, would be consumed within 100 years.
2. The fission breeder, fusion (if feasible), or solar energy, *in principle*, could carry this energy budget essentially forever.
3. The climatic changes caused by so large a release of energy cannot be predicted with our present knowledge of climatology, although the *average* increase in world temperature, assuming no changes in albedo, would be about 0.1 °C. Whether the effect of such energy output on climate will ever be predictable, *even in principle*, remains a moot question.
4. It seems likely that if fission breeders are the basis for the Age of Substitutability, large institutional changes will be called for so that man can live comfortably with fission—on the scale envisioned here. Thus, if the entire 75×10^9 kw were provided by fission breeders, with each reactor operating at 5×10^6 kw (of heat), the world would have to accommodate 15×10^3 reactors. If each reactor lasted 30 years, then 500 reactors would, in the asymptotic state, be retired each year. This amounts to about 10 reactors being built each week. Is this credible? Will we have the land, the waste disposal areas, and the capacity to deal with diversion and with transport of radioactivity on such a large scale?

Though technological fixes [13] will ease these problems, i.e., cluster siting would make diversion more difficult [14], one can see the point of those such as Mesarovic and Pestel who argue that a much more appropriate asymptotic per capita energy demand is the value that now characterizes much of Western Europe: about 4 kw/person. However, in Stage III of the Age of Substitutability, we estimate (Table 7) that loss of reduced CH_x would require an additional 2 kw/person for Demandite and agriculture. Since transport, space heating, and cooling would also require more energy in Stage III, we guess that to achieve the Pestel–Mesarovic world might require perhaps 7 to 8 kw/person, and that the problems enumerated above can hardly be avoided.

Mine Wastes in the Age of Substitutability [12]

The catastrophists often point out that as we exhaust our higher grade resources and mine lower grade rock, mine wastes will lead to disaster. But our analysis points to a rather different outcome. Since CH_x is by far the largest component of Demandite, most of the waste from mining is associated with extraction of coal—8 tons of spoil per ton of coal mined in the United States. Thus in Stage III of the Age of Substitutability, when we no longer mine coal, the mine spoil per person associated with other sources of energy (breeders, fusion, and solar) will be much less than the mine spoil now associated with our energy system.

As for mine wastes from Avalloy, these can never increase much more than threefold per ton because Avalloy is dominated by iron, and high-grade taconite iron ore is only three times as rich as infinite laterite ore. Moreover, the total waste from mining Avalloy is only 5% of that from coal mining: a threefold increase in Avalloy mining waste would still be small compared to the waste from coal mining.

We present in Table 8 our estimates for mining spoil at present and in the Age of Substitutability, for the U.S. and the world. These estimates are justified in Appendix 4.

From the table, one sees that at a per capita energy budget of 7.5 kw in Stage III the yearly mine spoil per person is 2.4 tons compared to the present 7.4 tons. Thus, one can argue that the world population could increase threefold before the mine wastes equal present mine wastes.

We have not attempted to estimate other pollutants in Stage III. However, since the bulk of air pollution is the result of burning of fossil fuel (CO_2, CO, SO_2, trace elements, fly ash), air pollution in Stage III ought to be *less* than present pollution.

Though our estimates are reassuring on the average, we are reluctant to leave an impression of facile optimism. The environmental impacts in specific places and specific situations might well be more serious than we have implied.

Table 8

Annual Fuels and Metals Ore Spoils Today and in The Age of Substitutability

	Stage I			
	Present[a]	Peak[b]	Stage II[c]	Stage III[d]
United States				
Population, 10^6 persons	215	250	275	300
Energy, kw/person	11	13	14	15
Spoil from Coal, tons/person	25	100	10	–
Spoil from Breeders, tons/person	–	–	–	0.6
Spoils from Avalloy, tons/person	1.3	~2	~3	4
Total Spoils, tons/person	26.3	102	13	4.6
Total Spoils, 10^9 tons/year	5.6	26	3.6	1.4
World				
Population, 10^6 persons	4000	6000	8000	10,000
Energy, kw/person	1.5	3	5	7.5
Spoil from Coal, tons/person	7	35	5	–
Spoil from Breeders, tons/person	–	–	–	0.3
Spoils from Avalloy, tons/person	0.4	~1	~1.5	2
Total Spoils, tons/person	7.4	36	6.5	2.3
Total Spoils, 10^9 tons/year	29.6	216	44	23

[a]Coal, oil, and gas available; half of coal strip-mined, half mined underground.
[b]Oil and gas gone, 80% of coal strip-mined, 20% mined underground.
[c]Strippable coal gone; all coal mined underground.
[d]All coal gone; all energy from non-fossil fuel resources.

Conclusions

Most of what we contend is speculation. Yet, there is one aspect of the future that seems to be scenario-proof: contrary to the assertions of the Neo-Malthusians, depletion of mineral resources *per se* need not create catastrophe, provided man finds an inexhaustible, non-polluting source of energy. The main problem is how to go from our present state, Stage I—with ample CH_x and other resource materials—to Stage III, the true Age of Substitutability, without incurring drastic social instabilities. Will we have the capacity and foresight to plan and execute the transition without such instabilities?

It is well to say that *in principle* substitutes can be found, that even without CH_x our per capita energy budget will not be much greater than it now is. But factors even of two in energy budget or price, though they seem small in the long run, can over the short term or in local situations cause great social dislocation.

What is at issue is the effectiveness of the marketplace in forcing a rational resource policy. To be sure, over the long term, the marketplace forces substitutions and use of lower grade ores. But the marketplace has a high discount rate: in technological changes that require many years, the marketplace as it now operates invariably seeks out paths that optimize short-term advantage. Such paths may waste resources in the long run. The situation is well illustrated by the light water breeder reactor (LWBR) and the light water reactor (LWR). Over a 30-year period, a 1000–MWe LWR with no recycle requires about 5000 tons of uranium; an LWBR of the same size may require perhaps 1500 tons of uranium. On the other hand, almost all of the LWBR inventory must be invested during the first few years, whereas the LWR uses uranium rather uniformly over its lifetime. Thus if one judges the relative economic costs on a very short write-off, the LWR wins; if the write-off is 30 years or longer, the LWBR wins. From the viewpoint of husbanding resources, over the long term LWBR is better than LWR; over the short term LWR is better.

Recycle faces the same dilemma. How can one use the marketplace to encourage recycle when, in the short run, it pays not to recycle? Is it possible for the marketplace to be modified, perhaps by government fiat, to reduce its discount rate, to take a longer, resource-dominated position?

Our *technical* message is clear: dwindling mineral resources in the aggregate, with the exception of reduced carbon and hydrogen, are *per se* unlikely to cause Malthusian catastrophe. But the exception is critically important; man *must* develop an alternative energy source. Moreover, the incentive to keep the price of prime energy as low as possible is immense. In the Age of Substitutability energy *is* the ultimate raw material. The living standard in the Age of Substitutability will almost surely depend primarily on the cost of prime energy. We therefore urge moving as vigorously as possible, not only to develop satisfactory inexhaustible energy sources—the breeder, fusion, solar, geothermal—but to keep the program sufficiently broad so that we can determine, perhaps within 50 years, the cheapest inexhaustible energy source.

Our *social* message is less clear and certainly less optimistic. Though we see during the Age of Substitutability no insuperable technical bars to living a decent—rather than a brutish—life, whether in fact this will happen is far from certain. As Robert Heilbroner [15] has pointed out, local shortages that in the course of history are destined to be viewed as transitory can and do cause large social and political instability. Heilbroner's Wars of Redistribution, pitting the overpopulated have-nots against the underpopulated haves, could collapse the society long before our carbon runs out. That the Age of Substitutability is *in principle* autarkic, since mankind no longer depends on reduced carbon (which is located in a few places), is little solace for governments or peoples today who look upon shortages of coming decades as threatening our entire social structure. We do not argue that the Age of Substitutability is an easily achieved technological heaven-on-earth. Rather we urge attention to those institutional deficiencies that now prevent us from passing through Stage II of our human voyage without causing the boat to capsize. The landfall, if we arrive at Stage III, should be surprisingly better than the catastrophists have predicted.

Appendix 1

This appendix provides: (1) an explanation of how the Demandite and Avalloy formulae were developed for Tables 1 and 2; (2) subsidiary formulae for Avrefract, etc; and (3) further detail about development of the crust, seawater, and air formulae of Table 3 and the "best resources" data of Table 4.

DEMANDITE

Determination today of a highly accurate formula for Demandite would be a formidable task and might even be impossible because of inadequate statistics. Nevertheless, since the first four materials—fossil fuels, sand and gravel, limestone, and iron—account for over 97% by weight of the non-renewable materials used by society, we believe the formulae of Table 1 are within a few percent of the ultimate formulae. Carbon and hydrogen have been combined because both of them have a primary common use: fuels. The different ratio of H/C for the U.S. and the world reflects the greater use of oil and gas (versus coal) in the U.S. compared to the world. The amount of carbon in limestone as opposed to that in fossil fuels is quite small (5 percent for U.S., 11 percent for the world). Of only academic interest are the very large amounts of oxygen associated with silica and limestone. Finally, formulae for Demandite show how small metals really are (1.3 mol % in U.S. and 1.6 mol % in the world) relative to all non-renewable materials used by society, and how the metals group is dominated by iron. Although the use of magnesium is large, less than 4% is used for production of metal.

AVALLOY

Determination of formulae for Avalloy is much simpler than for Demandite because statistics are more reliable. Note, however, that the formulae embrace two opposing "errors". First, many metallic elements include non-metal uses, and these are included (lead additives in gasoline, zinc and lead compounds in pigments, etc.). Second, metal recycle amounts to 1%–50% for specific metals; but since we are concerned with resource depletion, recycled amounts have been excluded.

SUBSIDIARY FORMULAE

In addition to Avalloy, Demandite can also be subdivided into other groups of elements having rather common use. Four such groups are: Avfert (average of agricultural chemicals), Avgas (average of industrial gases), Avchem (average of major non-metal chemicals and metals used in oxidized form), and Avrefract (average of refractory elements).

Avfert (including sulfur and hydrogen–NH_3–intermediates)
United States. $N_{.1803}$ $P_{.0284}$ $K_{.0029}$ $(S_{.0404}H_{.7280})$
World. $N_{.2638}$ $P_{.0391}$ $K_{.0384}$ $(S_{.0634}H_{.5953})$

Avgas
United States. $H_{.8138}$ $N_{.0483}$ $O_{.1026}$ $Cl_{.0342}$
$Ar_{.0008}$ $He_{.0003}$ $(Ne, Kr, Xe)_{<<.0001}$

World. $H_{.7779}$ $N_{.0613}$ $O_{.1306}$ $Cl_{.0294}$ $Ar_{.0004}$ $He_{.00001}$ $(Ne, Kr, Xe)_{<<.0001}$

Avchem
United States. $Na_{.7760}$ $S_{.1707}$ $Ba_{.0060}$ $F_{.0365}$ $B_{.0085}$ $Br_{.0020}$ $Other_{.0003}$
World. $Na_{.7634}$ $S_{.1930}$ $Ba_{.0050}$ $F_{.0305}$ $B_{.0071}$ $Br_{.0008}$ $Other_{.0002}$

Avrefract

United States. $Mg_{.8180}$ $Ti_{.1674}$ $Zr_{.0136}$ $(Be\ and\ other)_{.0010}$

World. $Mg_{.8722}$ $Ti_{.1196}$ $Zr_{.0072}$ $(Be\ and\ Other)_{.0010}$

AVERAGE ABUNDANCE OF ELEMENTS IN THE EARTH'S CRUST, SEA, AND ATMOSPHERE

The average composition of the lithosphere was calculated for all rocks to 1 kilometer depth, based on the assumption that in this thin layer on the continents 30% of the rocks are igneous, principally granitic, and that 70% are rocks or metamorphics derived from sedimentary formations. We assumed that 81.5% of the latter are shale, 12% sandstone, 6% limestone and dolomite, and 0.5% evaporites. In Table 3 the crustal composition is therefore a weighted average of the abundances of the various elements in these diverse type rocks.

Seawater and air are highly homogeneous. To provide a more satisfactory formula for seawater, the water content (96.5%) was omitted, so that the formula gives only the solid composition. This formula shows that seawater solids are heavily dominated by the first six elements which constitute 99.96 mol % of all the solids. The residual 0.04% consists principally of non-metallic elements such as boron and bromine and soluble metallic elements like lithium and rubidium. Table 3 shows that the common metals are extremely sparse—at least a thousandfold less prevalent than in rocks. Thus, terrestrial resources appear much more desirable sources of these elements unless very specific ion exchange, adsorption, or similar processes are developed for removal of a single element or groups of such elements from seawater. Even then, pumping costs would be enormous.

The formula for air deserves only one comment, namely that the concentration of helium in air (5 ppm by volume or 0.65 ppm by weight) is low compared to concentrations in natural gas so that recovery of helium from air will cost 50 or more times as much as recovery from most of the world's natural gas; unfortunately, the latter will have been consumed in less than 50 years.

HIGH-QUALITY "INFINITE" RESOURCES FOR WIDELY USED ELEMENTS

Table 4 gives "preferred" resources for the most commonly used elements in essentially infinite supply and demonstrates that at least for these elements recourse to less desirable, silicate-bearing igneous rocks, or "dirt", will never be required. The maximum concentration in the preferred resource, the total supply, and a resource-to-demand (R/D) ratio are given for each element. The R/D ratios are academic because of their large values, but the ratios do indicate that only the best and most accessible of these resources need be used for a very long time.

In determining quantities for the resource column of Table 4, the topmost kilometer of the earth's crust, all seawater in the oceans, and the entire atmosphere were assumed available for ultimate exploitation. This assumption seems acceptable because the rates of exploitation are so low ($<10^{-6}$/ year), and because over such a long time frame most of the exploited materials would return to the seas and the air.

Specifications for the world resource column of Table 4 are: The topmost kilometer of crust contains about 3.8×10^{17} tons of rock of which 1.15×10^{17} tons are igneous (including 2×10^{16} tons of basalt and 3×10^{14} tons of ultramafic rock rich in Ni, Cr, etc.), 2.2×10^{17} tons are shale (we assume only 10% is high alumina shale), 3×10^{16} tons are sand and sandstone, 1.6×10^{15} tons are limestone and dolomite, and 1.5×10^{15} tons are evaporites. The mass of the oceans is 1.5×10^{18} tons, containing about 5×10^{16} tons of solids; the mass of the atmosphere is 6×10^{15} tons. These great resources are assumed to be completely available for recovering all the elements in Table 4 except reduced carbon, phosphorus, and manganese; for these we use present estimates of reserves and hypothetical and speculative resources.

When all fossil fuels are depleted, recourse to the infinite carbon dioxide in the air and seas and in limestone may be required (as explained in Appendix 3) as a source of carbon for petrochemicals but not for fuels. Similarly, when CH_x is depleted we must acquire hydrogen from other sources for use as a metal ore reductant, for petrochemical production, and for ammonia synthesis; we assume this supply of hydrogen will come from electrolysis of water, which incidentally provides large amounts of by-product oxygen. As shown in Table 3, resources of "unextractable" CH_x are about 200 times that of coal, oil, and gas. Certainly, most of this resource will remain unexploited because more energy is needed to extract CH_x than is released by burning the CH_x. However, it may be used, in principle, as a raw material for petrochemicals; further study is required to compare CH_x from this source with CH_x from CO_2, and H_2 produced by water electrolysis.

Table 4 shows that some elements are recoverable from more than one infinite resource, e.g., salt and gypsum from both the lithosphere and the oceans.

Appendix 2
THE REMAINING ELEMENTS

To complete an appraisal of the elements, we consider briefly the non-metals (including the commonest metal oxides), refractory metals, and the non-ferrous by-product elements. Among the non-metals, 11 of the 12 most widely used in Demandite (iron being the sole exception) are either non-metals or metals used in their oxidized state (C, Si, Ca, Na, N, S, O, Cl, H, K, and P); only CH_x and P in this group have non-infinite resources. Together, these 11 non-metals or oxidized metals represent 94.6% of all non-renewable materials currently used by the world with iron accounting for an additional 4.8%. The other 70 stable elements comprise the remaining 0.6% of materials used by man. In 1968 in the pre-oil-embargo era the major non-metals, oxidized metals, and iron represented 86% of the value of all non-renewable resources; today this value is some 92% because fossil fuel prices have tripled.

Of the other non-metals or metals used predominantly in oxidized form (Ba, F, Ar, B, Br, Sr, the Rare Earths, I, He, Li, Y, Sc, Rb, Ne, Kr, and Xe), at least five (Ar, Ne, Kr, Xe, and Br) are infinite. The entire group accounts for a scant 0.05% of all non-renewable resources used by man: of these, 84% are barium and fluorine; 15% argon, boron, and bromine; and 1% the others. Even very large price increases for any or all of these elements are unlikely to appreciably affect the world's total resource bill.

The major refractory metallic elements used by society are titanium and zirconium; in addition, over 95% of all magnesium is used as the oxide, as is a small fraction of consumed aluminum. The quantity and value of this group of elements are about the same as for the minor non-metals. The resources of magnesium and aluminum are infinite and of titanium and zirconium large to very large; therefore, costs for this group should never increase more than severalfold. The major present use (>90%) of titanium is for titania pigment used largely to improve the appearance of paints, and society would hardly collapse if such goals were unfulfilled.

The final group, the non-ferrous metal by-product elements (As, Sb, Cd, Ge, Se, Te, Ga, In, Tl, and Bi), are obtained partly or wholly as by-products of copper, zinc, and lead. As a group they constitute only 0.0015% of all non-renewable substances used by society and less than 0.1% of the value of such materials; the first three account for 99% of the total. However, a number of the other non-ferrous metal by-products have a very important use: as semiconductors in communications and computers. Since copper, lead, and zinc are metals with limited resources which will be largely replaced during the Age of Substitutability by iron, aluminum, and magnesium, it seems prudent to recover and store these by-product elements.

Appendix 3
ENERGY DEMAND IN THE AGE OF SUBSTITUTABILITY

Annual energy consumption in the U.S. in 1968 was 19×10^{12} kwh (versus about 21×10^{12} kwh in 1972). This consumption is broken down into 25% for transportation, 34% for home and commercial use, and 41% for industry. Generation of electricity accounted for 22.5% of this total energy consumption. Use of energy by industry for five major industrial sectors is shown in Table A1. The first four sectors (metals, inorganic chemicals, cement-stone-clay-glass, fossil fuels), are based on non-renewable resources; the last (other manufacturing) is based mainly on output from the first four.

According to Eric Hirst [16], the energy expenditure per kilocalorie of food production is now nearly 7 kilocalories. Thus to grow, harvest, process, and distribute the 2,500 kilocalories of food/day necessary to feed the average American requires about 17,500 kilocalories daily, or 7,600 kwh/person/year (0.87 kw/person). This represents approximately 10% of total U.S. energy consumption. (To grow this food requires about 8,000 liters of water/person/day).

The purpose of this appendix is to examine how energy use by industry and agriculture may change during transition to and in the final stage of the Age of Substitutability, taking into consideration future resource availability and mix. As we will show, energy increases per ton of product in the inorganic chemical and cement, stone, clay, and glass sectors will be small, and that for

Table A1

Present (1968) U.S. and World Energy Demands

for Non-Renewable Resource Industries and for Agriculture

	Total Quantity 10^6 tons	Total Energy 10^9 kwh	kwh/ton	kwh/capita	kw/capita
United States					
Industry					
Metals, Primary	85.6	1,050	12,300	5,250	0.60
Metals, Total	132.0	1,595	12,080	7,975	0.91
Inorganic Chemicals	84.1	1,508	17,930	7,540	0.86
Cement, Stone, Clay, Glass	1,642	368	225	1,840	0.21
Fossil Fuels	1,500	837	560	4,185	0.48
Subtotal, new metal	3,312	3,763	1,136	18,815	2.14
Subtotal, total metal	3,358	4,308	1,280	21,540	2.46
Agriculture	–	1,520	–	7,600	0.87
Total, new metal	–	5,283	–	26,415	3.01
Total, total metal	–	5,828	–	29,140	3.33
Other Manufacturing	–	2,431	–	12,150	1.38
Total, total metal	–	8,260	–	41,300	4.71
Other Energy Use	–	11,040	–	55,200	6.3
Total, all energy	–	19,300	–	96,500	10 (11 in 1973)
World					
Industry					
Metals, Primary	424	4,700	11,100	1,305	0.15
Metals, Total	636	6,360	10,000	2,120	0.24
Inorganic Chemicals	221	3,962	17,930	1,100	0.13
Cement, Stone, Clay, Glass	11,670	2,626	225	730	0.08
Fossil Fuels	4,760	2,380	500	661	0.07
Subtotal, new metal	17,075	13,668	800	3,797	0.43
Subtotal, total metal	17,287	16,600	690	4,611	0.53
Agriculture	21,600	–	–	6,000	0.68
Total, new metal	38,675	–	–	9,797	1.12
Total, total metal	38,887	–	–	10,600	1.21

metals somewhat more, but certainly less than a factor of two. The most complex changes are in the petroleum refining and related industries sector.

The major resource problem in the future is the eventual depletion of natural supplies of reduced carbon and hydrogen (CH_x) as a source of energy, a reductant for metal ores and minerals, and as a raw material for petrochemical products, and ammonia for fertilizer. We envision two stages en route to Stage III of the Age of Substitutability. The first (present) stage should last about 30 to 50 years during which world petroleum and natural gas supplies will be depleted; however, Stage I may be extended if oil from high-grade shale can be massively exploited. The second stage, when coal is the sole remaining fossil fuel, may last several hundred years. In this stage, coke for making iron and steel will still be available, but petrochemicals, synthetic oil and gas, and hydrogen for ammonia synthesis must come from coal. Finally, in the true Age of Substitutability when *all* fossil fuels have been used, new energy systems–breeder reactors, fusion, solar–must be fully developed and widely deployed to meet world energy needs. The major metal ore reductant and source of hydrogen for ammonia will undoubtedly be hydrogen from water electrolysis. The source of carbon for petrochemicals, unless substitutes displace them—and some would be very hard to do without—must become carbon dioxide from the calcination of limestone, or conceivably CH_x from shale that is too dilute to mine as a source of energy.

With regard to future sources of high-temperature energy for industry, there now seem to be two options: (1) use of very high-temperature gas-cooled reactors with thermodynamic efficiencies approaching that available with the combustion of fossil fuels, and (2) electricity or electrolytic hydrogen with consideably lower overall thermodynamic efficiencies (30% to 40%). The former would not materially increase the input heat requirements of industry; the latter might double present requirements.

Reduction of metal ore concentrates and production of reduced carbon and ammonia will both require very large quantities of hydrogen during the Age of Substitutability. This hydrogen will probably come from water electrolysis because it does not appear that the overall energy budget for alternative chemical systems is much better than for electrolytic decomposition of water. The free energy change for water electrolysis is 32,700 kwh/ton of H_2 (or 1,840 kwh/ton of iron reduced from Fe_2O_3 with hydrogen). An advanced electrolysis cell [17] should produce hydrogen, using 48,500 kwhe/ton (121,000 kwht/ton at 40% thermodynamic efficiency for production of electricity). We consider use of hydrogen as a metal ore reductant in the discussions of metals, and use as a reductant for carbon dioxide below.

The basic step in producing petrochemicals from carbon dioxide is to produce methane by the following reaction:

$$CO_2 + 4H_2 \rightarrow CH_4 + 2H_2O.$$

This step is exothermic, releasing −39 kilocalories of enthalpy per mole and −27 kilocalories per mole of free energy. However, beginning with the electrolysis of water, the overall reaction is

$$2H_2O + CO_2 \rightarrow CH_4 + 2O_2$$

which is highly endothermic and requires +191 kilocalories per mole of free energy. Methane is then converted to ethylene ($\Delta F^\circ = 18.35$ kilocalories) or acetylene ($\Delta F^\circ = 37.62$ kilocalories), the more usual precursors of many plastics, solvents, etc. These reactions increase the overall change in free energies to 209 and 229 kilocalories per mole, respectively, but reduce the quantity of electrolytic hydrogen needed. As with metals production (Table 7), the total energy required in these processes is generally much greater than the free energy changes.

Although the energetics of such processes need further study, we estimate petrochemicals from air, water, and energy would require approximately two to five times[6] more raw energy than do plastics from natural gas. The energy we now use for plastics in the U.S. is 0.2% of our total consumption: in the Age of Substitutability this might increase to 1% or about 0.1 kw/person.

The metal industry has been discussed in the text. There we noted (Table 6) that energy for producing metals with abundant resources (Fe, Al, Mg, and Ti) would increase not more than 30% (63% for titanium) when utilization of lower grade infinite resources is required but before all fossil fuels have been used. When fossil fuels are gone and hydrogen becomes the main metals ore reductant, energy for extracting metals during the Age of Substitutability should still be less than twice the present requirement. Table 7 also indicates that the mining and beneficiation energy are only small parts of the total energy required. As recycle of metals improves, the relative amount of new metal from ores will decrease, and therefore the energy demand, because recycled metal requires less energy per ton. Conversely, as use of low-energy, but scarce, non-ferrous metals—Cu, Zn, Pb, and Sn—is replaced by the energy-intensive, but abundant substitute metals—Al, Mg, Ti, and Fe—the energy demand will increase. However, because Avalloy is dominated by iron and steel, iron strongly buffers the energy demand. For example, if all Cu were replaced by Al, the energy per ton of modified Avalloy would increase only about 10%. In any event, it is hard to see how Avalloy can ever require more than twice the present energy used per ton of metal.

Energy requirements for metals in Tables A1 and A2 are given for new metal production from ores and for total production, including recycle. It should be noted that the kwh/ton for Avalloy is almost

[6] The factor of two would apply to polyethylene; the factor of five, to polyvinyl-chloride because of the energy-intensive chlorine manufacture step.

Table A2

Possible U.S. and World Energy Demands for Non-Renewable Resources and for Agriculture in The Age of Substitutability[a]

	Total Quantity 10^6 tons	Total Energy 10^9 kwh	kwh/ton	kwh/capita	kw/capita
United States					
Industry					
Metals, Primary	128.4	3,150	24,600	10,500	1.20
Inorganic Chemicals	126.2	2,940	23,300	9,800	1.12
Cement, Stone, Clay, Glass	2,463	717	291	2,390	0.27
Carbon and Hydrogen	150	630	4,200	2,100	0.24
Subtotal, Industry	2,867	7,437	2,594	24,790	2.83
Agriculture					
Without Desalted Water	–	2,280	–	7,600	0.90
With Desalted Water	–	4.470	–	14,900	1.70
Total, Without Desalted Water	–	9,717	–	32,390	3.73
Total, With Desalted Water	–	11,907	–	39,690	4.53
World					
Industry					
Metals, Primary	2,134	52,500	24,600	5,250	0.60
Inorganic Chemicals	2,103	49,000	23,300	4,900	0.56
Cement, Stone, Clay, Glass	41,100	11,950	291	1,195	0.14
Carbon and Hydrogen	2,500	10,500	4,200	1,050	0.12
Subtotal, Industry	47,837	123,950	2,594	12,395	1.42
Agriculture					
Without Desalted Water	–	76,000	–	7,600	0.90
With Desalted Water	–	149,000	–	14,900	1.70
Total, Without Desalted Water	–	200,000	–	20,000	2.32
Total, With Desalted Water	–	273,000	–	27,300	3.12

[a]Populations: U.S., 300×10^6; World, $10,000 \times 10^6$

the same for both cases. The kwh/ton for new metal should be higher than recycled metal; however, the value for total metal may include many finishing operations not included in estimates for new metal.

The inorganic chemicals, and the cement, stone, clay, and glass groups are based on infinite resources for nearly all the major elements or materials of these groups. Thus, as long as fossil fuels last, it is difficult to see how the energy/ton for these materials can increase significantly. When fossil fuels are gone, however, energy requirements will increase–but not double–especially since much of the energy is already used as electricity that is now generated at about 30% efficiency. In our study we have estimated that the energy/ton of these materials in the Age of Substitutability might increase by 30%.

In the cement, stone, clay, and glass group, nearly half the energy consumed is for producing Portland cement from clay and limestone; the remainder is used for quarrying sand and gravel, crushed stone, and clay, and for manufacturing glass, bricks, and other clay products. A major increase in energy requirements will occur for increased transportation of these low-unit cost commodities as source-to-market distances increase, but we have not estimated this increase.

The inorganic chemicals sector is dominated by the major non-metals and oxidized metals—the seven elements (N, O, Na, Cl, S, P, and K) on line 3 in the formulae of Demandite (Table 1)—all of which have infinite resources with the possible exception of phosphorus. They are used principally in fertilizers, industrial gases, and basic chemicals such as caustic soda, soda ash, and sulfuric acid.

Energy demand for agriculture will also be higher during the Age of Substitutability. The extreme case would be agriculture based exclusively on desalted water and on use of fertilizers—fixed nitrogen, phosphates, and potash. If water must be won from the sea, since 6 kwh is the minimum work required to desalt 8,000 liters of water (which in turn is the average required to grow 2,500 kilocalories/person/day of food), then 6 kwh of *work*, or about 20 kwh of heat, is the thermodynamic daily minimum/person for desalting water now required for agriculture. This number must be multiplied by about 20 because desalting cannot be done reversibly; but the figure could be reduced twentyfold because with advanced farming methods we might grow a man's food with only 400 liters of water/day. Desalted water for agriculture might therefore require 20 kwh/person/day or 0.8 kw/person (7,300 kwh/person/year); and the total energy required to conduct agriculture in an era of water scarcity might be about 1.7 kw/person. This is almost surely an overestimate, since it is unlikely that most of the world will depend on desalted water for agriculture. Adding this to the 1.7 kw/person for industrial products[7] we estimate the *total* per capita energy in the Age of Substitutability for recovering non-renewables, and for agriculture based on artificial fertilizer and desalted water, to be some 3.5 kw/person.

The data on which these estimates are based are summarized in Table A2. Assumptions include a world population of 10×10^9 persons (300×10^6 for the U.S.), world use of energy at 7.5 kw/person (versus 15 for the U.S.), a world agricultural energy demand equal to U.S. demand, but a world industrial per capita energy use only one-half that of the U.S. Table 8 shows that such an estimate indicates an increase, for industry and agriculture, of 1.5 kw/person in the U.S. and a worldwide increase of 2 kw/person.

Appendix 4

POLLUTION IN THE AGE OF SUBSTITUTABILITY

The primary environmental insult in Stage II of the Age of Substitutability will result from mining associated with use of carbon and to a much smaller extent of iron and copper. In addition, society consumes very large quantities of sand and gravel, crushed and dimension stone, gypsum, pumice, and clay. The aggregate of these materials is 1.6 billion tons/year in the United States or 8 tons/capita. However, the bulk of these materials is quarried or dredged from a myriad of small rock quarries and sand pits, used almost as removed from the ground mainly for construction, with a minimum environmental impact other than leaving many holes in the ground. Probably the greatest environmental problem associated with the use of such materials is the local dust, smoke, and thermal pollution resulting from the manufacture of Portland cement. Because of such minor environmental impact, we will not discuss this class of materials further.

In Stage III of the Age of Substitutability, when all the reduced carbon has disappeared, the following estimate shows the environmental insult caused by mining should be *less* than today.

Each person in the U.S. now uses, annually, about 3 tons of coal, and about 0.4 tons of Avalloy, or a ratio of nearly 8:1. The spoils from mining coal usually dominate all others: about 8 tons of rock/ton of coal (50% strip, 50% deep mining) or about 25 tons/person/year. In 30 to 50 years, when oil and gas are gone, we will be stripping even more extensively, and spoils will increase considerably to as much as 100 tons/person/year. However, when all the strippable coal is depleted, we will undoubtedly return exclusively to underground coal mining, and the rock spoils will be reduced to perhaps 10 tons/person/year.

We presently mine 3 tons of rock per ton of Avalloy used: thus, the rock associated with Avalloy mining is about 1.3 tons/person/year. Even when we go to infinite sources of iron—10% instead of 30% ores—rock associated with Avalloy (4 tons/person/year) is still only one-sixth the rock now

[7] This becomes 1.4 kw/person when only new metal is considered.

associated with extraction of coal, and about 40% of the coal spoils when the strippable coal is no longer available.

When reduced carbon has been consumed, and we have to depend on fission, fusion, or solar energy, the rock moved per capita will be decreased. The worst case would be energy from fission breeders, though fusion based on lithium might be comparable. Consider energy from fission breeders based on rock bearing 10 ppm of uranium and thorium. Then our projected U.S. budget of 15 kw/person corresponds to burning 6 grams of uranium and thorium/person/year; this corresponds, at 10 ppm, to mining 0.6 ton of rock/person/year, rather than the 25 tons/person/year we now devote to energy. Thus, a striking characteristic of Stage III of the Age of Substitutability, as far as the environment is concerned, is that the amount of waste rock handled for fuel and metals will be *reduced* from about 30 tons/person/year to some 5 tons/person/year.

For the world, in the Age of Substitutability, we again assume 7.5 kw/person/year; for Avalloy, 0.2 ton/person/year (one-half the current U.S. figure). Then as long as strippable coal is available, the mine spoil would be 12 tons/person/year from coal-mining, and (at 10% ore) 2 tons of rock/person/year from Avalloy. When mining for coal has ceased, the rock per capita to extract Avalloy, on a world-wide basis, may be as little as 15% of the rock handled when carbon extraction was at its maximum. The *total* rock handled by the whole world, assuming a population of 10^{10}, comes to 2.5×10^{10} tons/year. This is about the same amount of rock as that now produced by coal mining throughout the world. It would seem that the total mining activity in the Age of Substitutability may be the same as now; however, in the interim it may increase and then decline as coal resources approach depletion.

References

1. Donella H. Meadows, Dennis L. Meadows, Jørgen Randers, and William W. Behrens III, *The Limits to Growth*, A Report for the Club of Rome's Project on the Predicament of Mankind, Universe Books Publishers, New York, 1972.
2. Mihajlo Mesarovic and Eduard Pestel, *Mankind at the Turning Point*, The Second Report to the Club of Rome, E. P. Dutton and Company, Inc./Reader's Digest Press, New York, 1974.
3. Charles Galton Darwin, *The Next Million Years*, Doubleday, Garden City, New York, 1953.
4. H. G. Wells, *The World Set Free: A Story of Mankind*, Dutton and Company, New York, 1914.
5. Meadows, et al., *The Limits to Growth, op. cit.*, p. 71.
6. Dean F. Frasché, *Mineral Resources*, A Report to the NAS–NRC Committee on Natural Resources, National Academy of Sciences-National Research Council Publication 1000–C, Washington, D.C., 1963, p. 18.
7. H. E. Goeller, "An Optimistic Outlook for Mineral Resources", presented before the Minnesota Forum on Scarcity and Growth, sponsored by University of Minnesota, Bloomington, Minnesota, June 22, 1972.
8. H. G. Wells, Julian S. Huxley, and G. P. Wells, "The Spectacle of Life", Book Six, pp. 1031–1032, *The Science of Life*, Volume Three, Doubleday, Doran & Company, Inc., Garden City, New York, 1931.
9. W. Fulkerson and H. E. Goeller (ed.), "Cadmium, The Dissipated Element", ORNL–NSF–EP–21, Oak Ridge National Laboratory, Oak Ridge, Tennessee, January 1973.
10. H. E. Goeller, "Summary of Current World Resources, Production and Uses of Mercury and Possibilities for Recycle and Substitution", presented before Ad Hoc Committee on the Rational Use of Potentially Scarce Metals, Scientific Affairs Division, North Atlantic Treaty Organization, London, April 17–18, 1975. (To be published 1976).
11. J. C. Bravard, H. B. Flora II, and C. Portal, "Energy Expenditures Associated with the Production and Recycle of Metals", ORNL–NSF–EP–24, Oak Ridge National Laboratory, Oak Ridge, Tennessee, 1972.
12. Alvin M. Weinberg and R. P. Hammond, "Limits to the Use of Energy", American Scientist **5(4)**, 414–418 (July/August 1970).
13. Alvin M. Weinberg, "Can Technology Replace Social Engineering?", *Bull. Atomic Scientists* **XXII**, 4–8 (December 1965).
14. Alvin M. Weinberg, "Moral Imperatives of Nuclear Energy", *Nuclear News* **14**, 33–37 (December 1971).

15. Robert L. Heilbroner, *An Inquiry into the Human Prospect*, W. W. Norton and Company, New York, 1974.
16. Eric Hirst, "Energy Use for Food in the United States", ORNL–NSF–EP–57, Oak Ridge National Laboratory, Oak Ridge, Tennessee, October 1973.
17. J. E. Mrochek and H. E. Goeller, "Generalized Capital and Operating Cost for Power Intensive and Allied Industries", ORNL–4296, Oak Ridge National Laboratory, Oak Ridge, Tennessee, December 1969.

Implications and Challenges

HAROLD A. LINSTONE

1. Leibniz and Substitution

In their categorization of technological forecasting methods according to Churchman's typology of Inquiring Systems, Mitroff and Turoff classified substitution as Leibnizian:

> The Leibnizian character of these models can rather easily be seen by spelling out a number of buried assumptions that underlie their applicability . . . It seems to be an implicit assumption that the reason why the forecasts can be relied on to predict the future is because the models reveal or embody a fundamental, enduring, structural feature of reality . . . A second assumption is that the models can be widely applied, again because the models supposedly embody a characteristic process that underlies a wide range of technical and social processes . . . [Thus it is assumed that] the models make possible the data that is fitted to the models, and that the data do *not* make possible the models. *The models are really prior to the data in the sense that they are used to uncover the kind of data that can be fitted to the models* . . . [hence] they are always 'true' by definition. [1]

The validity therefore rests on the "understanding" of the model builder. This description may fit the early substitution work, such as that of Fisher and Pry. It is clear that the developments presented in this book have made considerable use of other inquiring systems (IS). Martino and Conver are data-based rather than model-based (i.e., Lockean IS). Ayres and Shapanka as well as Sharif and Kabir use multiple models–including system dynamics and input-output–to improve the forecast (i.e., Kantian IS). Goeller and Weinberg are in another mode, virtually redefining contextual reality to facilitate the generation of new options and create the impetus for technical and social change (i.e., Merleau-Ponty IS) [2]. This diversity of philosophical underpinnings is probably a good indicator of future research in this area.

2. The Discounting Dilemma

The bane of forecasters is the inherent tendency of the rest of the world to apply a discount rate to future (and past) problems. The consequence of this form of popular optimism is to ignore the need for substitution until it assumes a gravity apparent to the most myopic, i.e., a problem readily manageable with ample lead time is left untouched until it looms as a staggering task of crisis management [3]. Goeller and Weinberg are well aware of the high discount rate with which our economy operates (see p. 262). An illustration of the impact of discounting on substitution-related decisions is the question of the development of solar energy versus coal. Solar energy incurs heavy initial costs and requires low maintenance later; coal has modest initial costs but high continuing costs. Without discounting, a solar energy program appears to be the preferred (lower cost) choice. With any significant discount rate the high future operating costs of coal are perceived as minimal and the initial high solar energy installation costs are forbidding. In other words, coal is the cheaper alternative and preferences are reversed. In the recurring debate between economists and environmentalists the discounting process invariably favors the former. Short term sacrifices (e.g., shutdown of a plant) furnish more telling ammunition than long term gains (e.g., environmental restoration). Significant improvement of this situation requires value changes and metapolicies which appear nowhere in sight in the Western societies.

3. New Kinds of Substitution

It has been a basic assumption of technological forecasters that substitution refers to the replacement of one technology by another that is more advanced. In the next generation this assumption may prove misleading at times. We may consider substitutions such as the following:

(a) *Substitution by a Less Sophisticated Technology.* E. F. Schumacher has popularized the concept of intermediate technology ("small is beautiful") in connection with the transfer of technology to developing areas. A bicycle may be more useful than an automobile, a small machine more effective than a large one designed for a highly industrialized society. Consider the case of eggs in Zambia. A need for egg cartons led to the search for a carton manufacturing source. The initial response of industry was that the limited annual quantity of cartons required made the stamping machines being produced in the West uneconomical for Zambia. However, it did not take long for technological entrepreneurs to have a better idea—development of a less sophisticated system to produce egg cartons economically under the given constraints. Ironically this development has subsequently led to orders for the new system from developed countries such as Canada! Proceeding in his familiar tracks the technological forecaster may completely miss this possibility—and profitable new lines of business!

(b) *Substitution by a Non-Technological Alternative.* The high rate of change of technological systems has not been matched by a corresponding pace in societal systems. The slowness of change of social institutions has created serious gaps. Many adverse technological impacts today are the result of misuse, particularly overuse, of available technology.

With embarrassing frequency, the excesses could, in principle, be alleviated or eliminated far more easily by regulatory or organizational change than by demanding a technological substitution. Consider the case of solid waste management, specifically residential or household waste in American cities. The depletion of resources provides an impetus for substituting new means of collecting and processing solid waste to facilitate recycling and reuse of materials. The obvious technological response is the development of sophisticated (and expensive) waste separation equipment for use by municipalities. A valuable non-technological solution is the separation of waste at home with separate collection [4]. Clearly, if waste is not mixed at the start of the collection process, energy is not required to separate it again later.

The history of the U.S. involvement in Vietnam is replete with examples of the drive to substitute new technological systems for existing concepts when non-technological alternatives were far more suitable. The design of special aircraft and sophisticated weaponry were often exercises in futility brought about by obsessive concentration on technological options.

(c) *Substitution at a Higher Level.* We are accustomed to technological substitution of one system for another in aircraft, computers, engines, manufacturing processes, and consumer products. In the future pervasive resource problems may lead to technological substitution at a higher, more strategic level. For example, fossil fuel shortages may, for some time period, lead to the substitution of communications for transportation. It takes far less energy to move ideas than people. This suggests that the horizon of the forecaster must expand. Not only should he consider the envelope formed by a sequence of logistic curves, but the envelope produced by a series of such envelopes. It has been recognized that forecasts of railroad technology miss the jump to aircraft and that an envelope based on all modes of transport is more useful. It may now be necessary to place communication and transportation on the same graph.

4. Abortion of Technological Substitution

The impact of technology on the environment is now well publicized. The seriousness of potential adverse effects is leading repeatedly to abrupt stoppages in technological substitution. The SST development was halted by governmental action, invalidating a number of substitution forecasts. Current concerns regarding genetic, biomedical, and behavioral research may lead to legislative or regulatory constraints and cessation of research and development that would normally culminate in substitution. The anticipated shift to nuclear energy (breeder reactors, fusion) faces similar threats. These possibilities point up the importance for technology assessment in determining potential substitutions.

5. Limits to Growth

The logistic curve has become the symbol of "limits to growth" and this concept has become the rallying cry of many intellectuals in the West. It may be a dangerous strategy. In terms of human evolution a steady state or no-growth future may prove as deadly as nuclear annihilation–slow death rather than quick death, drop by drop water torture rather than a single apocalyptic wave. Continued human evolution implies continued growth, but not necessarily growth as commonly defined in an industrial society. Confronted with a logistic economic growth curve as that in the Introduction to the book (p. xiv) the technological forecaster with a historical perspective is likely to insist that man's development is more correctly viewed as a series of ascending logistic curves similar to those that he finds useful in his own work (Fig. 1). It would seem far more meaningful to consider each logistic curve in a different dimension. GNP per capita is a suitable

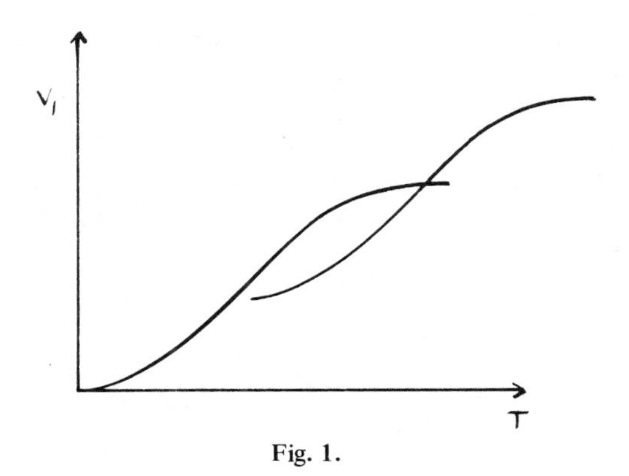

Fig. 1.

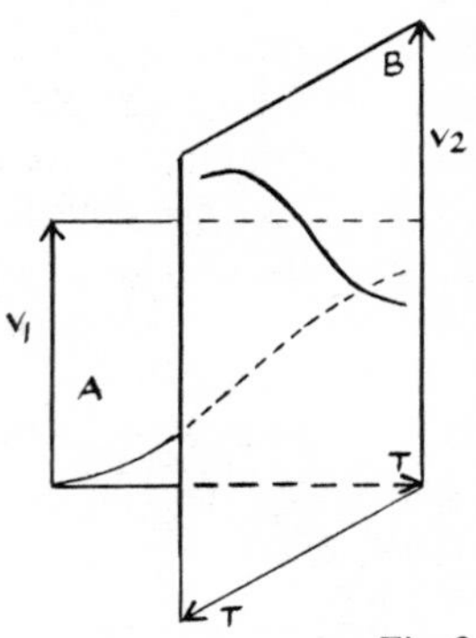

Fig. 2.

measure of growth in an industrial society, but not in either an agricultural or a post-industrial society. As growth along the logistic curve in the present plane tapers off, it begins in a new plane (Fig. 2). If we accept this schema then it becomes evident that the debate between the steady state and continued economic growth proponents is misdirected. The question is not "to grow or not to grow." If evolution is to proceed the question should be: what is the next direction in the ascent of man?

In this context we can envision a two-fold role for technology:

(1) Facilitate the slowing growth process following the inflection point on our current logistic curve (e.g., energy conservation technology, intermediate technology);

(2) Assist in the search for the next growth dimension, the process of design for evolution [5]. (Is the new frontier inner space? outer space?) And herein lies the most inspiring challenge to the virtuoso in technological substitution.

References

1. I. I. Mitroff and M. Turoff, Technological Forecasting and Assessment: Science and/or Mythology?, *Technological Forecasting and Social Change* **5** (1973), American Elsevier Publishing Co., N.Y., p. 117.
2. D. S. Scheele, Reality Construction as a Product of Delphi Interaction, in H. A. Linstone and M. Turoff (eds.), *The Delphi Method*, Addison-Wesley Publishing Co., Reading, Mass., 1975, p. 43.
3. Cf. H. Linstone, On Discounting the Future, *Technological Forecasting and Social Change* **4** (1973), American Elsevier Publishing Co., N.Y., pp. 335–338, or H. Linstone and M. Turoff, *The Delphi Method*, Addison-Wesley Publishing Co., Reading, Mass., 1975, pp. 574–578.
4. Duncan, R., The Oregon Plan for Recycling Household Solid Waste, *Compost Science* (Jan.–Feb. 1975).
5. E. Jantsch, *Design for Evolution*, Braziller, New York, 1975.

Technological Substitution: A Selected Bibliography

Of the more than 2,000 published studies of diffusion of innovations, the great majority are the work of sociologists. A comprehensive bibliography of these works can be found in E. Rogers and F. Shoemaker, *Communication of Innovations: A Cross Cultural Approach*, New York, The Free Press, 1971. In the following bibliography, however, we concentrate on the more recent studies, especially those relevant to forecasting technological substitution and those that are particularly illustrative of the major traditions of research in this area.

Babcock, J. M., 1962, "Adoption of hybrid corn: A comment", *Rural Sociology*, **27**, pp. 332–38.

Bailey, N. J. T., 1957. *The Mathematical Theory of Epidemics*. New York: John Wiley.

Bain, A. D., 1963, "The growth of demand for new commodities", *Journal of The Royal Statistical Society*, **126**, Series A.

——, 1962, "The growth of television ownership in the United Kingdom", *International Economic Review*, **3**, pp. 145–67.

——, 1964. *"The Growth of Television Ownership in the United Kingdom", A Lognormal Model.* Cambridge: University Press.

Bartholomew, D. J., 1967. *Stochastic Models for Social Processes.* New York: John Wiley.

Bass, Frank, M., 1969, "A new product growth model for consumer durables", *Management Science*, **15**, pp. 215–27.

Belcher, J., 1958, "Acceptance of the Salk polio vaccine", *Rural Sociology*, **23**, pp. 158–70.

Bernhardt, I., 1970, "Diffusion of catalytic techniques through a population of medium-size petroleum refining firms", *The Journal of Industrial Economics*, **19**, pp. 50–64.

——, and K. D. MacKenzie, 1972, "Some problems in using diffusion models for new products", *Management Science*, **19** (2), pp. 187–200.

Blickwede, M., 1972. "Development and implementation of new technologies in the steel industry", in *Proceedings, Third National Seminar: Steel Industry Economics*. New York:

Boff, Du R. B., 1964, "A note on the substitution of inanimate for animal power", *Journal of Political Economy*, **LXXII**, pp. 196–97.

Bonus, H., 1973, "Quasi-Engel curves, diffusion, and the ownership of major consumer durables", *Journal of Political Economy*, **81** (3), pp. 655–77.

Brandes, O., 1971. *Supply Models, An Empirical Study of Adaptation and Innovation in the Firm.* Gothenburg Studies in Business Administrations.

Brittain, J. E., 1976, "The international diffusion of electrical power technology, 1870–1920", *The Journal of Economic History*, **36**, pp. 108–30.

Brown, L. A., 1968. *Diffusion Dynamics: A Review and Revision of the Quantitative Theory of the Spatial Diffusion of Innovation*. Lund: Gleerup, Lund Studies in Geography.

——, 1969, "Diffusion of innovation: A macro view", *Economic Development and Cultural Change*, **17**, pp. 189–211.

——, 1974. "The Market and Infrastructure Context of Adoption: A Perspective on the Spatial Diffusion of Innovation". Studies in the Diffusion of Innovation Discussion Paper Series, Department of Geography, Ohio State University.

——, and K. R. Cox, 1971, "Empirical regularities in the diffusion of innovation", *Annals of the Assoc. of American Geographers*, **61**, pp. 551–59.

Capron, W. M., ed., 1971. *Technological Change in Regulated Industries.* Washington, D.C.: The Brookings Institute.

Carlson, J. W., 1967. "Aspects of the Diffusion of Technology in the U.S.". Paper prepared for the 5th Meeting of Senior Economic Advisers, United Nations, Economics Commission for Europe, Geneva, October 2, 1967.

Casetti, E., 1969, "Why do diffusion processes conform to logistic trends?", *Geographical Analysis,* **1**, pp. 101–5.

——, and R. K. Semple, 1969, "Concerning and testing of spatial diffusion hypothesis", *Geographical Analysis*, **1**, pp. 254–59.

Chow, G. C., 1967, "Technological change and the demand for computers", *American Economic Review*, **57**.

Colman, G. P., 1968, "Innovation and Diffusion in Agriculture", *Agricultural History*, **42** (3).

Coleman, J., E. Katz, and H. Menzel, 1966. *Medical Innovation: A Diffusion Study.* Indianapolis: Bobbs-Merrill.

Crain, R. L., 1966, "Fluoridation: The diffusion of an innovation among cities", *Social Forces,* **44**, pp. 467–76.

David, P. A., 1966, "The mechanization of reaping in the ante-bellum midwest", Ch. 1 in *Industrialization in Two Systems* (H. Rosovsky, ed.). New York: Wiley.

David, P. A., 1969, *A Contribution to the Theory of Diffusion.* Memorandum 71, Research Center in Economic Growth, Stanford University.

Dodd, S. C., 1956, "Testing message diffusion via harmonic logistic curves", *Psychometrica,* **21**, pp. 191–205.

Evenson, R., 1974, "International diffusion of agrarian technology", *Journal of Economic History*, **34**, pp. 51–90.

Fisher, J., and R. Pry, 1971, "A simple substitution model of technological change", *Technological Forecasting and Social Change,* **3**, pp. 75–88.

Fliegel, F. C., and J. E. Kivlin, 1966, "Attributes of innovations as factors in diffusion", *The American Journal of Sociology*, **72**, pp. 235–48.

Fourt, L. A., and J. W. Woodlock, 1960, "Early prediction of market success for new grocery products", *Journal of Marketing*, **XXV** (2), 31–8.

Friedman, L., 1973. "Innovation and diffusion in non-markets: Case studies in criminal justice". Yale University Ph.D. Dissertation, unpublished.

Gersho, A., and D. Mitra, 1975, "A simple growth model for the diffusion of new communication service", *IEEE Transactions on Systems, Man and Cybernetics*, **5**, pp. 209–16.

Gold, B., 1969. "The framework of decision for major technological innovation", in *Values and the Future* (Baier, K., and Rescher, N., eds.), New York: The Free Press.

——, W. S. Peirce, and G. Rosegger, 1970, "Diffusion of major technological innovations in U.S. iron and steel manufacturing", *Journal of Industrial Economics*, **18**.

Gray, V., 1973, "Innovation in the States: A diffusion study", *American Political Science Review,* **67**, pp. 1174–85.

Green, P. E., and H. F. Sieber, Jr., 1967. "Discriminant techniques in adoption patterns for a new product, Part four", in *Promotion Decisions Using Mathematical Models* (P. J. Robinson, ed.). Boston, Mass.: Allyn and Bacon, Inc.

Griliches, Zvi, 1957, "Hybrid corn: An exploration in the economics of technological change", *Econometrica*, **25**, pp. 501–22.

——, 1962, "Profitability versus interaction: Another false dichotomy", *Rural Sociology*, **27**, pp. 327–30.

Hägerstrand, T., 1967. *Innovation Diffusion as a Spatial Process.* Chicago: University of Chicago Press.

____, 1965, "A Monte Carlo approach to diffusion", *European Journal of Sociology,* **6**, pp. 43–67.

____, 1952. *The Propagation of Innovation Waves.* Lund: Gleerup, Lund Studies in Human Geography.

Haines, G. H., Jr., 1964, "A theory of market behaviour after innovation", *Management Science*, **10**, pp. 634–58.

Hayani, Y., 1974, Conditions for the diffusion of agriculture technology, An Asian perspective, *The Journal of Economic History*, **34**, pp. 131–48.

Haynes, K. E., F. Y. Phillips, and G. M. White, 1972. "Information Theoretic Approaches to Spatial Interaction". Department of Geography, The University of Texas, Austin, Mimeograph.

——, V. Mahajan, G. M. White, 1975. "Innovation Diffusion: A Deterministic Model of Space-Time Integration With Physical Analog". Dept. of Geography, The University of Texas, Austin, Mimeograph.

Hildebrand, P., and E. Partenheimer, 1958, "Socio-economic characteristics of innovators", *Journal of Farm Economics*, **60**.

Hirshberg, A., and R. Schoen, 1974, Barriers to the widespread utilization of residential solar energy: The prospects for solar energy in the U.S. housing industry, *Policy Sciences*, December.

Hudson, J., 1972. *Geographical Diffusion Theory*. Evanston, Ill.: Northwestern University Press, Studies in Geography, Number 19.

——, 1971, "Some properties of basic classes of spatial diffusion models", in *Perspectives in Geography* (H. McConnell and D. W. Yaseen, eds.). DeKalb, Illinois: Northern Illinois University Press.

Katz, E., 1935. "The diffusion of new ideas and practices", in *The Science of Human Communication* (W. Schramm, ed.). New York: Basic Books, pp. 77–93.

——, M. L. Levin, and H. Hamilton, 1963, "Traditions of research on the diffusion of innovation", *American Sociological Review*, **28**, pp. 237–52.

Kurz, Mordecai, and Alan S. Manne, 1963, Engineering estimates of capital-labor substitution in metal machining, *The American Economic Review*, **53**, pp. 662–81.

Kennedy, C., and A. P. Thirlwall, 1972, "Surveys in applied economics: Technical progress", *Economic Journal*, **82**.

Knoerr, A. W., 1963, "Role of literature in the diffusion of technological change", *Special Libraries*, **54** (May/June).

Korea Institute of Science and Technology, 1973. Final report on a Study of the Scope for Capital-Labor Substitution in Mechanical Engineering Sector. Seoul, Korea.

Kuehn, A., 1962, "Consumer brand choice–A learning process?", *Quantitative Techniques in Marketing Analysis* (Frank, R. E., Kuehn, A. A., and Massy, W. F., eds.). Homewood, Ill.: Richard D. Irwin, Inc., pp. 390–403.

Landau, H. G., and A. Rapoport, 1953, "Contributions to the mathematical theory of contagion and spread of information", *Bulletin of Mathematical Biophysics*, **15**, p. 173.

Lekvall, P., and C. Wahlbin, 1973, "A study of some assumptions underlying innovation diffusion functions", *Swedish Journal of Economics.*

Lynn, F., 1966. "An investigation of the rate of development and diffusion of technology in our modern industrial society", in *Studies Prepared for the National Commission on Technology, Automation, and Economic Progress*, Appendix Vol. II. Washington, D.C.: Technology and the American Economy.

Mansfield, E., 1963, "Intrafirm rates of diffusion of an innovation", *Review of Economics and Statistics*, **45**, November.

——, 1963, "The speed of response of firms to new techniques", *The Quarterly Journal of Economics*, **77** (2).

——, 1966. "Technological change: measurement, determinants, and diffusion", *Report to the President by the National Commission on Technology, Automation, and Economic Progress*, Appendix Vol. II. Washington, DC, Feb. 1966.

——, 1968. *Industrial Research and Technological Innovation*. New York: Norton.

——, 1961, "Technical change and the rate of imitation", *Econometrica*, **29**.

——, et al., 1971. *Research and Innovation in the Modern Corporation.* New York: Norton.

——, 1971. "Determinants of the Speed of Application of New Technology". Paper delivered at the meeting of the International Economic Association in St. Anton, Austria, August 27–September 2, 1971.

Markowitz, H. M., and A. J. Rowe, 1963. "A machine tool substitution analysis", in *Studies in Process Analysis* (A. Manne and H. Markowitz, eds.). New York: Wiley.

Mohr., L., 1969, "Determinants of innovation in organization", *American Political Science Review*, **63**, pp. 111–26.

Mason, R., and A. Halter, 1968, "The application of a system of simultaneous equations to an innovation diffusion model", *Social Forces*, **47**, pp. 183–95.

Massy, W., D. Montgomery, and D. Morrison, 1969. *Stochastic Models in Marketing.* Cambridge, Mass: MIT Press.

Morrill, R. L., 1968, "Waves of spatial diffusion". *Journal of Regional Science*, **8**, pp. 1–18.

Nabseth, L., and G. F. Ray, 1974. *The Diffusion of New Industrial Processes.* Cambridge: University Press.

National Institutes of Health (NIH), 1972. Conference Papers for National Institutes of Health Conference on the Diffusion of Medical Innovations, September 24–7, 1972, working papers.

Nelson, R. R., 1968, "A 'diffusion' model of international productivity differences in manufacturing industry", *American Economic Review*, **58**.

——, M. Peck, and E. Kalachek, 1967. *Technology, Economic Growth, and Public Policy.* Washington: The Brookings Institution.

Nicosia, F. M., 1966. *Consumer Decision Processes.* Englewood Cliffs, NJ: Prentice-Hall.

Ozga, S., 1960, "Imperfect markets through lack of knowledge", *Quarterly Journal of Economics*, **74**, pp. 29–52.

Peirce, W. S., 1974. "Factors Affecting Responsiveness to Technological Innovations in Coal Mining". Working Paper No. 54. Research Program in Industrial Economics, Case Western Reserve University.

Pemberton, H. E., 1936, "The curve of culture diffusion rate", *American Sociology Review*, **1** (4), pp. 547–56.

Perry, A., G. A. Sullivan, R. J. Dolan, and G. P. Marsh, 1967, "The adoption process: S-curve or J-curve?", *Rural Sociology*, **32** (2), 220–22.

Pitts, F., 1963, "Problems in computer simulation of diffusion", *Papers and Proceedings of the Regional Science Association*, **11**, pp. 111–19.

Rapoport, A., 1956, "The diffusion problem in mass behaviour", *General Systems Yearbook*, **1**, pp. 48–55.

——, and L. I. Rebhuhn, 1952, "On the mathematical theory of rumor spread", *Bulletin of Mathematical Biophysics*, **14**, p. 375.

Ray, G. F., 1969, "The diffusion of new technology: A study of ten processes in nine industries", *National Institute Economic Review*, **48**.

Renshaw, E. F., 1963, "The substitution of inanimate energy for animal power", *Journal of Political Economy*, pp. 284–92.

Rhee, Y. W., and L. E. Nestphal, 1973. Planning Future Import Substitution and Export Expansion in Korea's Mechanical Engineering Industries. Paper Presented at Joint Korea Development Institute, Development Advisory Service, Harvard University. Conference in Seoul, Korea, October 10–12.

Robertson, T. S., 1967, "The process of innovation and the diffusion of innovation", *Journal of Marketing*, **31**, pp. 14–19.

Robinson, E. H., 1974, "The early diffusion of steam power", *The Journal of Economic History*, **34**, pp. 91–107.

Rogers, E. M., 1962, "Rejoinder to Griliches' another False Dichotomy", *Rural Sociology*, **27**, pp. 168–70.

——, and E. Havens, 1961, "Adoption of hybrid corn: Profitability and the interaction effect", *Rural Sociology*, **26**, pp. 409–14.

——, 1962. *Diffusion of Innovations.* New York: The Free Press.

——, and F. Shoemaker, 1971. *Communication of Innovations: A Cross-Cultural Approach.* New York: The Free Press.

Rohlfs, J., 1974, "A theory of interdependent demand for a communications service", *Bell J. Econ. Management Science*, **5** (1), pp. 16–37.

Rosegger, G., 1976, "Diffusion research in the industrial setting: Some conceptual clarifications", *Technological Forecasting and Social Change*, **9**.

Rosenberg, N., 1972, "Factors affecting the diffusion of technology", *Explorations in Economic History*, **10**.

Ryan, B., and M. Gross, 1943, "The diffusion of hybrid seed corn in two Iowa communities", *Rural Sociology*, **8**, pp. 15–24.

Saxenhouse, G., 1974, "A tale of Japanese technological diffusion in Meiji period", *The Journal of Economic History*, **34**, pp. 149–65.

Schumpeter, J., 1969. *The Theory of Economic Development.* New York: Oxford University Press.

Sharif, M. N., and G. A. Uddin, 1975, "A procedure for adapting technological forecasting models", *Technological Forecasting and Social Change*, **7** (1), pp. 99–105.

Shaw, S. J., 1965, "Behavioural science offers fresh insights on new product acceptance", *Journal of Marketing*, **29**, pp. 9–13.

Sutherland, A., 1959, "Diffusion of an innovation in cotton spinning", *Journal of Industrial Economics*, March issue.

Wahlbin, C., 1973. Stochastic Models of Innovation Diffusion Processes". Working Paper 73–10. European Institute for Advanced Studies in Management, Brussels.

Walker, J., 1969, "The diffusion of innovations among the American States", *American Political Science Review*, **63**, pp. 880–99.

____, 1971. "Innovations in state politics", in *Politics in the American States–A Comparative Analysis*, 2nd ed. (H. Jacob and K. Vines, eds.). Boston: Little Brown.

Warner, K. (no date). The Diffusion of Leukemia Chemotherapy: A Study in the Nonmarket Economics of Medical Care. Yale University Ph.D. Dissertation, in progress.

____, 1976, "The need for some innovative concepts of innovation: An examination of research on diffusion of innovations", *Policy Sciences*, **5** (4), pp. 433–51.

Watson, A. M., 1974, "The Arab agricultural revolution and its diffusion 700–1100", *The Journal of Economic History*, **34**, pp. 8–35.

Wilkins, M., 1976, "The role of private business in the international diffusion of technology", *The Journal of Economic History*, **34**, pp. 166–93.

Wilson, L. O., 1972, "An epidemic model involving a threshold", *Mathematical Biosciences*, **15**, pp. 109–21.

Utterback, J., 1974, "Innovation in industry and the diffusion of technology", *Science,* **183**, February 15, pp. 620–26.

Zaltman, G., R. Duncan, and J. Holbek, 1973. *Innovations and Organizations.* New York: Wiley.

Index

Italic numbers in entries refer to figures